Marine Policy

This textbook provides the reader with a foundation in policy development and analysis and describes how policy, including legal mechanisms, is applied to marine environments around the world. It offers a systematic treatment of all aspects of marine policy, including environmental protection, fisheries, transportation, energy, mining and climate change.

It starts with a biophysical overview of the structure and function of the marine environment with a particular emphasis on the challenges and opportunities of managing the marine environment. An overview of the creation and function of international law is then provided with a focus on international marine law. It explores the geographic and jurisdictional dimensions of marine policy, as well the current and anticipated challenges facing marine systems, including climate change-related impacts and resource over-exploitation. The book should appeal to senior undergraduate and graduate students and form a core part of the curriculum for marine affairs, science and policy courses. It will also provide supplementary reading for students taking a course in the law of the oceans, but is not aimed at legal specialists.

Mark Zacharias is an Assistant Deputy Minister in the British Columbia Ministry of the Environment, Canada. He is also Adjunct Professor at the University of British Columbia Fisheries Centre and Associate Professor in the Department of Geography at the University of Victoria, Canada. He is co-author of the book *Marine Conservation Ecology* (Earthscan, 2011).

Marine Policy

An introduction to governance and
international law of the oceans

Mark Zacharias

Routledge
Taylor & Francis Group

LONDON AND NEW YORK

from Routledge

First published 2014
by Routledge
2 Park Square, Milton Park, Abingdon, Oxon OX14 4RN

and by Routledge
711 Third Avenue, New York, NY 10017

Routledge is an imprint of the Taylor & Francis Group, an informa business

British Library Cataloguing-in-Publication Data
A catalogue record for this book is available from the British Library

Library of Congress Cataloging-in-Publication Data
Zacharias, Mark, 1967–
Marine policy : an introduction to governance and international law of the oceans / Mark
Zacharias.
pages cm
Includes bibliographical references and index.
ISBN 978-0-415-63307-9 (hbk : alk. paper) — ISBN 978-0-415-63308-6 (pbk : alk. paper)
— ISBN 978-0-203-09525-6 (ebk) 1. Law of the sea. I. Title.
KZA1145.Z33 2014
341.4'5—dc23
2013037660

ISBN: 978-0-415-63307-9 (hbk)
ISBN: 978-0-415-63308-6 (pbk)
ISBN: 978-0-203-09525-6 (ebk)

Typeset in Goudy
by FiSH Books Ltd, Enfield

Contents

List of illustrations

Figures

Tables

Boxes

List of acronyms and abbreviations

AAC	Arctic Athabaskan Council
ABNJ	areas beyond national jurisdiction
ACAP	Agreement on the Conservation of Albatrosses and Petrels
ACIA	Arctic Climate Impact Assessment
AEPS	Arctic Environmental Protection Strategy
AHP	analytic hierarchy process
AIA	Aleut International Association
AMAP	Arctic Monitoring and Assessment Programme
APFIC	Asia-Pacific Fisheries Commission
ARREST	International Convention on Arrest of Ships
ASBAO	Regional Convention on Fisheries Cooperation among African States Bordering the Atlantic Ocean
ASOC	Antarctic and Southern Ocean Coalition
AT	Antarctic Treaty
ATCM	Antarctic Treaty Consultative Meeting
ATS	Antarctic Treaty System
BUNKERS	International Convention on Civil Liability for Bunker Oil Pollution Damage
BWM	International Convention for the Control and Management of Ships' Ballast Water and Sediments
CAFF	Conservation of Arctic Flora and Fauna
CAM	coastal area management
CBD	Convention on Biological Diversity
CCAMLR	Convention on the Conservation of Antarctic Marine Living Resources
CCAS	Convention for the Conservation of Antarctic Seals
CCSBT	Commission for the Conservation of Southern Bluefin Tuna
CDM	Clean Development Mechanism
CDQ	community development quota
CECAF	Committee for the Eastern Central Atlantic Fisheries
CEP	Caribbean Environment Program
CER	corporate environmental responsibility
cif	cost, insurance and freight
CITES	Convention on International Trade in Endangered Species of Wild Fauna and Flora
CLC	International Convention on Civil Liability for Oil Pollution Damage

CLCS	Commission on the Limits of the Continental Shelf
CLEE	Convention on Civil Liability for Oil Pollution Damage Resulting from Exploration for and Exploitation of Seabed Mineral Resources
CLL	International Convention on Load Lines
CMI	Comité Maritime International
CMS	Convention on the Protection of Migratory Species of Wild Animals
COFI	Committee on Fisheries
COLREG	Convention on the International Regulations for Preventing Collisions at Sea
COLTO	Coalition of Legal Toothfish Operators
COMNAP	Council of Managers of National Antarctic Programs
COP	conference of the parties
CPPS	Permanent Commission on the South Pacific
CP	consultative party
CPUE	catch per unit effort
CRM	coastal resource management
CSI	Container Security Initiative
CZM	coastal zone management
DFO	Department of Fisheries and Oceans, Canada
DMC	dangerous maritime cargo
dwt	dead weight tonnes
EAF	ecosystem approach to fisheries
EAGGF	European Agricultural Guidance and Guarantee Fund
EAM	ecosystem approaches to management
EAP	Action Plan for the Protection, Management and Development of the Marine and Coastal Environment of the Eastern African Region
EbA	ecosystem-based adaptation
EEC	European Economic Community
EEDI	Energy Efficiency Design Index
EEZ	Exclusive Economic Zone
EFR	ecological fiscal reform
ELECRE	elimination and choice expressing reality
EPPR	Emergency Prevention, Preparedness and Response
ERFEN	Protocol on the Regional Program for the Study of the El Niño Phenomenon in the South-East Pacific
ESG	environmental, social and governance
ETR	ecological tax reform
EU	European Union
EVI	Economic Vulnerability Index
FAL	Convention on Facilitation of International Maritime Traffic
FAO	Food and Agriculture Organization
FCWC	Fishery Committee of the West Central Gulf of Guinea
FFA	South Pacific Forum Fisheries Agency
F_{MSY}	fishing mortality less than maximum sustainable yield
fob	free on board
FSC	flag state control
GBRMPA	Great Barrier Reef Marine Park Authority

GDP	gross domestic product
GDS	geographically disadvantaged state
GFCM	General Fisheries Council for the Mediterranean
GGI	Gwich'in Council International
GNI	gross national income
GPA	Global Programme of Action
GSP	green shipping practice
GT/G.T./gt	gross tonnage
GW	gigawatt
HAFS	International Convention on the Control of Harmful Anti-fouling Systems on Ships
HAI	Human Assets Index
HFC	hydrofluorocarbon
IAATO	International Association of Antarctic Tour Operators
I-ATTC	Inter-American Tropical Tuna Commission
IBSFC	International Baltic Sea Fisheries Commission
ICC	Inuit Circumpolar Council
ICCAT	International Commission for the Conservation of Atlantic Tunas
ICES	International Council for the Exploration of the Sea
ICJ	International Court of Justice
ICNAF	International Commission for the Northwest Atlantic Fisheries
ICRW	International Convention for the Regulation of Whaling
ICZM	integrated coastal zone management
IFQ	individual fishing quota
ILC	International Law Commission
ILO	International Labour Organization
IMO	International Maritime Organization
IOPC Funds	International Oil Pollution Compensation Funds
IOS	Indian Ocean Sanctuary
IOTC	Indian Ocean Tuna Commission
IPHC	International Pacific Halibut Commission
IPOA	international plans of action
IPOA-IUU	International Plan of Action to Prevent, Deter and Eliminate Illegal, Unreported and Unregulated Fishing
ISA	International Seabed Authority
ISMC	International Safety Management Code
ISO	International Organization for Standardization
ISPS	International Ship and Port Facility Security Code
ITLOS	International Tribunal for the Law of the Sea
ITQ	individual transferable quota
IUCN	International Union for the Conservation of Nature
IUU	illegal, unreported and unregulated
IVQ	individual vessel quota
IWC	International Whaling Commission
kg	kilogram
lb	pound
LBS	Protocol for the Protection of the Mediterranean Sea Against Pollution

	from Land-Based Sources and Activities
LDC	London Dumping Convention (Convention on the Prevention of Marine Pollution by Dumping of Wastes and Other Matter)
LDS	landlocked developing state
LLDC	landlocked developing country
LLMC	Convention on Limitation of Liability for Maritime Claims
LME	large marine ecosystem
LOSC	Law of the Sea Convention
MAP	Mediterranean Action Plan
MARPOL	International Convention for the Prevention of Pollution from Ships
MAUT	multi-attribute utility theory
MBI	market-based instrument
MBTA	Migratory Bird Treaty Act
MCDA	multi-criteria decision analysis
MEPC	Marine Environment Protection Committee
MLC	Maritime Labour Convention
MOU	memorandum of understanding
MPA	marine protected area
MSC	Maritime Safety Committee
MSP	marine spatial planning
MSY	maximum sustainable yield
MW	megawatt
NAFO	North Atlantic Fisheries Organization
NAMMCO	North Atlantic Marine Mammal Commission
NASCO	North Atlantic Salmon Conservation Organization
NEAFC	North East Atlantic Fisheries Commission
nmi	nautical mile
NOAA	National Oceanic and Atmospheric Administration
NPA	national programme of action
NPAFC	North Pacific Anadromous Fish Commission
NT/N.T./nt	net tonnage
OCA/PAC	Oceans and Coastal Areas Programme Activity Centre
OECD	Organization for Economic Co-operation and Development
OLDEPESCA	Latin American Organization for the Development of Fisheries
OPA	Oil Pollution Act
OPC	Oil Pollution Convention
OPRC	International Convention on Oil Pollution Preparedness, Response, and Co-operation
OPRC-HNS	Protocol on Preparedness, Response and Co-operation to Pollution Incidents by Hazardous and Noxious Substances
OSPAR	Convention for the Protection of the Marine Environment of the North East Atlantic
OSY	optimum sustained yield
P&I	protection and indemnity
PAL	Athens Convention relating to the Carriage of Passengers and their Luggage by Sea
PAME	Protection of the Arctic Marine Environment

PCA   Permanent Court of Arbitration
PCB   poly-chlorinated biphenol
PERSGA   Programme for the Environment of the Red Sea and Gulf of Aden
PFC   perfluorocarbon
PICES   North Pacific Marine Science Organization
PROMETHEE   preference ranking organizational method for enrichment evaluation
PSC   Pacific Salmon Commission
PSI   Proliferation Security Initiative
PSSA   particularly sensitive sea area
PSU   practical salinity units
PTBT   Partial Nuclear Test Ban Treaty
RAIPON   Russian Arctic Indigenous Peoples of the North
ReCAAP   Regional Cooperation Agreement on Combating Piracy and Armed
               Robbery against Ships in Asia
RFMO   regional fisheries management organization
RMP   Revised Management Procedure
RMS   Revised Management Scheme
ROPME   Regional Organization for the Protection of the Marine Environment
ro-ro   roll-on, roll-off
SAP BIO   Strategic Action Plan for the Conservation of Marine and Coastal
               Biodiversity in the Mediterranean
SAR   Convention International Convention on Maritime Search and Rescue
SC   Saami Council
SCAR   Scientific Committee on Antarctic Research
SDC   Seabed Disputes Chamber
SDR   Special Drawing Right
SDWG   Sustainable Development Working Group
SEAFO   South East Atlantic Fisheries Organization
SECA   sulphur emission control area
SEEMP   Ship Energy Efficiency Management Plan
SIDS   small island developing state
SIOFA   South Indian Ocean Fisheries Agreement
SOLAS   International Convention for the Safety of Life at Sea
SOS   Southern Ocean Sanctuary
SPA   Protocol Concerning Specially Protected Areas and Biological Diversity
               in the Mediterranean
SPLOS   State Parties to the Law of the Sea Convention
SPR   spawning potential ratio
SPREP   South Pacific Regional Environmental Programme
SPRFMO   South Pacific Regional Fisheries Management Organization
SRCF   Sub-Regional Commission on Fisheries
SSB   spawning stock biomass
SSBR   spawning stock biomass per recruit
STCW   International Convention on Standards of Training, Certification and
               Watchkeeping for Seafarers
SUA   Convention for the Suppression of Unlawful Acts Against the Safety of
               Maritime Navigation

TAC	total allowable catch
TBT	tributyltin
TEU	twenty-foot equivalent units
TURF	territorial user right in fisheries
UK	United Kingdom
ULCC	ultra large crude carrier
UN	United Nations
UNCED	United Nations Conference on Environment and Development
UNCHE	United Nations Conference on the Human Environment
UNCITRAL	United Nations Commission on International Trade Law
UNCLOS I	1958 United Nations Conference on the Law of the Sea
UNCLOS II	1960 United Nations Conference on the Law of the Sea
UNCLOS III	1982 United Nations Conference on the Law of the Sea
UNCLOS	United Nations Convention on the Law of the Sea
UNCTAD	United Nations Conference on Trade and Development
UNEP	United Nations Environment Programme
UNFCCC	United Nations Framework Convention on Climate Change
UNFSA	UN Fish Stocks Agreement
UNGA	United Nations General Assembly
UNGAR	United Nations General Assembly Resolution
UNODA	United Nations Office for Disarmament Affairs
UNSCR	United Nations Security Council Resolution
US	United States of America
VLBC	very large bulk carrier
VLCC	very large crude carrier
VLOC	very large ore carrier
WACAF	Abidjan Convention for Cooperation in the Protection and Development of the Marine and Coastal Environment of the West and Central African Region
WCED	World Commission on Environment and Development
WCPFC	Western and Central Pacific Fisheries Commission
WECAFC	Western Central Atlantic Fisheries Commission
WiP	with policy
WoP	without policy
WRC	Nairobi International Convention on the Removal of Wrecks
WSSD	World Summit on Sustainable Development
WTO	World Trade Organization
WWF	World Wildlife Fund

Acknowledgements

I cannot hope to thank everyone that shaped the development of this book. However, a number of key individuals were critical in its execution. Laura Feyrer, Kevin Knox, Meherzad Romer and Lori Waters were instrumental in assisting with research, editing and the development of figures. I am similarly grateful to Jeff Ardron, Rod Dobell, Lorne Kriwoken, Martha McConnell, Ted McDorman, Andrew Stainer and Rashid Sumaila for donating their time to review the various chapters. Above all, I would like to thank John Roff for his continued support over the years and for his permission to incorporate modified sections of our previous text, *Marine Conservation Ecology*, into this book.

Many colleagues and friends contributed indirectly to this endeavour and include: Jackie Alder, Hussein Alidina, Jeff Ardron, John Baxter, Rosaline Canessa, Villy Christensen, Chris Cogan, Phil Dearden, Jon Day, Dave Duffus, Zach Ferdana, Leah Gerber, Ed Gregr, Ben Halpern, Ellen Hines, Don Howes, David Hyrenbach, Sabine Jessen, David Kushner, Nancy Liesch, Josh Laughren, Olaf Niemann, Carol Ogborne, Charlie Short, Rashid Sumaila and Mark Taylor. I apologize to anyone I have forgotten.

Lastly, I would like to acknowledge the past contributions of the authors, editors and staff of *The International Journal of Marine and Coastal Law*, *Marine Policy*, *Ocean & Coastal Management* and *Ocean Development & International Law*, without which this book would not have been possible.

Preface

In his classic text *Unpopular Essays*, Bertrand Russell contended that humanity confronts two different problems: the first is mastering nature to provide for human needs, which is firmly in the domain of science and technology (Russell, 1950); the second is wisely utilizing the fruits of science and technology to improve the human condition. Prudence, however, has not always prevailed. Russell provided examples of technological advances throughout history that led to disastrous outcomes. In particular, he evoked the domestication of the horse (conquest), mechanization (slavery) and modern physics (nuclear weapons) as specific advances that were accompanied by unfortunate consequences. Russell stated that while science and technological achievements and skills are important, '...something more than skill is required, something which may perhaps be called "wisdom". This is something that must be learnt, if it can be learnt, by means of other studies than those required for scientific technique'.

Russell's observations are especially apropos to our relationship with the oceans. Evidence confirms that humans consumed seafood 164,000 years ago, and early pre-Western cultures developed sufficient technical expertise to overharvest species they could access (Mueller-Dombois and Wirawan, 2005; Marean *et al.*, 2007). The advent of industrialization created the opportunity to fish in remote oceans, extract oil and gas from continental shelves, transport goods over great distances and use the ocean as a repository for waste. Even as late as the early twentieth century, many thought the oceans immutable and immune to human activities.

Paralleling Russell's examples, scientific and technological advancements have outpaced humanity's ability to understand the consequences of our activities in the marine environment. Nor have we been able to envision how to regulate these activities. Even after evidence emerged on the effects of overfishing and pollution, responses took decades. This unfortunate delay is exasperated by the fact that, even to this day, most of the ocean is owned by no one and thus must be managed as a commons resource. This book explores what Russell termed the 'wisdom' necessary to ensure that humanity's scientific and technical achievements – and the consequences of these achievements – lead to what he termed 'contentment'.

The management of the oceans is incredibly complex. Even those employed in the field of ocean governance are rarely fluent in all its aspects. This ignorance is forgivable: at present there are over 150 multilateral international agreements that govern the use of the marine environment. In turn, many of these agreements establish dozens of multinational agencies, commissions, organizations and secretariats to oversee their implementation. These administrative bodies then create policies, guidelines and reporting requirements.

Increasing the confusion is that many of these agreements often nest hierarchically within other agreements or require other agreements to be fully implemented in order to operate successfully. Complicating the Byzantine maze is the fact that many agreements – and particularly the more important international conventions – are frequently amended. As if this were not enough, many geographic regions of the ocean are governed by multiple, separate regional agreements.

Unsurprisingly, very few people have a comprehensive understanding of how the world's oceans are governed and whether this governance is effective. Fewer still can competently gauge if our current approach to ocean governance is sufficient to address complex global issues such as climate change, sovereignty claims and supporting indigenous rights.

Purpose of this book

This intent of this book is to provide the reader with an understanding of how nation states work together to address socioeconomic, environmental and cultural issues with respect to the management of the world's oceans. Marine governance is an amalgam of historical practice and custom, economic and trade considerations, domestic law and policy, and regional and international agreements. This book provides a systematic overview of ocean governance with the intention of providing a foundational treatment of how humans manage the various activities undertaken on, in or under the world's oceans. How oceans are governed is captured here under the aegis of 'marine policy'. Broadly defined, marine policy is the various courses of action to influence decisions, actions and other ocean-related matters.

While this text introduces the legal context and international law that underpins ocean governance, it is not a law textbook nor is it an authoritative guide to any particular international law or policy. Dozens of textbooks have been written on each of the international conventions discussed here and the intent is not to duplicate these works. Instead, this book explores the legal, policy and state approaches to ocean governance to demonstrate the commonalities and differences in how various sectors (e.g. fishing, transportation) are governed and the opportunities to integrate governance approaches to improve socioeconomic and environmental outcomes. In particular, this book attempts to provoke the reader, who may be a specialist in one particular aspect of marine law or policy, towards thinking about the opportunities and benefits of thinking beyond a sector-based approach to marine management.

This text primarily focuses on the international aspects of marine management. For this reason, it is not a treatise on how individual nations manage their domestic waters. While each nation state (even landlocked states) has legal standing in ocean law and may have its own set of marine laws and policies, it is not the purpose of this book to preferentially focus on individual nations and their domestic marine agendas. Since there are many excellent 'coastal management' texts, domestic examples will be used only when they are illustrative of a particular system, approach, philosophy, success or failure.

Why this book now?

The current environmental condition of marine environments world-wide is steadily declining and these trends have been widely chronicled (Pauly et al., 2002; Roberts, 2013). Traditional, sector-based approaches to solving socioeconomic and environmental problems (such as managing fisheries or regulating transportation) have to date been unable to resolve the complex social and environmental predicaments facing humanity (see Box 0.1). Climate change, sovereignty claims and regional economic disparity are not addressed by simply improving the management of one part of the ocean environment or economy. Instead, complex, interrelated problems require innovative ways of integrating management actions towards common goals.

Box 0.1 Governance challenges resulting from the unique characteristics of the marine environment

- Sector-based, universal solutions are incompatible with the biophysical complexity of marine environments and interconnections with the economy and human communities.
- Socioeconomic and ecological interactions at multiple geographical scales are poorly understood.
- Management based on 'command and control' regulatory approaches are less effective in marine systems because of their dynamism and complexity.
- Attitudes and time frames are short term, compromising planning for long-term sustainable solutions.
- A 'frontier mentality' prevails that assumes few ecological, thermodynamic and economic constraints to resource development.
- Marine systems transcend national and regional boundaries, leaving management systems developed for property rights (ownership) inadequate.

Source: modified from Glavovic (2008)

Book outline

The discrete inner workings of the oceans are largely invisible to us and often counterintuitive to our terrestrial experience. As such, a basic knowledge of the biological and physical characteristics of marine systems is required to understand how the application of marine law and policies may impact marine environments. An 'Oceans 101' is provided in Chapter 1 to familiarize the reader with the structure, function and processes that operate in marine environments. It begins with an exploration of the differences between terrestrial and marine environments. The intent of this section is to evince how our 'land referenced' perspective instills a bias that must be overcome if we are to fully embrace how marine environments could be governed. The chapter then outlines the major biological, oceanographic and physiographic characteristics and features of marine environments and the relationship of these features to their astute management.

Chapter 2 introduces the history and operation of international law. Within this chapter is a discussion of the types of ocean ownership regimes, an introduction to the world's different legal systems, and a brief explanation of the workings of international law. The chapter then introduces international marine law and the differences between marine law and other legal systems. Finally, the chapter introduces the United Nations Convention on the Law of the Sea (LOSC). The LOSC is one of the most comprehensive international agreements created and tomes written solely on the LOSC and its application abound. This chapter provides an overview of the content and operation of the LOSC while subsequent chapters will delve further into the specific functions of the Convention.

An introduction to the field of policy encompasses Chapter 3. It explores the purpose of public policies, types of policies and the relationship between law and public policy. The unique characteristics of marine law and policy are then discussed with particular attention to areas beyond national jurisdiction (the high seas and deep seabed). The third chapter concludes with an exploration of the application of various types of policy instruments in marine environments. These include regulatory, economic, price-based, quantity-based, market enhancement, informational and voluntary instruments.

Chapter 4 reviews how policies are formed, the characteristics of policy development, considerations in selecting policies and methods for selecting and analysing policies. Qualitative versus quantitative policy analysis figures substantially in this analysis.

The primary international mechanisms that govern the protection and preservation of the oceans – along with the important regional conventions and examples of bilateral and multilateral agreements – are presented in Chapter 5. The chapter begins with a discussion on the primary threats to marine biodiversity: overharvesting, habitat loss, introduced species, global climate change and pollution. This is followed by a brief history of international environmental marine law and policy. The remainder of the chapter outlines the primary international and regional agreements governing the protection of marine environments.

The operation and regulation of marine fisheries is discussed in Chapter 6. The chapter provides an overview of marine fisheries management, focusing on divergent approaches (single-species, multi-species, ecosystem-based) that have been applied – sometimes successfully and sometimes not. A history of international fisheries law and policy is then provided along with a summary of the organizations that oversee fisheries management. The primary international agreements that govern fishing are then discussed.

The seventh chapter explores the operation and regulation of international marine transportation. First, it begins with an overview of contemporary shipping and introduces marine transportation law and policy. Organizations that regulate marine transportation are then discussed along with the key international conventions and agreements dealing with the ship, shipping company and seafarer. The chapter then addresses international laws and policies related to maritime security (including piracy) and the environmental effects of shipping.

The polar regions (Arctic and Antarctic) warrant a separate treatment in Chapter 8 since they have unique governance arrangements due to their remoteness, inhospitable climate and unique biological systems. While the polar regions are similar in climate, they are distinctive in almost all other respects, be they biological structure, governance or human use. Chapter 8 introduces the Antarctic and Southern Ocean and the Antarctic Treaty System that governs these regions. Governance of the Arctic regions is then discussed along with the key institutional arrangements that assist with Arctic management.

Chapter 9 introduces the international law and policy instruments that apply to ocean energy and mining activities. It first differentiates energy and mining opportunities between the continental shelves and the deep seabed; it then discusses the role of the LOSC in energy and mineral development. Organizations that oversee energy and mining activities are then reviewed. A brief summary of marine energy and mineral resources and their management in the continental shelf and deep sea areas ensues.

Integrated approaches to ocean management are elicited in Chapter 10. These approaches are loosely defined as tools to assist marine decision-making and the analysis of policies. Coastal management, ecosystem approaches to management, marine protected areas, marine spatial planning, cumulative effects and private rights are analysed.

Terminology used in this book

Lawyers, policy analysts, environmental planners, engineers, diplomats and bureaucrats have their own lexicons related to ocean governance. This text borrows from all of these disciplines and uses the following terms based on recent practice:

States are equivalent to countries, nations, nation states and economies (e.g. Taiwan).
Marine and *oceans* are used interchangeably. However, 'maritime' has specific connotations with the marine transportation sector and is used only in this context.
Law may either connote domestic or international law.
Policy is a plan or course of action to influence and determine decisions, actions and other matters.
Governance is the act of enacting or implementing laws or policies.
Convention is synonymous with an international agreement or treaty. When a particular convention is discussed it is capitalized.
Law of the Sea Convention (LOSC) is used to abbreviate the United Nations Convention on the Law of the Sea as per common practice in the international law literature. Other texts may use the term 'UNCLOS', which is synonymous with 'LOSC'.
More developed and *less developed states* are used to differentiate 'developed' (or 'first world') from 'developing' (or 'third world') states.

References

Glavovic, B. (2008) 'Ocean and coastal governance for sustainability: Imperatives for integrating ecology and economics', in M. Patterson and B. Glavovic (eds) *Ecological Economics of the Oceans and Coasts*, Edward Elgar, Cheltenham.

Marean, C. W., Bar-Matthews, M., Bernatchez, J., Fisher, E., Goldberg, P., Herries, A. I., Jacobs, Z., Jerardino, A., Karkanas, P., Minichillo, T., Nilssen, P. J., Thompson, E., Watts, I. and Williams, H. M. (2007) 'Early human use of marine resources and pigment in South Africa during the Middle Pleistocene', *Nature*, 449 (7164), 793–794.

Mueller-Dombois, D. and Wirawan, N. (2005) 'The Kahana Valley Ahupua`a, a PABITRA study site on O`ahu, Hawaiian Islands', *Pacific Science*, 59 (2), 293–314.

Pauly, D., Christensen, V., Guenette, S., Pitcher, T. J., Sumaila, U. R., Walters, C. J., Watson, R. and Zeller, D. (2002) 'Towards sustainability in world fisheries', *Nature*, 418 (6898), 689–695.

Roberts, C. (2013) *The Ocean of Life: The Fate of Man and the Sea*, Penguin Books, New York.

Russell, B. (1950) *Unpopular Essays*, Routledge, New York.

Chapter 1

An introduction to the marine environment

The structure and function of the world's oceans

Introduction

Appreciating the ocean's contributions to life on earth and human needs is essential. Effective marine management requires an understanding of the structure and function of the oceans (Boxes 1.1 and 1.2). For example, an understanding of ocean temperature and ocean chemistry is fundamental to comprehending the impacts of greenhouse gas emissions on marine environments (e.g. sea level rise, ocean acidification). This knowledge also contributes to evaluating opportunities to either mitigate (reduce) greenhouse gas emissions or adapt (respond to) to climate change. Another example is the need to understand how deep sea mining may affect marine environments. This requires knowledge of the oceanographic, physiographic and biological structures and processes that contribute to the ecological health of the deep sea and associated communities.

The unique structure and function of the ocean, compared to the terrestrial realm, suggest that approaches to terrestrial management may not be successful when applied in the marine environment. For example, management approaches to emissions (e.g. pollution, noise) from terrestrial transportation activities (e.g. aircraft, vehicles) have no marine

---

Box 1.1 Importance of the oceans

Globally, the oceans are the:

- main reservoir of water: 71 per cent of the earth's surface is covered by oceans; less than 0.5 per cent is freshwater;
- main place for organisms to live; they comprise over 99 per cent of the inhabitable volume of the 'earth';
- main planetary reservoir of O_2;
- possible main planetary producer of O_2 from phytoplankton;
- planetary thermal reservoir and regulator;
- medium for longitudinal heat transfer and circulation;
- major reservoir of CO_2 especially in HCO_3-, CO_3 = forms;
- habitat for enormous diversity of living organisms, from bacteria to whales;
- reservoirs of enormous resource potential, both renewable and non-renewable, oil, minerals, etc.; also, the location of half the planet's carbon fixation.

Source: after Roff and Zacharias (2011)

Box 1.2 Marine goods and services that support humanity

Production services

Food provision: Marine organisms for human consumption.
Raw materials: Marine organisms for non-consumptive purposes.

Cultural services

Identity/cultural heritage: Value associated with the marine environment, e.g. for religion, folklore, painting, cultural and spiritual traditions.
Leisure and recreation: Revitalizing the human body and mind through the observation of and engagement with marine organisms in their natural environment.
Cognitive value: Cognitive development, including education and research.

Option use value

Future unknown and speculative benefits: Currently unknown potential future uses of the marine environment and associated biodiversity.

Regulation services

Gas and climate regulation: Balance and maintenance of the chemical composition of the atmosphere and oceans by marine living organisms.
Flood and storm protection: Dampening of environmental disturbances by biogenic structures.
Bioremediation of waste: Removal of pollutants through storage, dilution, transformation and burial.

Supporting services

Nutrient cycling: Storage, cycling and maintenance of nutrients by living marine organisms.
Resilience/resistance: Extent to which ecosystems can absorb recurrent natural and human perturbations and continue to regenerate without degrading or unexpectedly switching to alternate states.
Biologically mediated habitat: Habitat that is provided by marine organisms.

Source: Beaumont et al. (2007); Ruiz-Frau et al. (2011)

analogue. Noise can have far reaching implications for marine mammal communities, while pollution from ballast water discharges can have significant economic and environmental impacts if species transported in ship's ballast become established outside of their natural range. Another example is the difference in complexity between marine and terrestrial food webs. Terrestrial approaches to wildlife management utilizing models with few trophic levels may have limited utility in marine environments that are comprised of five to six trophic levels, which shift according to annual and decadal oceanographic regimes.

This chapter provides a short overview on the structure and function of marine environments, and in doing so informs discussions on approaches to marine management in following chapters. Knowledge of the workings of marine environments is necessary to understand how marine management can overcome humanity's biased 'land referenced' perspective. The chapter is not a comprehensive treatise on oceanography or marine biology and readers interested in a more detailed treatment should refer to, among others, Valiela (1995), Barnes and Hughes (1999), Bertnes *et al.* (2001), Knox (2001), Sverdrup *et al.* (2003) and Mann and Lazier (2005).

Similarities and differences between marine and terrestrial environments

To reiterate a crucial point, marine systems differ significantly from terrestrial systems. Consequently, any attempts to manage marine environments simply by applying terrestrial theories, paradigms and concepts are likely to fail. For this reason, this section provides a brief overview of the similarities and differences between terrestrial and marine ecosystems (adapted from Day and Roff, 2000; Roff and Zacharias, 2011).

Similarities between terrestrial and marine ecosystems

At a very broad conceptual level, marine and terrestrial systems do have some similarities:

- Both are composed of interacting physical and biological components with energy from the sun driving almost all ecosystems.
- Both are complex patchworks of differing environments and habitats that are occupied by different communities and species.
- Both marine and terrestrial species show gradients in diversity (number of species) with latitude – generally species diversity increases with decreasing latitude.
- In both types of ecosystem, the primary zones of biological activity tend to be concentrated nearer the surface (i.e. the sea–air or land–air interface).

Differences between terrestrial and marine ecosystems

Difference in size

The oceans are far larger in area than all the land masses combined, covering 71 per cent of the earth's surface. The Pacific Ocean alone could easily contain all the continents. Even more marked is the difference in vertical extent and volumes between marine and terrestrial habitats. Life on land generally extends from a few metres underground to the tops of trees – a vertical extent of perhaps no more than 30–40 metres. Although birds, bats, insects and bacteria may periodically rise above these heights, the atmosphere above the trees is a medium of temporary dispersal only. Since species must return to the terrestrial environment for resources, reproduction and shelter, it is not considered a habitat. In contrast, the average depth of the oceans is 3,800 metres and contains life at all depths. The sea's habitable volume is therefore hundreds of times greater than that of the land.

productivity vs
area vs
depth

Difference in physical properties

Terrestrial and marine ecosystems differ markedly in their physical properties. For example, the water overlying the seabed is 60 times more viscous than air and has greater surface tension. Water is also about 850 times denser than air; this provides buoyancy and allows organisms to survive without the need for powerful supporting structures. Buoyancy enables organisms to exist in the sea that are morphologically and anatomically very different from those on land. Seawater's buoyancy and viscosity keep food particles suspended and result in an environment – the pelagic realm (Box 1.3) – without an analogue in the terrestrial environment.

Box 1.3 Divisions in the marine environment

The ocean is composed of two fundamentally distinct realms: the pelagic environment and the benthic environment. A secondary division is the intertidal or estuarine realm that exhibits characteristics of both the pelagic and benthic realms along with contributions (inputs) from terrestrial environments.

Pelagic realm

The pelagic realm (pelagos meaning 'open sea' in ancient Greek) is the water column itself and all the organisms that inhabit it. The pelagic zone consists of a volume of $1{,}330 \times 10^6 km^3$, with an average depth of 3.68km and a maximum depth of 11km (Charette and Smith, 2010). A fully three-dimensional world, the pelagic realm is capable of functioning independently from either terrestrial or benthic environments. It generates its own primary production from planktonic organisms that, in turn, often engender complex food webs with birds, sharks and marine mammals as top predators.

Benthic realm

The benthic realm (benthos meaning 'depths of the sea' in Greek) is loosely defined as the seabed (or lakebed), sediments on the seabed and adjacent waters and their associated biological community. Benthic environments are primarily structured by depth, substratum (bottom type) and nutrient inputs from the water column. These inputs are driven by abiotic mechanisms (e.g. water motion) at larger scales and biological interactions (e.g. competition, predation) at smaller scales. Nearshore benthic environments generally exhibit higher spatial and temporal variability, while offshore environments exhibit lower variability.

Intertidal and estuarine environments

Intertidal environments, while fundamentally a type of benthic environment, are those coastal marine areas that are exposed during low tides. As such, they are structured by atmospheric, marine and terrestrial (e.g. runoff) processes and exhibit characteristics of both marine and terrestrial environments. Estuarine environments may be intertidal environments, but they are defined as areas at the confluence of freshwater and marine systems. Estuarine environments are characterized by low salinities, high biological productivity and low species diversity.

Differences in temperatures

The strong seasonal and inter-annual fluctuations in the terrestrial climate contrast with the much more moderate fluctuations in the marine environment. Seawater has much greater heat capacity than air; therefore temperatures change more slowly than they do on land. The higher viscosity of seawater also causes it to circulate more slowly than air.

Light and vertical gradients

Compared to the terrestrial environment, a far greater proportion of the ocean is light-limited. Due to changes in availability of light and nutrients, only the upper 50–200 metres of the oceans – the *photic zone* – will support the growth of primary producers (Figure 1.1). Light is necessary for primary production through photosynthesis, both within the water column and in the benthos. The exceptions are a few unique communities such as hydrothermal vents that are dependent on their own chemosynthetic producers. Given these exceptions, essentially no organic matter is produced within the vast unlit depths of

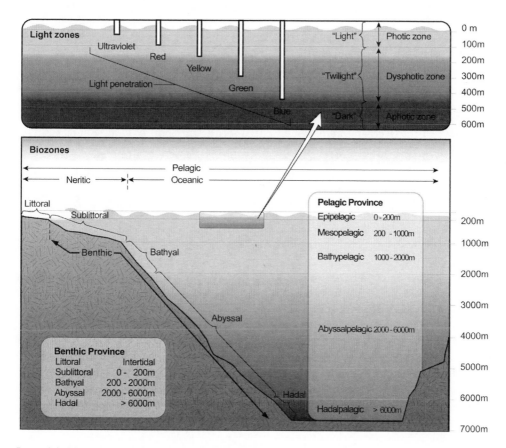

Figure 1.1 Diagram of the pelagic and benthic realms of the marine environment, showing generally recognized vertical depth and light zones

the ocean (the dysphotic and aphotic zones). In these dark areas, the entire biological economy of the seas is dependent on the flux of detrital organic matter from the productive surface layer.

Mobile and fluid nature of water

The fluid nature of the marine environment means that most marine species are widely dispersed, and individuals can be far ranging. In addition to enhancing cross fertilization and dispersal of larvae and other propagules, water motion also enhances the migration and aggregation of marine species (especially those in pelagic systems). Water also dissolves and circulates nutrients. Even marine species that can be considered static as mature benthic forms (e.g. many molluscs and seaweeds) usually have highly mobile larval or dispersive reproductive phases within the plankton of the pelagic realm. Seabed sediments are often mobile to a degree that makes land erosion cycles seem slow and mild.

Circulation differences

Although both terrestrial and aquatic environments exhibit patterns of circulation (i.e. circulation of the water in the oceans and of the atmosphere above the land and water), the two are not strictly comparable. In the oceans, the medium – water – contains the organisms themselves; aquatic organisms live in the medium, flow with it and are subject to its physics and chemistry. In contrast, the presence of organisms in the atmosphere is strictly temporary. Thus in the oceans, species and communities can be passively transported long distances, whereas on land, colonization of distant areas generally requires active migration and passive dispersal (National Research Council, 1994). Even for benthic communities, movements of the overlying waters are essential for the transport of food resources; this is generally not the function of the atmosphere for terrestrial communities.

Differences in primary production

Perhaps the most obvious biological difference between terrestrial and marine ecosystems is the type and source of primary production (Steele, 1991). In terrestrial ecosystems, primary producers (mainly vascular plants such as trees and grass) constitute the great majority of biomass, and individual organisms are often large. In contrast, the dominant primary producers in marine ecosystems – the phytoplankton – are generally microscopic and can reproduce rapidly. Consequently, they have much higher turnover rates compared to the forests or prairies of terrestrial ecosystems. In terrestrial ecosystems, the biomass of primary producers tends to be highly conserved (e.g. in woody plants). In contrast, in marine ecosystems the biomass of primary producers is rapidly processed, either by consumers or reducers (e.g. bacteria). As a consequence, marine sediments are typically much lower in organic carbon than terrestrial soils, and become progressively lower still with increasing depth. Even marine macrophytic plants (e.g. algae and seagrass angiosperms), though often perennial, have short generation times compared to terrestrial plants. No marine plants have the longevity of terrestrial gymnosperms and angiosperms. At lower latitudes the partially seawater-adapted mangroves (emergents) are recent re-invaders of marine waters.

Taxonomic differences

In the oceans, there is a high diversity of divisions among the algae, comprising many phyla of microscopic unicellular or multicellular forms (see below and Tables 1.1 to 1.4). This in fact represents a much greater diversity at the level of division (phylum) and therefore genetic diversity, than in the terrestrial environment. This is often not recognized because of our bias and perception of the overwhelming diversity of flowering plants on land – which nevertheless all belong to a single division – the Angiosperms (or Anthophyta). Bryophytes, Pteridophytes and Gymnosperms are all absent in marine waters, and Angiosperms are significantly under-represented, mainly by a few species of seagrasses and mangroves.

Of the animal phyla, all are represented in the oceans in one way or another. On land, however, several of the major phyla are completely absent, having failed to adapt to its more demanding climate. Although the greatest species-diversity undoubtedly lies within the terrestrial insects, the greatest phyletic (and therefore genetic) diversity is clearly among the flora and fauna of the oceans.

Temporal and spatial scales

Other important differences between terrestrial and marine ecosystems result from – or are associated with – the temporal and spatial scales of ecological responses to changes in the physical environment. In the oceans, these changes are related to the fundamental differences in the communities of primary producers. The high storage of biomass in terrestrial organisms and organic detritus tends to decouple biological and physical processes. In the oceans, the space and timescales of physical and biological processes are nearly coincident (at least in the pelagic realm), such that biological communities can respond rapidly to physical processes. Spatial variation and distributions of primary producers on land are largely related to topography and soil, and at the same scale, they change slowly compared to marine primary producers. On land, largely as a function of the longer generation times of larger organisms, population and community cycles may depend more on biological processes than on immediate physical processes (Steele *et al.*, 1993).

Relatively short response times to environmental perturbations

As a result of the mobility, fluidity and interconnectedness of marine waters, environmental perturbations (e.g. oil spills, the introduction of toxic substances, toxic algal blooms) can be rapidly dispersed throughout the ocean environment. Depending on the properties of a pollutant, this can readily occur in two or three dimensions.

Boundary differences

Compared to their marine counterparts, terrestrial environments have more pronounced physical boundaries between ecosystems. Especially at finer scales, it may be difficult to identify distinct boundaries in marine systems. This is due to the dynamism of marine systems, and because pelagic and benthic realms require separate consideration. This does not mean that there are no distinct marine ecosystems, but, generally, their boundaries are more 'fuzzy' or transitional. Thus the concept is less useful in marine than in freshwater or terrestrial environments.

Longitudinal diversity gradients

In addition to the latitudinal diversity gradients observed in both terrestrial and marine communities, there is also a longitudinal diversity gradient in the marine environment. Species diversity decreases from west to east in both the Atlantic and Pacific oceans. Furthermore, faunas in the Pacific Ocean (e.g. coral reefs) are on the whole more diverse than in the Atlantic Ocean (Thorne-Miller and Earle, 1999) (Figures 1.2 and 1.3). The greater diversity in the western part of the basins can be related to the eccentric circulation pattern of the two oceans, whereby the west receives waters from low latitudes (with higher species diversity) at high velocity (e.g. the Gulf Stream and Kuroshio), while the eastern part of the basin receives lower velocity water from higher latitude (with lower species diversity). The greater overall species diversity of the Pacific is generally explained in terms of its greater geological age.

Diversity at higher taxonomic levels

At higher taxonomic levels, the diversity of marine fauna is much greater than the diversity of terrestrial fauna (Ray and McCormick-Ray, 1992). All phyla of animals are represented in the oceans. Some taxa, the fishes for example, are extraordinarily diverse while others are less diverse (although in the 'lower' phyla many species remain to be described). Most marine communities are also highly patchy and variable in species composition.

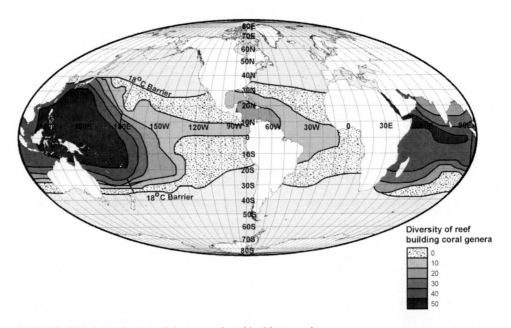

Figure 1.2 Global distribution of diversity of reef-building coral genera
Source: Adapted from Stehli *et al.* (1967)

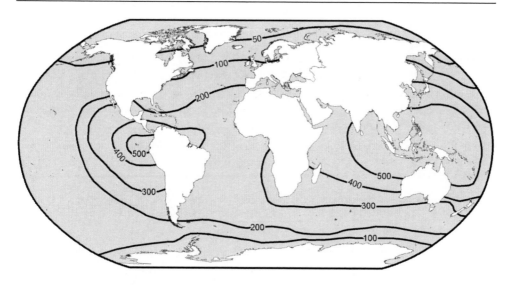

Figure 1.3 Global species richness of bivalve mollusks
Source: Adapted from Stehli *et al.* (1967)

Physiographic characteristics of the marine environment

Physiographic characteristics are those features broadly recognized as 'marine landforms' – in essence relate to the topography and substrates of the seafloor. The physiography of the shoreline and seabed determines the broad character of benthic biological communities (along with the overlying geological, biogeographical and oceanographic contexts). Primary physiographic characteristics of the marine environment are discussed below and in Box 1.4.

Horizontal divisions and bathymetry

In the horizontal dimension, the marine environment is generally regarded as divisible into several major provinces (Figure 1.1). Coastal waters comprising the *coastal zone* (generally defined as less than 30m in depth) extend seaward from the high water level and include the *littoral zone*. Near to shore are various kinds of *inlets* including estuaries, bays and coves and associated wetlands. *Estuaries* are the meeting place of freshwaters with the ocean, where the salinity is measurably diluted by freshwater runoff (see above). *Bays* and *coves* are shoreline concavities of the ocean where salinity may not be diluted (unless they are bays *within* estuaries). The coastal regions run into the *neritic province* of *sub-littoral waters*, which is the region of the ocean that lies above the *continental shelf* out to depths of 200m. Although (in some places) the edge of the continental shelf corresponds to the *Exclusive Economic Zone* (EEZ) in some locales there is no relationship between the two. The neritic province then merges into the *oceanic province* at the *shelf edge* or the *shelf break*. The oceanic province comprises those vast areas of the oceans that physically lie beyond the edge of the continental shelves and whose waters exceed depths greater than 200m.

Box 1.4 Other physiographic characteristics of the marine environment

Basin morphology: The general topography or morphometry of a region (e.g. an estuary, inlet, basin) can significantly affect the character of a coastal region.

Substrate type and particle size: Substrate particle size is a dominant influence on marine communities and often delineated into various descriptive categories, such as rock and boulders, gravel, coarse sand, fine sand or mud/silt.

Relief: The slope, rate of change of slope, rugosity or benthic complexity of an area. Areas of higher relief tend to have a greater range of habitat types and thus high biodiversity.

Geology and rock type: In the intertidal and sub-tidal zones, rock type may be sedimentary, metamorphic or igneous, which, due to various erosion rates, may affect the types of habitat and associated benthic organisms.

Depth, light and pressure

The next major division of the oceans is made in terms of depth, where a variety of terms are used to describe the habitats and their biological communities. The conventional descriptive divisions of the oceans with respect to depth, which have long been recognized, are shown for both the pelagic and benthic realms in Figure 1.1. Depth is an important factor in both the pelagic and benthic realms. In combination with temperature, salinity, light and pressure (with which it co-varies) it defines the distributions of major community types (e.g. Glenmarec, 1973).

Somewhat arbitrarily, the oceans are vertically subdivided into the *epipelagic* (down to 200m), the *mesopelagic* (200 to 1000m), the *bathypelagic* (1000 to 2000m) and the *abyssal/hadal* zones (>2000m). Similar terms are applied to the benthic realm (see Figure 1.1). In the oceans, light intensity diminishes exponentially with depth. In the vertical dimension, and fundamentally for the photosynthetic organisms, the oceans can therefore be subdivided as follows: the *euphotic* (= photic, or well lighted) zone is the region in which sufficient light penetrates to allow net photosynthesis and plant growth to occur; below this is the *dysphotic* (or poorly lighted) zone where light is still present, but its intensity is too low to support plant growth; and below the dysphotic zone, the great majority of the oceans' depths lie within the *aphotic* zone, where no light penetrates (Figure 1.1).

The *compensation point* occurs at the bottom of the euphotic zone, a depth below which the rate of photosynthesis exceeds the rate of respiration. The actual depth of the euphotic zone increases with water depth itself, from the coast towards the edge of the shelf and into oceanic waters, and it also varies at different times of the year. For example, in estuaries, the euphotic zone may be less than 2m in depth, in average coastal waters it approximates 30–50m, while in oceanic waters it may exceed 200m. In contrast, in the Arctic Ocean, the euphotic zone may exceed 100m during the spring, and suddenly decrease to only a few metres during the summer phytoplankton bloom.

Depth is also a surrogate for *pressure*. The increase in pressure with depth has a significant impact on organisms. With every 10 metres of depth, the water pressure increases by approximately one atmosphere (with the greatest change from 0 to 1 atmosphere occurring in the top 10 metres). Additional physical, chemical and biological changes lead to a decrease of dissolved oxygen and increase of dissolved carbon dioxide (see below). Organisms that live in the deeper regions of the oceans are adapted to physical conditions of high pressure, low temperatures and dilute resource concentrations and rarely move into the epipelagic region.

Temperature also decreases with depth, from ambient surface values to a nearly constant 0–4°C in deepest oceanic waters. Conversely, salinity typically increases with depth. Concentrations of particulate organic carbon (the detrital flux from the euphotic zone) also decrease exponentially with depth, while oxygen concentrations decline and carbon dioxide concentrations increase with depth. Depth is therefore an index of a variety of concurrently changing physical and chemical conditions, which collectively influence the nature of biological communities.

Oceanographic characteristics of the marine environment

Oceanographic structures

Temperature

Temperature is often regarded as the most critical factor governing the distribution and behaviour of organisms on earth (Gunter, 1957). There is considerable scientific literature on the effects of temperature on the distribution, physiology and behaviour of aquatic organisms. All marine organisms show some pattern of limitation of their distribution according to temperature, and very few species have geographical ranges covering the entire range of observed aquatic temperatures.

Temperatures are relatively stable within the oceans, and there is far less seasonal fluctuation than in terrestrial systems. Seasonal variation in temperature is generally low at both high and low latitudes, but fluctuations may be higher at mid-latitudes – for example from 0 to +20°C. The temperature in the oceans varies from -2°C in polar oceans to +32°C in tropical waters. The lower temperature is set by the freezing point of seawater. Due to its salt content, the freezing point of seawater is depressed below that of freshwater (by definition -0°C) and the temperature of freezing is proportional to the salinity. The temperature of maximum density lies below the freezing point for all open ocean seawaters; it is not +4°C as for freshwaters. This means that seawater of a given salinity will continue to sink below warmer waters of the same salinity as it cools.

The oceans are vertically thermally stratified by a 'permanent' thermocline. To a first approximation, ocean waters can be divided into an upper wind-mixed layer – from the surface to a depth of 600 to 1000m. Here temperatures range from +8°C to +32°C. Temperatures of the lower layers – from 600m to the bottom – are generally well below those at the surface, averaging between 0°C and +4°C. The temperature of most marine waters therefore lies between 0°C and +4°C. Water temperatures in the deeper layers are also more constant than in surface waters, and may be significantly influenced by ocean currents. Bottom topography also has an influence on water temperature, with pockets of warmer (but more saline) waters often being trapped for periods of time in large basins. Deeper colder waters can also be transported towards the surface in upwellings.

With its impact on growth and its complex relations to biodiversity and nutrient supply (see below), it is no wonder, then, that temperature is a major determinant of marine communities.

Ice cover and scour

Ice cover may be permanent, seasonal or absent. Its extent and development is a substantial influence on community types over broad geographical areas. Not surprisingly, the presence of permanent ice greatly restricts marine productivity. Through a variety of mechanisms, seasonal ice has a major impact in seasonally eliminating or enhancing production. The physical effects of ice scour and glaciers have the greatest impact on shallow water marine communities, especially in intertidal regions, primarily by reducing community diversity or by restricting some organisms to crevices. Except for inshore areas – or in polar oceans where effects of rafted ice can reach to considerable depths – ice does not have an impact on the sub-tidal benthos.

Temperature gradients and anomalies

Temperature anomalies are well documented as being associated with several types of regions. Upwelling regions, carrying nutrient-rich waters to the surface, also carry a signature of surface temperature lower than surrounding waters. While the geographic location of upwelling events is predictable, the timing may not be, and may depend on unpredictable meteorological events. Upwellings may be seasonally or annually variable in development; the water column is typically vertically mixed, and stratification is weak or absent.

Salinity

The salt content of virtually all parts of the oceans lies between 32 and 39psu (*practical salinity units* or parts per thousand – also written as ‰, or ppt) and the chemical composition of the principal ions of seawater remains virtually constant. This is because the rate of input and export of salts to and from the oceans is low compared to the rate of global mixing. The salinity of surface waters around the globe correlates strongly to the difference between latitudinal variations in precipitation and evaporation, and to the rate of mixing with subsurface waters. Salinity variations with depth also contribute to vertical stratification and stability of the water column (see below).

Variations in salt content over the range of 32 to 39psu have little discernible effect on marine organisms. However, the majority of marine species are *stenohaline* (adapted to constant salinity), and are intolerant of salinity changes greater than this. Although, in the past, considerable attention has been devoted to the relationships between distributions of organisms and salinity, other factors such as temperature and the movements of water masses dictate the distributions of marine organisms.

Only in estuaries, estuarine bays or sheltered mountainous coastal regions experiencing high rainfall – where salinity varies from close to zero to over 33psu does salinity have major effects on the distribution of aquatic species. This is because of its significance in osmotic and ionic regulation. The majority of aquatic species are adapted either to a life in freshwaters at low but relatively constant salt content, or to the much higher, nearly constant salinities of the oceans. Relatively few species (*euryhaline*) are well adapted to life in estuaries at intermediate and fluctuating salt content. In estuaries, salinity exhibits complex behaviour, both physically and in its impact on the distribution and abundance of organisms (see below).

Composition of seawater and nutrients

Seawater is a complex solution containing virtually every element of the periodic table. This is a testament to the high dissolving power of water. Despite variations in total salt content (the salinity), the composition of seawater (that is the ratio of the major elements to one another) remains virtually constant. This is occasioned by the fact that the concentrations of most elements are high relative to their rate of use by the biota. The exception to the rule is for elements (or ionic compounds) called nutrients.

Nutrients in marine environments are those inorganic substances (PO_4, NO_3, NH_4, Fe and Si) that are required for, and can control the rate of, aquatic plant growth. Only nutrients within the euphotic zone are significant to our understanding of processes for marine management purposes. Nutrient concentrations below the euphotic are much higher than within it. Not until these nutrients reach the surface waters in upwellings can they be exploited by primary producers.

Nutrients in aquatic environments are derived from the atmosphere, from land drainage, by recirculation of deep waters to the surface and from internal regeneration. Various components of the marine environment obtain their nutrients by different processes and from different predominant sources.

In the open-water pelagic realm, nutrient inputs are predominantly a function of the mixing regime and/or internal regeneration by the food web. Atmospheric inputs are assumed to be uniform over macroclimatic regions. Terrestrial inputs to the oceans are low compared to internal recycling, except in estuaries and the nearshore coastal zone. Except in the immediate coastal zone, internal regeneration and recirculation of nutrients within the ocean thus far outweigh the significance of external inputs.

A major (internal) source of nutrients to coastal areas is upwelling or entrainment of subsurface water. While the geographic location of these events is largely predictable, the timing may depend on unpredictable meteorological events. Such water mixing regimes may be evident from local low-temperature anomalies.

Within water masses, the nutrient regime is a function of whether the water column is stratified or mixed, as described by the stratification parameter (see below).

Oxygen and other dissolved gases

Within the neritic and pelagic realms of the oceans (and larger lakes), dissolved gases are normally at or close to saturation in surface waters. This is because physical processes that cause exchange of gases between water and atmosphere typically overshadow the biological processes of photosynthesis and respiration in magnitude. The wide diel variations of gas content, typical of smaller bodies of water, are generally not observed. Even below a thermocline, oxygen rarely becomes naturally seasonally depleted within the water column. The exceptions to this are stratified inlets and estuaries where there are major inputs of organic matter and the occurrence of summer stratification. In oceanic tropical and subtropical waters, an oxygen minimum layer develops at a depth of between 600 and 1000m in both the Atlantic and Pacific oceans (more highly developed in the latter).

Low oxygen concentrations – often engendered by human actions – are prompting considerable concern for several regions around the globe. These anoxic areas (also termed 'dead zones') are caused by the inputs of dissolved nutrients and suspended solids from land runoff (Diaz and Rosenberg, 2008). Perhaps the largest of these areas lies in the Gulf of

Mexico. Here, large areas to the south and west of the Mississippi River have become anoxic following algal blooms caused by terrestrial and freshwater nutrient inputs into coastal waters.

Water masses and density

The combination of temperature and salinity determines the density of seawater. Higher temperature reduces density and higher salinity increases it. Outside of estuarine environments, density itself is of little direct biological consequence in the oceans because it varies only within narrow limits. Density is, however, of major global significance for physical oceanographic processes, as a determinant of ocean thermohaline circulation (see below).

Certain 'usual' or 'habitual' combinations of temperature and salinity (defining the water masses) persistently occur, and the distribution of these combinations can be traced over wide swaths of the oceans, frequently for thousands of square kilometres. Water masses can in some respects be considered as analogues of the major climatic regions of terrestrial environments, and they define the extent and influence of major ocean currents (Figures 1.4 and 1.5). Both temperature and salinity need to be quantified in order to define the origins, movements and extent of the distribution of water masses. A water mass will seasonally change its temperature (salinity is far less seasonally variable) because of atmospheric interactions. Temperature by itself, therefore, is not ideal as a measure or descriptor of water masses or origin of water masses.

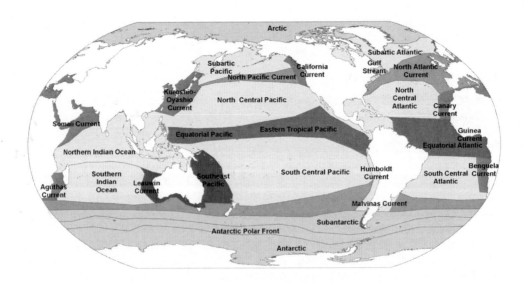

Figure 1.4 Global map of pelagic provinces
Source: UNESCO (2009)

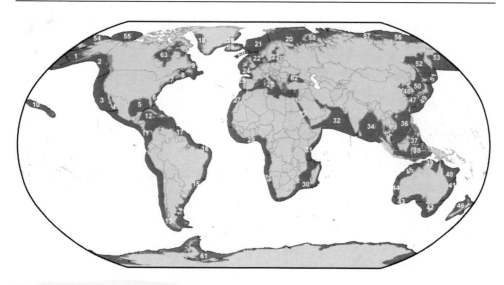

Figure 1.5 Global map of large marine ecosystems

Note: Large marine ecosystems (LMEs) are ocean areas approximately 200,000 km² or greater, adjacent to the continents in coastal waters where primary productivity is generally higher than in open ocean areas. The 64 LMEs are delineated based on bathymetry, hydrography, productivity, trophic relationships. LME names are as follows:

1 East Bering Sea	23 Baltic Sea	44 West-Central Australian Shelf
2 Gulf of Alaska	24 Celtic-Biscay Shelf	45 Northwest Australian Shelf
3 California Current	25 Iberian Coastal	46 New Zealand Shelf
4 Gulf of California	26 Mediterranean Sea	47 East China Sea
5 Gulf of Mexico	27 Canary Current	48 Yellow Sea
6 Southeast U.S. Continental Shelf	28 Guinea Current	49 Kuroshio Current
7 Northeast U.S. Continental Shelf	29 Benguela Current	50 Sea of Japan
8 Scotian Shelf	30 Agulhas Current	51 Oyashio Current
9 Newfoundland-Labrador Shelf	31 Somali Coastal Current	52 Okhotsk Sea
10 Insular Pacific-Hawaiian	32 Arabian Sea	53 West Bering Sea
11 Pacific Central-American Coastal	33 Red Sea	54 Chukchi Sea
12 Caribbean Sea	34 Bay of Bengal	55 Beaufort Sea
13 Humboldt Current	35 Gulf of Thailand	56 East Siberian Sea
14 Patagonian Shelf	36 South China Sea	57 Laptev Sea
15 South Brazil Shelf	37 Sulu-Celebes Sea	58 Kara Sea
16 East Brazil Shelf	38 Indonesian Sea	59 Iceland Shelf
17 North Brazil Shelf	39 North Australian Shelf	60 Faroe Plateau
18 West Greenland Shelf	40 Northeast Australian Shelf –	61 Antarctic
19 East Greenland Shelf	Great Barrier Reef	62 Black Sea
20 Barents Sea	41 East-Central Australian Shelf	63 Hudson Bay
21 Norwegian Shelf	42 Southeast Australian Shelf	64 Arctic Ocean
22 North Sea	43 Southwest Australian Shelf	

Source: Sherman *et al.* (1992); lme.noaa.gov

Oceanographic processes

Water motions

Natural bodies of water show many types of movements at all spatial and temporal scales. Water motion is an essential feature of the oceans, and is essential for all life. It follows that a proper perspective on water movements is imperative to understanding aquatic ecosystem function. Ultimately, three 'forces' cause water motion on the earth: the sun's radiation (causing heating and evaporation/precipitation); the earth's rotation on its own axis (a modifying action rather than a true force); and the gravitational effects of the sun and moon. A minor fourth factor is geological (tectonic), which causes tsunamis. Proximately, water motion may be generated by winds, tides or density differences.

Water motion sustains life by replenishing and stimulating the production of resources for organisms of every trophic level. Motion provides the passive transport for the dispersal of organisms (or their larvae) in the broad-scale or in more local patterns of ocean connectivity. The junctions between the gyral systems and major ocean currents contribute to the barriers that separate species and higher taxa. Essentially, these intersections define the major biogeographic provinces and regions of the oceans.

Surface ocean circulation

Water motion in the surface waters of the oceans (down to a depth of about 600m) is driven primarily by winds and tides. Below approximately 600m, water motion and horizontal currents are much slower and driven primarily by density differences. The frequency and strength of water movements at the ocean's surface have profound effects on biological communities.

At the global scale, the basic patterns of surface ocean circulation are the same in the Pacific and Atlantic oceans. These patterns consist of sets of currents which together comprise a sub-polar gyre and a subtropical gyre (Figure 1.6). The counterparts of each of these currents in the North Atlantic and North Pacific, and in the South Atlantic and South Pacific are very clear. For example, the counterparts of the sub-polar Labrador current and the subtropic Gulf Stream in the North Atlantic Ocean are the Kurile and Kuroshio currents in the North Pacific Ocean. In all oceans, circulation is eccentric resulting in a westward intensified current. It is the effect of the earth's rotation (described in terms of the

Figure 1.6 Surface ocean circulation and the major surface ocean currents

Coriolis 'force' or parameter) that produces these various eccentric ocean gyres with their westward, intensified currents.

Each gyre revolves around a central but eccentric hub, simply referred to as a 'central gyre'. The only gyre named specifically is the Sargasso Sea of the North Atlantic. Each subtropical gyre encloses a relatively warm, salty, clear blue sea, poor in nutrients and of low productivity. Production is dependent primarily on internally recycled NH_4. These are essentially the deep, desert areas of the ocean. Although the best-known example is the Sargasso Sea, other analogous central gyre areas are found in the North and South Pacific oceans.

In the tropical areas of the oceans, both a north equatorial current and a south equatorial current can be recognized flowing from east to west. In the Atlantic and Pacific oceans, an equatorial countercurrent is found between these two, running from west to east. The Pacific equatorial countercurrent is 8000 nautical miles (nmi) in length and about 250nmi in width. The Atlantic equatorial countercurrent is more variable; it becomes stronger in summer, extending its influence westward towards South America.

Of all the ocean currents, the most pronounced and persistent is the Antarctic circumpolar current (West Wind Drift). It subsumes the sub-polar gyres of the southern oceans and completely circles the globe running from west to east. Meanwhile, the most variable surface ocean currents are found in the Indian Ocean. Here, even the major ocean currents show great variability in velocity, direction and latitudinal and longitudinal position. For example, the Somali current flows south in winter, but north in summer at high velocity. The north equatorial current reverses direction under monsoon influence; indeed, while flowing east in summer it is dubbed the monsoon current. The equatorial countercurrent is only observed in winter, and disappears in summer when the monsoon current occurs. The major force that causes these changes is the variable winds of the monsoon season, clearly showing the dependence of surface ocean circulation on the atmospheric winds.

Oceanic convergences and divergences

Where the various oceanic gyral systems meet, corresponding to regions of reversal of major atmospheric wind systems, water masses either converge or diverge. That is, at the surface, the circulation pattern induces a series of convergences and divergences where water either descends below the surface, or is brought up from below the surface. Where oceanic water masses meet and descend they are known as *convergences*; where they rise to the surface and disperse they are known as *divergences*. They are present throughout the oceans, but typically restricted to oceanic waters and more pronounced in the southern oceans. These are typically regions of much higher than average production, at all trophic levels. However, these regions are much more clearly defined in the tropics, subtropics and southern oceans than at northern latitudes.

It is important to realize the differences in the character of oceanic convergences and divergences. Convergences are where surface waters meet and biota may accumulate – thus biomass is increased. They may be areas of rich life that has physically accumulated – that is, it has been advected to the region, not produced there. Rich fish feeding areas may nevertheless correspond to convergence zones. A divergence is not usually so well defined, though it carries a signature of reduced surface temperature. At divergences, nutrient-rich water is brought to the surface stimulating new production. Perhaps the best-known example is the Antarctic divergence where extremely high levels of primary planktonic production are known to occur.

Coastal upwelling

Production is also high in regions of near-coastal upwelling and divergence, where production is dependant primarily on inputs of NO_3 from deeper waters. If the coast lies to the left of the wind direction, then surface water is transported to the right in the northern hemisphere. This is a result of deflection cause by the Coriolis parameter. This necessitates a deep replacement current moving onshore and upwards – that is, upwelling. This upwelling water brings with it much higher concentrations of nutrients that have been biologically regenerated in deep water. The Coriolis 'force' is thus directly linked to areas of high production. Examples include the Grand Banks of Newfoundland where the Gulf Stream has moved offshore.

In the southern hemisphere, the situation is reversed. The best known example here is the northward flowing Peru–Chile current that flows into the Humbolt current (Figure 1.6). It progresses along the west coast of South America, leads to upwelling of cold deep nutrient-rich water that, in turn, stimulates planktonic production. Coastal upwellings may be relatively constant, periodically interrupted (e.g. as in the El Niño Southern Oscillation System off Peru and Chile), or more episodic as in the case of the Benguela current off southwestern Africa.

Deep ocean circulation

Deep ocean currents generally move at a slower pace than surface waters, and are driven by small differences in density caused by both temperature and salinity variations (thermohaline circulation). Formation of the deep and bottom waters of the oceans occurs only in two ways. Saline water is transported to high latitudes and is subsequently cooled without significant dilution and then sinks. Cold surface water freezes and, as it does, it forms ice (above the eutectic point where solvent alone freezes and eliminates the salt solutes), so that the salinity of the remaining water is increased and consequently sinks. The first process is of greatest importance in the North Atlantic, between Iceland and Norway. The second is around Antarctica primarily in the Weddell Sea region. These are the only two regions of the world where deep and bottom waters are formed in any quantities.

Water sinking in the North Atlantic makes its way south towards the Antarctic. Here it meets water sinking from the Weddell Sea where it is re-cooled and moves off to the Indian Ocean, the North Pacific, and then the South Pacific. As deduced from carbon-14 dating techniques, this entire journey takes some 1500 years. Despite the long time that these waters have been out of contact with the atmosphere, they still retain significant levels of oxygen. This indicates the sparse nature of life in the depths of the oceans, and its low metabolic rates at low temperatures.

Tidal amplitude and currents

Tides are caused by the gravitational effects of the moon and the sun. In the absence of land, the world's oceans would experience semi-diurnal tides of a periodicity of 12.25 and 24.50 hours, and a spring-neap cycle of variable amplitude depending on conjunction of moon and sun. Regionally, both tidal periodicity and tidal amplitude depend on the physical dimensions of the ocean basin. Some locations in the world experience only one tide per day (diurnal) and others up to four per day (quadridiurnal). Which type occurs depends on the location of the amphidromic points around which tides circulate, and which of the sun's and moon's gravitational influences most nearly reinforce the natural period of oscillation of the regional ocean basin.

Tidal amplitude is a consequence of the natural period of oscillation of a regional ocean basin, sea or bay. Canada's Bay of Fundy, and its extension, the Minas Basin, experience the highest tides in the world because the natural period of oscillation of the Bay of Fundy is approximately 13 hours. This corresponds closely to the moon's gravitational period (12.25 hours), creating a natural resonance with the waters in the Bay of Fundy that induces spring tide amplitudes up to 16m. In contrast, the natural periods of oscillation of the basins in the Mediterranean Sea are about 9.30 hours. Here, there is interference between gravitational period and natural period of oscillation. The result is tides of only a few centimetres.

Currents in the oceans can originate from several processes, including general ocean circulation, density-driven origins of water masses, tides and winds. In coastal waters, currents due to tides are generally the most predictable and best documented. Tidal amplitude, together with bottom slope, determines the vertical and horizontal extent of the intertidal zone. Tidal currents and bottom slope determine, in part, the substrate characteristics that, in turn, determine the extent and type of benthic communities. A major distinction between substrate types and associated benthic communities can be recognized as follows: erosional areas, where current speeds are high and material is removed from the substrate resulting in hard bottoms (rock, boulders, gravel etc.); and depositional areas, where water movement is sufficiently slow that particles sediment out, resulting in soft bottoms (mud, silt etc.).

Stratification and mixing regime

All temperate marine and fresh waters typically mix vertically in spring and autumn, and may or may not become stratified in the summer. The seasonal cycle and timing of stratification has major implications for the productivity regime of a region. Two forces cause near-surface water mixing: tides and winds. The effects of wind mixing are difficult to formulate, but they are more uniformly distributed than those of tides. In the oceans, when the water column depth exceeds ~50m, winds cannot prevent the water column from stratifying during the annual heating cycle; however, tidal action can. In freshwaters, tides are insignificant, and temperate lakes over approximately 50m in depth always stratify.

Polar waters typically mix in early summer, and then become stratified. Mainly, this results from vertical salinity differences. In sub-Arctic waters, stratification can be due to a combination of salinity and temperature effects. In temperate and subtropical waters, stratification of the water column is primarily due to temperature; however, it may also result from a change in salinity. In coastal waters this may be associated with seasonal freshwater runoff from land, and it can also be related to the cycle of primary production.

Tsunamis, storm surges, hurricanes and water spouts

These massively destructive atmospheric events are now more-or-less predictable in terms of location of impact and scale of effect, at least on the timescale of days. Although considerable local destruction to coastal marine communities may result from these phenomena, over a period of years (for most intertidal communities) to decades (for coral reefs, mangroves) the local biological communities are generally reestablished. Such phenomena, although with profound local effects in shaping marine communities, should be thought of as resetting the ecological 'clock of succession'.

The marine biological environment

The flora and fauna of the oceans range in taxonomy and size from minute marine viruses to large marine mammals and are generally organized into a series of more-or-less definable communities. Tables 1.1 to 1.4 summarize the major animal and plant divisions in the oceans, freshwaters and on land. Some key differences can be seen: first, the lower groups of bacteria, fungi and protists are widely distributed among all environments; however, the lower plants (the algae, both microscopic single-cell forms and multicellular macrophytic forms) are found in both marine and freshwaters, but not on land. Second, the higher multi-cellular plants, with few exceptions, are limited to the terrestrial environment. The great majority of phyla of multicellular animals are found only in the marine environment, with fewer phyla in freshwaters and fewer still on land. Third, the great majority of phyla of multicellular animals are found only in the marine environment, with fewer phyla in fresh-waters and fewer still on land.

Tables 1.5 and 1.6 present two other ways in which to examine the biology of the marine environment. Table 1.5 classifies species according to their ecological taxonomic type (also called 'ecospecies' or 'trophospecies'), which allows similar species to be presented in the context of their habitats and life cycle. Table 1.6 groups the dominant biological commu-nity types in their relevant biological realms and major community units.

This section discusses the major biological community types within the pelagic and benthic realms.

Marine biology of the pelagic realm

The pelagic realm supports a diversity of organisms and life forms – collectively, the plank-ton and nekton – that have no dimensional or community counterparts on land. Pelagic

Table 1.1 The major higher taxonomic groups of marine, freshwater and terrestrial flora and fauna: predominantly unicellular forms

Kingdom or phylum	Marine	Freshwater	Terrestrial
Archaea. Two main phyla of 'bacteria'	Y	Y	Y
Eubacteria. Several phyla of true bacteria	Y	Y	Y
Fungi. Four to six major phyla	Y	Y	Y
'Slime moulds'. A polyphyletic group	Y	Y	
Protista. Unicellular (and some multicellular) eukaryotes 30–40 separate phyla	Y	Y	

Note: The terms 'phylum' and 'division' (for animals and plants respectively) are equivalent. The taxonomy of all these groups is in a state of flux.

Table 1.2 The major higher taxonomic groups of marine, freshwater and terrestrial flora and fauna: the major divisions of algae

Division	Marine	Freshwater	Terrestrial
Bacilliarophyta	Y	Y	
Charophyta	Y	Y	
Chloroarachniophyta	Y	Y	
Chlorophyta	Y	Y	
Chrysophyta	Y	Y	
Cryptomonads	Y	Y	
Cryptophyta	Y	Y	
Cyanophyta	Y	Y	
Dinophyta	Y	Y	
Euglenophyta	Y	Y	
Eustigmatophyta	Y	Y	
Glaucophyta	Y	Y	
Haptophyta	Y	Y	
Phaeophyta	Y	Y	
Prasinophyta	Y	Y	
Rhodophyta	Y	Y	
Rhaphidophyta	Y	Y	
Xanthophyta	Y	Y	

Note: The Chlorophyta, Phaeophyta and Rhodophyta are the major groups of macrophytic algae. Most of the other groups in this table are either unicellular forms or undifferentiated multicellular forms.

Table 1.3 The major higher taxonomic groups of marine, freshwater and terrestrial flora and fauna: the major divisions of 'higher' plants

Division	Marine	Freshwater	Terrestrial
Anthocerotophyta		Y	Y
Bryophyta		Y	Y
Marchantiophyta			Y
Lycopodiophyta			Y
Pteridophyta			Y
Pteridospermatophyta			Y
Coniferophyta			Y
Cycadophyta			Y
Ginkgophyta			Y
Gnetophyta			Y
Anthophyta (or Magnoliophyta)			Y

Note: These are the familiar, almost exclusively terrestrial macrophytic plants, the mosses, liverworts, ferns, conifers and flowering plants

organisms inhabit a fully three-dimensional realm; most of its species inhabit it permanently (among the planktonic organisms – the holoplankton) or some temporarily as larvae (e.g. the meroplankton). Within the realm conditions vary dramatically with depth, changing the array of organisms present and community structure. Pelagic species may also occupy different trophic levels and different stages of development. The 'fringing communities' of the benthic realm – intertidal and sub-tidal communities within the euphotic zone – which

Table 1.4 The major higher taxonomic groups of marine, freshwater and terrestrial flora and fauna: the major phyla of multicellular animals

Phylum	Marine	Freshwater	Terrestrial
Acanthocephala	Y	Y	
Acoelomorpha	Y		
Annelida	Y	Y	Y
Arthropoda	Y	Y	Y
Brachiopoda	Y	Y	
Bryozoa	Y	Y	
Chaetognatha	Y		
Chordata	Y	Y	Y
Cnidaria	Y	Y	
Ctenophora	Y		
Cycliophora	Y		
Echinodermata	Y		
Echiura	Y		
Entoprocta	Y		
Gastrotricha	Y	Y	
Gnathostomulida	Y		
Hemichordata	Y		
Kinorhyncha	Y		
Loricifera	Y		
Micrognathozoa		Y	
Mollusca	Y	Y	Y
Nematoda	Y	Y	Y
Nematomorpha		Y	
Nemertea	Y	Y	Y
Onychophora			Y
Orthonectida	Y		
Phoronida	Y		
Placozoa	Y		
Platyhelminthes	Y	Y	
Porifera	Y	Y	
Priapulida	Y		
Rhombozoa	Y		
Rotifera	Y	Y	
Sipuncula	Y		
Tardigrada	Y		Y
Xenoturbellida	Y		

contain various photosynthetic plants, could be considered the functional equivalent of terrestrial communities, where species either live on the ground or are dependent on it for habitat and resources. However, a significant difference between the benthic realm and terrestrial environments is that most of the benthic realm lies below the photic zone and is therefore devoid of its own primary producers. It must therefore rely on detrital resources settling from the photic zone and hence has no terrestrial community counterpart. To a first approximation, terrestrial and benthic ecosystems can be considered as two dimensional. Certainly at larger spatial scales and for management purposes, they are typically considered in this way. Marine ecosystems, therefore, can be considered as comprising five dimensions, two of the benthos and three of the pelagic realm.

Table 1.5 Ecological taxonomic types or 'ecospecies'

VIRUSES
ARCHEBACTERIA
EUBACTERIA
FUNGI
PLANTS

Ecological type	Habitat	Lifecycle/reproduction
Microphytic algae planktonic	Water column	Asexual fission; some sexual
Microphytic algae benthic	Hard substrates encrusting	Asexual fission; some sexual
Microphytic algae benthic	Soft substrates adherent	Asexual fission; some sexual
Microphytic algae – symbiotic	E.g. corals	Symbionts with coral
Macrophytic algae benthic	Hard substrates attached	Asexual; Sexual motile spores
Angiosperms mangroves	Estuaries, bays	Sexual; Dispersing Propagules
Angiosperms seagrasses	Sandy bottom	Asexual; Sexual seed dispersal
Angiosperms saltmarsh	Saltmarshes	Asexual; Sexual seed dispersal
PROTISTS		
Protists planktonic	Water column	Asexual fission; some sexual
Protists benthic	Encrusting or burrowing	Asexual fission; some sexual
ANIMALS		
Marine mammals	Pelagic water column migratory	Sexual; live birth
Marine mammals	Shore referenced	Sexual; live birth
Marine birds	Oceanic	Sexual; eggs nesting
Marine birds	Shore referenced	Sexual; eggs nesting
Marine reptiles	Pelagic water column migratory and shore referenced	Sexual; eggs nesting
Fish	Pelagic highly migratory	Sexual; free larvae
Fish	Anadromous	Sexual; benthic larvae
Fish	Catadromous	Sexual; free larvae
Fish	Pelagic regionally defined	Sexual; larvae
Fish	Demersal regionally defined	Sexual; larvae
Fish	Territorial	Sexual; protected larvae
Nekton	Water column regionally defined	Sexual; free larvae
Zooplankton	Water column regionally defined	Sexual; free larvae
Zoobenthos	Regionally migratory/mobile	Sexual; free larvae
Zoobenthos corals	Epibenthic attached	Asexual fragmentation; sexual planula larva
Zoobenthos	Epibenthic attached	Sexual; free larvae or 'direct'
Zoobenthos	Burrowing infauna	Sexual; free larvae or 'direct'

Not only do pelagic species demonstrate a great diversity of body size, form and taxonomic type, they evince a high diversity of both plant and animal phyla (Table 1.1). In fact, a single plankton collection can contain a higher diversity of either plant divisions or animal phyla (depending upon the sampling apparatus) than any other single collection on the planet. Nevertheless, at the species level, diversity in the pelagic realm tends to be lower than in the benthos (Gray, 1997), perhaps as a function of greater environmental heterogeneity in the benthos.

For descriptive purposes the pelagic realm can be subdivided into the *planktonic* community of smaller organisms, and the *nektonic* assemblage of larger, more mobile species. The

Table 1.6 Summary of some major marine communities

Biological realm	Major community 'units'	Community type
Fringing Communities	Estuaries Non biogenic communities	 Rocky shores Tidal flats Sub-tidal soft bottoms Shorebirds
	Biogenic communities	Biogenic reefs Corals Kelp bed macrophytic algae Mangroves Seagrasses Salt marshes
Pelagic	Pelagic communities	Phytoplankton Holozooplankton Meroplankton Fish Diadromous fish Seabirds/waterfowl Marine reptiles Marine mammals
Benthic	Benthic communities	Demersal fish Territorial fish Epi-benthic zoobenthos In fauna burrowing zoobenthos Cold seeps
Deep Sea		Deep sea corals Sponge beds Soft corals (Gorgonians)
	Seamounts Hydrothermal vents Abyssal plains Trenches	

plankton comprises all those aquatic organisms that are 'suspended' freely in the water mass. They drift passively in the water currents, and their powers of locomotion are insufficient to enable them to move against the horizontal motion of the water. However, many planktonic species may make extensive vertical movements. One specialized component of the plankton is the *neuston* that comprises a diversity of organisms inhabiting the immediate water surface.

The nekton is composed of those actively swimming consumers that constitute the middle and upper trophic levels of marine ecosystems. Among these, the dominant organisms are many types of fish (e.g. herring, tuna). They increase in size up to, and including, marine reptiles and mammals. Some species of fish, reptiles and marine mammals make extensive seasonal migrations, while other fish species make regional movements. The smaller components of the nekton, which may also make seasonal migrations, comprise invertebrates such as euphausiids. The chief feature separating nekton from plankton is their relative mobility, and their ability to swim or migrate contrary to ocean currents. The distinction between plankton and nekton is by no means absolute. For example, while being nektonic as adults, fish larvae are members of the plankton.

Because of their ubiquity, plankton has generally not been considered worthy of management attention until recently. Nevertheless, an appreciation of their global significance is warranted. As in other communities, the plankton comprises primary producers, primary and secondary consumers and decomposers. The primary producers, the *phytoplankton*, are unicellular or chain-forming microalgae, ranging from <1μm to >1mm in diameter. Functionally, they are often divided into *picoplankton* (<2μm), *nannoplankton* (2–20μm) and *netplankton* (>20μm). Taxonomically, the phytoplankton comprise some 12–18+ divisions of plant life (Table 1.1), a far greater phyletic diversity than on land. These are the dominant photosynthetic organisms on the planet, yet they remain virtually unknown by most humans.

Within the oceans themselves the phytoplankton are responsible for >95 per cent of the annual primary production. In temperate waters, the seasonal cycle of primary production within the plankton is typically highly pulsed. It typically peaks in the spring with an outburst of diatom growth, as the water column becomes *stratified*. Seasonal cycles are generally more extreme at high latitudes (e.g. in the Arctic) and in coastal waters. Cycles are less modulated in the tropics and in offshore oceanic waters. Within the pelagic realm, the rate of primary production is controlled by the availability of light and nutrients (typically: nitrogen, phosphorus, iron, silica).

Net carbon fixation can only occur within the euphotic zone, where the rate of photosynthesis exceeds the rate of respiration. The depth of this zone varies greatly, even among natural aquatic environments. In those affected by human activities, the depth of the euphotic zone is typically decreased, due to elevated plant production (*eutrophication*). Some examples of the depths of euphotic zones would be: smaller lakes 5–10m; larger lakes 10–30m; estuaries 5–15m; coastal and neritic waters 10–60m; oceanic waters, up to 200m. The biomass and productivity of phytoplankton can be estimated *in situ* from the uptake rate of radioisotopes such as ^{14}C, or from remotely sensed ocean colour and water clarity (e.g. Bukata *et al.*, 1995).

The presence of floating ice at high latitudes also affords new habitat for microalgae. A major new community of primary producers – the *epontic* or ice-algae community develops seasonally in the bottom few centimetres of ice.

The zooplankton comprises a wide diversity of phyla, body forms and modes of life. *Holoplankton*, the predominant members, spend their entire lifecycle within the water column. They are distinguished from the *meroplankton*, the larval forms of benthos, which recolonize the benthos after a period in the plankton. A category of *ichthyoplankton*, comprising the larvae of many fish species, is also often recognized. Individuals may be herbivores, omnivores or strict carnivores, or function ontogenically at different 'trophic levels'. Every phylum of the animal world is represented somewhere in the marine plankton; however, they typically do not exhibit a high diversity of species, compared to benthic or terrestrial communities. *Zooplankton* are the most abundant animals (metazoa) in the world, yet like phytoplankton, they are largely unknown to the public. The most abundant zooplankton are the copepods. Their nauplius larva is the most common and abundant type of animal body plan on our planet.

Zooplankton range in size from <50μm (nauplii) to over 2m (lion's mane jellyfish *Cyanea spp*). Despite being at the mercy of horizontal water motions, many zooplankton can make extensive vertical migrations. The extent of these migrations generally increases with increasing body size and with depth. In neritic and oceanic waters, this typically leads to larger species inhabiting deeper levels of the water column and overlapping in their migrations with smaller species living closer to the surface.

The phytoplankton are grazed by organisms of two major food webs. The *classical food web* consists of the larger phytoplankton (nanno- and net-plankton), grazed by zooplankton, which, in turn, are consumed by the nekton (euphausiids, fish, marine mammals etc.). This food web has been recognized for many years since the major commercial fisheries of the world depend on it. The classical food web, fuelled largely by nutrients from seasonally mixed or upwelling waters, is most highly developed in coastal (neritic) regions; it becomes progressively diluted in oceanic waters. A second pelagic food web, the *microbial food web*, has more recently been recognized (e.g. Azam, 1998), though in evolutionary terms it is much more ancient. It is spatially less variable from neritic to oceanic waters, and ecologically it dominates in oceanic waters. It consists of the smaller primary producers (pico- and nanno-plankton), bacteria, flagellates and ciliates. The ubiquitous nauplii may be an important interface between these two food webs (Turner and Roff, 1993).

The planktonic community is preeminently a region where organic production is directly available to grazing consumers. Since there is no equivalent to the resistant woody tissues of terrestrial plants a large proportion of the production of the phytoplankton is consumed immediately. Ungrazed organic material and fecal pellets sink towards the ocean floor. Depending on depth, seasonal timing of cycles and so on, this material may be efficiently recycled several times within the water column before the residue finally becomes available to the zoobenthos.

Both pelagic and benthic communities are strongly related to depth in the water column, as a function of light and pressure (Figure 1.1). The epi-pelagic lies within the uppermost part of the water column, with characteristic communities of plankton and fish. Organisms that live in the meso-pelagic and bathy-pelagic regions are adapted to these physical conditions and dilute resource concentrations, having evolved in response to the enduring factor of pressure. However, some fish and invertebrates in the bathy-pelagic zone (such as the sergestid prawns and myctophid fishes) migrate into the upper two layers at night. Where bathy-pelagic zones are occupied by specially adapted fish or invertebrates, the actual populations may be low (except in areas where food is more abundant such as around sea mounts). Vertical changes in these biological communities with depth are real, but especially in the pelagic realm, where organisms form a series of overlapping distributions and vertical migration ranges, divisions should not be regarded as discrete.

Marine biology of the benthic realm

The benthic realm is sometimes conceived as a fluid extension of the terrestrial environment. In fact, in evolutionary terms and from a global view, the opposite perspective is more appropriate; the terrestrial environment is really an extension of the benthic realm – above water.

For practical purposes, the benthic realm can be considered two dimensional. While it can be argued that benthic organisms use their space three dimensionally (e.g. plants extend well into the water column, animals burrow into the substrate), their mode of life and physical adaptations lie in sharp contrast to organisms of the pelagic realm. Benthic plants and animals are bottom referenced and their distributions are also strongly related to water depth.

Benthic communities are sometimes further subdivided (e.g. Hedgpeth, 1957) into:

- fringing communities (comprising the primary producers and consumers of the euphotic zones) and
- benthic proper communities (those below the euphotic zone dependent on a detrital economy).

There are pros and cons to recognizing such a distinction. Here fringing and benthic communities are considered together, in order to avoid a facile differentiation between producers and their attendant consumers.

Fringing communities

In the shallowest euphotic regions of aquatic habitats, photosynthetic fixation of carbon is performed by several types of fringing floras. Despite their diversity and their potential high biomasses, compared to phytoplankton their overall contribution to primary production of marine ecosystems is low, except in shallow coastal waters. Rates of primary production for all types of marine communities can be found in Mann (2000).

Several community types, many of them familiar to us because of superficial similarities to terrestrial forms, can be recognized:

- Dense stands of sub-tidal (and therefore submergent) macrophytic angiosperm vegetation (basically of terrestrial ancestry). In marine habitats these are represented by species such as the truly marine seagrasses *Spartina* and *Thalassia*.
- Salt marshes around the ocean margins are intertidal (rarely extending below mean tide level) and essentially semi-terrestrial in nature.
- In wetlands and marshes both submergent and emergent macrophyte plant communities develop. These communities are a mixture of truly aquatic and semi-terrestrial species.
- The seaweed communities are a dominant feature of rocky marine shores. High biomasses of these macrophytic algae may develop both intertidally where they are temporarily exposed to the air during the phases of the tide, and sub-tidally within the euphotic zone.
- The intertidal and sub-tidal microalgae are often forgotten in comparison with the more visually striking macrophytic algae of the shorelines. These microscopic algal communities (single cells, chains or mats) are often dominated by the benthic diatoms that form mats on the bottom sediments or rock or may form epiphytic growths on the macroalgae themselves.
- The hermatypic corals forming reef structures. Although the corals themselves are animals, their symbiotic microalgae (zooxanthellae) are responsible for high rates of photosynthesis. Reef forming and other hard corals are not represented at higher latitudes although non-photosynthetic 'soft corals' extend into the polar regions.
- Mangrove swamps typically develop in hot wet regions of the ocean margins, where tidal amplitude is low.

The rate of primary production in these fringing communities may be extremely high. It is usually much higher per unit area per unit time than that of the planktonic algae. However, the greater vertical extent and horizontal coverage of the phytoplankton more than compensates in terms of overall production. A major distinction between planktonic and benthic producers is that the macrophytic plants tend to conserve their biomass. The carbon fixed by them tends to enter detritus pathways (as is generally the case in terrestrial ecosystems) rather than being directly grazed (Mann, 2000).

Associated with the plants of the fringing communities is a wide diversity of bottom-dwelling invertebrates. These are considered next under zoobenthos communities.

Zoobenthos communities

Marine life extends from the highest limit of the high tides to the abyssal depths of the oceans. Yet, unlike the pelagic realm, life is either attached to or roams the substrate, or is buried only a few centimetres within it.

In the shallow waters of the euphotic zone, benthic and nektonic grazers can feed directly on a diet of macrophytic or microphytic algae. Within the intertidal zone, the upper limits to distributions are set by tolerance to atmospheric desiccation, while lower extents are often set by less physically tolerant predators. There are major differences between the communities that develop on rocky shores versus those consisting of silts and muds.

In the immediate sub-tidal, plants are still present within the euphotic zone. However, in deeper waters the living plants are absent below the photic zone, and crucial changes occur in the animal communities. Below the euphotic zone, the substrate has generally changed to a relatively uniform sand/silt/mud in composition. The benthos in these regions can only be supported by the rain of detrital material that is derived from the other photosynthetic aquatic communities. The benthos in the dysphotic and aphotic zones is essentially a region of detritus-based food webs and of heterotrophic bacterial activity. It is also a dominant site of nutrient regeneration. Overall, detritus-based food chains are extremely significant; in fact, as mentioned earlier, they are the dominant food webs of the ocean.

Specialized members of the fish community also exist at various depths within the benthos. For example, flatfishes, skates, dogfish and many other species are essentially bottom dependent, or at least bottom referenced. They may be relatively static, or highly localized and territorial. Some bottom-dwelling species are periodically mobile or migratory (e.g. crabs and lobsters). Others are either attached permanently to a substrate or burrow within it.

Pelagic–benthic coupling

Pelagic and benthic communities are often spoken of as being 'coupled'. However, it is perhaps better to refer to them as 'interdependent', since the degree of coupling varies in space and time. Typical pelagic and benthic communities do not share adult species members, except temporarily; however, there are significant two-way interactions between the communities.

Below the euphotic zone, benthic communities are entirely dependent on the rain of detrital material from the pelagic realm. The only exception to this is found in the hydrothermal vent communities around the oceanic ridges. These are essentially independent of resources from the overlying water column, and are self-reliant on a chemosynthetic sulphur economy. The two major realms might best be considered to interact in three ways:

1   by exchange of energy;
2   by recruitment of larvae;
3   by temporary exchange of members.

The rate of sedimentation of primary energy (ungrazed algal cells) and recycled organic matter (fecal pellets and other detritus) to the benthos decreases with depth (e.g. Suess, 1980). Pelagic–benthic interactions therefore 'weaken' with depth because the flux of organic matter diminishes and the benthos must become food limited. In shallow waters however, the benthos can graze the phytoplankton of unstratified water columns directly.

The flux of organic matter to the benthos is also highly seasonal, being greatest during and immediately following the spring diatom increase, and lowest during the summer stratified period. Losses from the plankton to the benthos are probably proportionately highest in the Arctic and lowest in the tropics (Roff and Legendre, 1986). This is because the greater the degree of modulation of the seasonal cycle, the less efficiently energy is used within the water column, and a greater proportion sinks to the benthos. At lower latitudes at higher temperatures in the benthos a greater proportion of this energy must be lost to bacterial decomposition. The net effect of these ecological differences and seasonal and latitudinal changes in processes is that the planktonic realm tends to be proportionately more productive than the benthos at lower latitudes and in systems whose production is not highly modulated, while the reverse is true (i.e. the benthos tends to be proportionately more productive) at high latitudes or in strongly pulsed ecosystems (e.g. Longhurst, 1995).

This is by no means the end of the complexities of pelagic–benthic interactions. Lateral transport of energy can also dominate in a region where tidal currents are strong. For example, the planktonic and benthic systems in the outer Bay of Fundy have been described as 'uncoupled', since production of the zoobenthos is heavily dependent on tidal transport of detritus (Emerson et al., 1986).

The planktonic world is also heavily, but temporarily, colonized by the larvae of benthic forms. In their larval forms, lobsters, crabs, molluscs and such are all temporary members of the meroplankton. In this form they have direct access to the richer pelagic food resources, and may use this realm as a means of dispersal. The proportion of benthic species having meroplanktonic larvae tends to decrease with depth (because of greater distance to and from recruitment), and towards higher latitudes (because of shorter and more pulsed seasonal production cycles).

Lastly, the pelagic and benthic realms may temporarily exchange members. Several species may inhabit the water column – often at night – but become bottom referenced during the day. However, they may predominantly inhabit the water column, but use benthic resources, or vice-versa. Mysids and euphausiids and several fish species do this. Such invertebrate species are often referred to as *epibenthic*, while such fish may be referred to as *demersal*.

Chapter summary

* Temperature governs the distribution of marine biological communities more than any other factor.
* Water masses are defined by homogenous temperature and salinity regimes and are analogous to the major terrestrial regions.
* High biological productivity is found in areas where water masses converge or diverge.

- The great majority of phyla of multicellular animals are found only in the marine environment.
- Phytoplankton are responsible for >95 per cent of the annual primary production in the ocean.

References

Azam, F. (1998) 'Microbial control of oceanic carbon flux: The plot thickens', *Science*, 280 (5364), 694–696.

Barnes, R. S. K. and Hughes, R. N. (1999) *An Introduction to Marine Ecology*, Blackwell Scientific Publications, Oxford.

Beaumont, N. J., Austen, M. C., Atkins, J. P., Burdon, Degraer, D. S., Dentinho, T. P., Derous, S., Holm, P., Horton, T., van Ierland, E., Marboe, A. H., Starkey, D. J., Townsend, M. and Zarzycki, T. (2007) 'Identification, definition and quantification of goods and services provided by marine biodiversity: Implications for the ecosystem approach', *Marine Pollution* Bulletin, 54, 253–265.

Bertnes, M. D., Gaines, S. D. and Hay, M. E. (2001) *Marine Community Ecology*, Sinauer Associates, Sunderland, MA.

Bukata, R. P., Jerome, J. H., Kondratyev, K. Y. and Pzdnyakov, D. V. (1995) *Optical Properties and Remote Sensing of Inland and Coastal Waters*, CRC Press, Boca Raton, FL.

Charette, M. A. and Smith, W. H. F. (2010) 'The volume of earth's ocean', *Oceanography*, 23 (2), 112-114.

Day, J. C. and Roff, J. C. (2000) *Planning for Representative Marine Protected Areas: A Framework for Canada's Oceans*, World Wildlife Fund Canada, Toronto.

Diaz, R. J. and Rosenberg, R. (2008) 'Spreading dead zones and consequences for marine ecosystems', *Science*, 321 (5891), 926–929.

Emerson, C. W., Roff, J. C. and Wildish, D. J. (1986) 'Pelagic benthic coupling at the mouth of the Bay of Fundy, Atlantic Canada', *Ophelia*, 26, 165–180.

Glemarec, M. (1973) 'The benthic communities of the European north Atlantic continental shelf', *Oceanography and Marine Biology Annual Review*, 11, 263–289.

Gray, J. S. (1997) 'Marine biodiversity: Patterns, threats and conservation needs', *Biodiversity and Conservation*, 6, 153–175.

Gunter, G. (1957) 'Temperature', *Memorandum of the Geological Society of America*, 67 (1), 159–184.

Hedgpeth, J. W. (1957) 'Marine biogeography', in J. W. Hedgpeth (ed.) *Treatise on Marine Ecology and Paleoecology. I. Ecology.*, Geological Society of America 67, New York.

Knox, G. A. (2001) *The Ecology of Seashores*, CRC Press, London.

Longhurst, A. R. (1995) 'Seasonal cycles of pelagic production and consumption', *Progress in Oceanography*, 22, 47–123.

Mann, K. H. (2000) *Ecology of Coastal Waters*, Blackwell Science, Oxford.

Mann, K. H. and Lazier, J. R. N. (2005) *Dynamics of Marine Ecosystems: Biological-physical Interactions in the Oceans*, Blackwell Science, Cambridge, MA.

National Research Council (1994) *Priorities for Coastal Ecosystem Science*, National Academy Press, Washington, DC.

Ray, G. C. and McCormick-Ray, M. G. (1992) *Marine and Estuarine Protected Areas: A Strategy for a National Representative System within Australian Coastal and Marine Environments*, Australian National Parks and Wildlife Service, Canberra, Australia.

Roff, J. C. and Legendre, L. (1986) 'Physico-chemical and biological oceanography of Hudson Bay', in I. P. Martini (ed.) *Canadian Inland Seas*, Elsevier, New York.

Roff, J. C. and Zacharias, M. A. (2011) *Marine Conservation Ecology*, Earthscan, London.

Ruiz-Frau, A., Edward-Jones, G. and Kaiser, M. J. (2011) 'Mapping stakeholder values for coastal zone management', *Marine Ecology Progress Series*, 434, 239–249.

Sherman, K., Alexander, L. M. and Gold, B. D. (1992) *Large Marine Ecosystems: Patterns, Processes and Yields*, AAAS Press, Washington, DC.

Steele, J. H. (1991) 'Ecological Explanations in Marine and Terrestrial Systems', in T. Machline and T. Nemoto (eds) *Marine Biology: Its Accomplishment and Future Prospect*, Elsevier, Amsterdam, Netherlands.

Steele, J. H., Carpenter, S. R., Cohen, J. E., Dayton, P. K. and Ricklefs, R. E. (1993) 'Comparing Terrestrial and Marine Ecological Systems', in S. A. Levin, T. M. Powell and J. H. Steele (eds) *Patch Dynamics*, Springer-Verlag, New York.

Stehli, F. G., McAlester, A. L. and Heisley, C. E. (1967) 'Taxonomic diversity of recent bivalves and some implications for geology', *Geological Society of America Bulletin*, 78, 455–466.

Suess, E. (1980) 'Particulate organic carbon flux in the oceans–surface productivity and oxygen utilization', *Nature*, 288, 260–263.

Sverdrup, K. A., Duxbury, A. C. and Duxbury, A. B. (2003) *An Introduction to the World's Oceans*, McGraw-Hill, NY.

Thorne-Miller, B. and Earle, S. A. (1999) *The Living Ocean: Understanding and Protecting Marine Biodiversity*, Island Press, Washington, DC.

Turner, J. T. and Roff, J. C. (1993) 'Trophic levels and trophospecies in marine plankton: Lessons from the microbial food web', *Marine Microbial Food Webs*, 7 (2), 225–248.

Valiela, I. (1995) *Marine Ecological Processes*, Springer, New York, NY.

UNESCO (2009) *Global Open Oceans and Deep Seabed (GOODS) – Biogeographic Classification*, UNESCO, Paris.

Chapter 2

An introduction to international law

International marine law and the United Nations Convention on the Law of the Sea

Introduction

Human use of the marine environment began approximately 164,000 years ago (Marean *et al.*, 2007). Since this time 6000–8000 distinct coastal indigenous groups have used the oceans as a means of transport, trade, and as a source of protein and materials (Hattendorf, 2007). To provide uninterrupted benefits from marine areas early cultures established practices to assign rights, restrict access, allocate resources and defend the goods and services provided by their coastal and marine resources. Compliance with these practices was realized through a combination of rules, spiritual beliefs, myth and experience.

However, it was not until the rise of secular sovereign states in seventeenth-century Western Europe that modern ocean governance regimes and jurisprudence emerged to address the socioeconomic, political and environmental challenges created by population growth and technological advancement. Currently, over 500 major multilateral instruments have been deposited with the Secretary-General of the United Nations (UN). They cover a range of subject matter, including human rights, conflict, trade, refugees, the environment and the law of the sea. Considering only the environmental treaties between two or more countries, there are currently 464 international agreements.

There are many comprehensive marine law texts that dissect the various international agreements and the cases that arise from them (e.g. Anand, 1983; Brown, 1994; Rothwell and Stephens, 2010, Tanaka, 2012) and it is not the intent of this chapter to duplicate these efforts. Instead, this chapter will focus on introducing marine law and policy for the purposes of gaining a better understanding on how marine environments are governed.

This chapter begins by tracing the development of international law and international marine law through an exploration of types of ownership. This is followed by a brief summary of the primary international treaties, conventions and agreements and their associated institutions. Lastly, this chapter briefly introduces regional marine agreements and their operation and examines some of the more important regional agreements.

Types of ownership regimes in the ocean

Before discussing the legal frameworks that govern the oceans, an understanding of the ways marine environments can be appropriated is necessary, for much of international marine law and policy is concerned with concepts of appropriation and ownership. There are four types of ownership regimes applicable to coastal and marine environments. The first is *open access* (*res nullis* meaning 'nobody's property'), where the oceans and resources within them are neither exclusive nor transferable and, as such, available to all. The concept, developed under Roman law, applies to things that cannot be appropriated, such as wild animals or the high seas and also in situations where there is an absence of the right to exclude. Under the open access regime users have no responsibility for the maintenance of the asset. Marine capture fisheries before the end of the nineteenth century were, for all intents and purposes, open to all and 'ownership' only occurred once a resource was appropriated – a caught fish for example. The 1893 Bering Sea Arbitration between Great Britain and the United States of America (US) on the harvest of seals was perhaps the first example of the repudiation of *res nullis* as a foundation of marine management. Another component of open access is the concept of *terra nullis* (land belonging to no one), which is a Roman law concept that describes territory (terrestrial or marine) over which no state has declared or exerted sovereignty. The concept has particular relevance to the Antarctic continent which still has unclaimed areas and high seas environments (discussed in Chapter 8).

Private ownership (*res privata*), confers exclusive, transferable rights (*abusus* in legal terms) to a resource with an associated duty to refrain from socially unacceptable uses. While relatively rare in marine environments, private ownership of rights to certain amounts of fish within certain fish stocks termed individual transferrable quotas is increasing (this is discussed in Chapters 6 and 10). Private ownership also occurs when a non-privately owned resource is captured or killed.

Common ownership (*res communes*) is a blended model between pure property rights and pure regulation where a defined group of independent users exert a collective right to a jointly held resource or geographic area. The model uses explicitly or implicitly understood rules about how the resource is allocated. Most common ownership resources are rarely exclusive and transferable. Common ownership in marine environments is usually associated with smaller coastal fisheries in developing countries or Aboriginal-controlled fisheries in developed countries.

Lastly, *public ownership* (*res publicae*) is where governments, on behalf of their citizens, control marine areas either through international agreements/treaties or through outright appropriation through force. States make laws or rules to control access and management of resources. Public ownership is exerted by nearly every coastal nation and a key objective of the various Law of the Sea conventions (discussed below) was to establish a common agreement on the geographical extent of public ownership. Under public ownership, use rights in marine environments are fundamentally a function of the relationship between states and persons (in the legal sense) regarding its use.

An introduction to law and types of legal systems

The term 'law' most likely originates from 'lagu', an Anglo-Saxon word generally meaning the rules that bind a community. In the contemporary context 'law' refers to the principles and rules that govern the behaviour of individuals and groups within a society as well as the institutions that oversee these rules. Law is typically administered through a political system comprised of three components: politicians (elected, appointed or otherwise) in senior positions of power that make laws; the bureaucracy, which administers laws; and the 'judiciary' (judicial system) comprised of courts where judges interpret laws and rule on questions and disputes.

In order to inform a discussion on the evolution of the development and application of international law, an understanding of domestic (national) law is useful. While there continues to be disagreement on how to categorize the world's legal systems, five types of legal systems are in common use throughout the world (Table 2.1). Importantly, these types are not mutually exclusive and many states utilize a combination of two or more systems. In addition, many sub-national entities (e.g. French territories) have different legal systems than their parent nation. With respect to religious law, Muslim law is the only recurring religious-based legal system in use across many states. Many other religion-based legal systems have been subsumed into customary law with the exception of Jewish law, used only in Israel.

Beginning with the most popular law system, *civil law* – used in some form by approximately 70 per cent of UN member states – is based on the Roman and Germanic law heritage where all laws are codified (written down). In civil law systems, statutes (written laws) take precedence over case law (law made by judges based on past precedent). Judges have limited authority to interpret law and use of juries is limited to serious criminal matters.

Common law – used in some form by approximately 40 per cent of UN member states – is the primary contrasting legal system to civil law. Common law systems, originally developed in England, are based on case law, or precedent, where law is based on prior court decisions rather than written statutes. Common law is based on the principle of *stare decisis*, which states that judges are obligated to respect past judicial decisions. This entails that case law is produced in common law jurisdictions. However, case law may be overturned by higher (appellate) courts and may not be recognized between sub-national jurisdictions.

Customary law, used in some form by approximately 23 per cent of UN member states, is a legal system based on accepted practice and behaviour (termed 'norms') where a legal practice (although not written down or developed by the judiciary) becomes accepted as law by all parties affected by the practice. Customary law is a foundational part of international law (see below) and is part of the legal systems of nearly 60 states, including China, Japan and states in Africa and the Persian Gulf.

Islamic law, also known as Muslim law or Sharia ('a path to the watering hole') law is a religious-based legal system used in approximately 33 states. Islamic law is a religious code

Table 2.1 Distribution of legal systems among UN member states

Legal system	Number of member states	Percentage of member states
Civil law	77	40.1
Civil law/customary law	25	13.0
Common law	23	12.0
Common law/customary law	14	7.3
Civil law/Muslim law	11	5.7
Civil law/common law	10	5.2
Civil law/common law/Muslim law	7	3.7
Common law/Muslim law/customary law	6	3.1
Civil law/common law/customary law	5	2.6
Muslim law/common law/civil law/customary law	4	2.1
Muslim law	3	1.6
Customary law	1	0.5
Civil law/common law/Jewish law/Muslim law	1	0.5
Muslim law/customary law	1	0.5

Source: www.juriglobe.ca/eng/syst-onu/index-alpha.php

that governs both the secular (e.g. crime, governance, finance) as well as the spiritual (e.g. diet, prayer). Islamic law is based on the Qur'ān and the rulings of Muhammad (the Sunnah). Many Muslim states have adopted Islamic law as a component of their legal systems and the law is interpreted by judges and religious leaders.

A primer on international law

International law is not new: evidence exists of treaties between states as far back as the Babylonian and Sumerian empires (>2200 years ago). Greek city states and the ancient kingdoms of India were known to have well-developed relationships with other political bodies that were codified in what we would term today as international law. Since its background is so rich and complex, this section provides only a brief overview of the history and development of international law, mainly focusing on the structure and processes that form international marine law and policy agreements (those interested in a more thorough treatment of this topic are encouraged to review Higgins (1995), Malanczuk (1997) and others).

International law, also known as the law of nations and language of international relations, is the framework for how states conduct themselves at the national level. It comprises both a system for securing values (e.g. freedom, security, material goods) and a method to resolve disputes between states (Higgins, 1995). The objective of international law is to establish acceptable practices or behavioural norms between states to regularize and make predictable international interactions.

The central differences between domestic and international law are: who makes it (legislatures versus states); to whom it applies (persons versus states); who interprets it (judges/bureaucracies versus states); the near absence of sanctions in the international system; and courts (largely irrelevant in international law). Given the absence of a central governing authority and judicial system, international law is often perceived as weak at best and non-existent at worst. The 'law' in international law is therefore more a collection of rules rather than a codified system of enforceable statutes.

States generally have legislatures that enact laws, a judiciary that enforces laws and an executive branch that oversees the legislature and judiciary. International law, by contrast, has no central governing authority; however, states use their executive and legislative branches to – often unilaterally – establish rules on how the state will interact with other states. The state may also unilaterally ignore the rules other states create for themselves without the threat of sanctions.

International law is sometimes further subdivided into public and private international law. *Public law* applies to the interactions between states and may be defined by agreements, customs, principles and rules. *Private law* applies to individuals and corporations who interact with and between states.

The terms *soft law* and *hard law* are also used to further differentiate international law. While many jurists and scholars consider all international law a type of soft law, soft law, in the international law context, generally consists of non-binding multilateral agreements. These include decisions/resolutions adopted at conferences, recommendations, codes of conduct, principles, action plans and resolutions such as those from the UN General Assembly (UNGA) (Juda, 2002). Soft law can be an alternative to creating treaties and, while non-binding, states often pay careful attention to the development of soft law as it can signal future developments of hard law (Boyle, 1999). Hard law and customary law consist of binding agreements between states in the form of treaties (conventions). Certain treaties may be appealed to the International Court of Justice (ICJ) or other convention-specific tribunals.

The state in international law

It is important to note that the state (see Preface) is the fundamental entity in international law. While international law affects sub-sovereigns, corporations and occasionally individuals (e.g. pirates), the state is the primary agent in international law. It is important also to note that governments are different from states. To be considered a state under international law (articulated in the Montevideo Convention on the Rights and Duties of States) the state must:

- demonstrate sovereignty (control) over a defined territory and its nationals;
- support a permanent human population;
- have a functioning government;
- have the capacity to engage with other states.

While recognition by other states is not a requirement for statehood, recognition is a key goal for newly formed states (e.g. South Sudan), proposed states (e.g. Palestine) and states where it is unclear who is the government (e.g. Somalia). Unrecognized states (e.g. Macedonia, a member of the UN) may act as states and undertake the duties of a recognized state. More recently, after the dissolution of the former Yugoslavia in the early 1990s the European Economic Community (EEC) added a number of other criteria (e.g. support for democracy, human rights, rule of law) before the EEC would recognize the newly formed Balkan states.

Other key aspects of statehood and international law include:

- The political existence of a state is independent of recognition by the other states.
- It is possible to recognize a state without recognizing its government.
- It is also possible to recognize a government but elect not to have any diplomatic ties to the government.
- A government in exile may not exert control of a state but the government can be recognized.
- An unrecognized state may function in the same manner as a recognized state.
- People have the right of self-determination but states may only be established to end colonialism or foreign subjugation.

While these criteria are straightforward for most states, many new states including South Sudan, Palestine and those states in the former Yugoslavia have difficulty meeting them.

International law also governs any entity that possesses 'international personality'. These 'non-state' entities include governmental organizations (e.g. the UN). Very rarely will multinational companies, individuals and non-governmental organizations be recognized as having 'international personality'. It is important to note that international law governs states, which in turn regulate the activities of their nationals and businesses. Thus many types of international laws – including human rights or trade – are enacted through domestic legislation.

Sources of international law

Unlike domestic law that is established by a legislative body, international law is much more fluid and draws on a range of sources to inform its development and application. While treaties and customary law are generally viewed as the primary sources of international law, the ICJ (Article 38.1) defines the sources of international law, which include:

- International conventions, whether general or particular, establishing rules expressly recognized by the contesting states.
- International custom, as evidence of a general practice accepted as law.
- The general principles of law recognized by civilized nations.
- Judicial decisions and the teachings of the most highly qualified publicists of the various nations, as subsidiary means for the determination of rules of law.

Treaties and other bilateral agreements to which states are parties that directly address the issue are the primary source of international law. These are often referred to as 'express international agreements'. Agreements – if given the authority by a treaty – may be between international organizations (e.g. World Bank) and states. It is important that while the ICJ statute refers to 'conventions', the terms 'treaties', 'agreements', 'declarations', 'accords', 'protocols' and so on are synonymous with 'conventions' although there are certain exceptions.

The second principal source is customary international law. It is best understood as a practice of states that becomes so accepted that it becomes legally binding. Any action by any state that is adopted (or accepted) by other states can become part of international law, which is also known as *opinion juris sive necessitates* (the belief by states that it is an international legal obligation). Stated differently, there are two elements that demonstrate whether a practice/action is part of customary international law: consistent action taken by numerous states over a period of time, and the belief by a state that a practice has become binding. Statements by government officials may be evidence of customary international law or demonstrate what a state considers as customary. However, customary international law is normally predicated on the actions of a state rather than the verbal or written intentions of states. No written agreement is required to develop customary international law and the routes to establishing customary international law are beyond the scope of this book; but it is important to note that international law may be created by actions of a single state that are adopted by other states.

Third, the general principles of mature legal systems can be used to create international law (i.e. the 'law recognized by civilized nations'). Although there has been much discussion on how to define mature legal systems, it has come to be understood that international law may be formed by 'borrowing' aspects of national law common across states and, where appropriate, expanding national law into the international realm.

Lastly, international law may be formed by accepting and incorporating rulings of state's legal systems (e.g. Supreme Court) into international law as well as the 'teachings' of highly qualified jurists (Malanczuk, 1997; Higgins, 1995).

There are other sources of international law not contemplated by the ICJ in the 1940s. These include the acts or resolutions of international organizations such as the UN General Assembly, which produce frequent resolutions that may become accepted as customary international law. The growth of international organizations in recent years has vastly expanded the number of agreements between nations that could be considered as customary international law.

Operation of international law

A foundation of international law is the Vienna Convention on the Law of Treaties (signed 23 May 1969 and entered into force 17 January 1980), which is generally accepted as the 'rules' related to treaties (United Nations, 2005a). For the Vienna Convention to be applicable, parties must be states (Article 1), parties must have agreed (Articles 11–13), agreements must be in writing and intended to be binding (Article 11). Moreover, the agreement must state that the governing law will be international law (Box 2.1).

Box 2.1 Key terms as defined by the Vienna Convention on the Law of Treaties

Treaty: An international agreement in written form forged between states and governed by international law, whether embodied in a single instrument or in two or more related instruments and whatever its particular designation (Article 2, paragraph 1[f]).

Party: A state which has consented to be bound by the treaty and for which the treaty is in force (Article 2, paragraph 1[g]).

Plenipotentiary: A person (usually a diplomat) invested with the full power to represent their government in negotiations.

Date of acceptance: The date at which a state becomes a party to a treaty. Treaties normally identify whether treaties are as 'signature subject to acceptance' (analogous to ratification) or by acceptance without prior signature (analogous to accession).

Date of accession: The date at which a state becomes a party to a treaty of which it is not a signatory. The right of accession is independent of the entry into force of the treaty; that is, a state may accede to a treaty which has not yet entered into force.

Date of adoption: The date when states participating in the negotiation of a treaty agree on its final form and content. This usually occurs before signature.

Date of denunciation: When a state announces that it is no longer willing to be bound by a treaty.

Date of entry into force: The date when a treaty binds the parties that have agreed to be bound by it. It is often triggered when a certain number of states have ratified the treaty.

Date of ratification: The date when states obtain internal approvals to be bound by a treaty. This usually occurs after signature.

Date of reservations: The date when a state excludes or modifies the legal effect of certain provisions of a treaty.

Date of signature: The date when a state consents to be bound by a treaty but prior to ratification when internal approvals have been obtained.

Date of succession: The date when newly formed states (e.g. Russia) agree to be bound by treaties signed by predecessor states (e.g. Soviet Union).

Source: United Nations (2005a)

Under the Vienna Convention states have to agree to be bound to any aspect of international law. The only exception to this rule is if an action violates a 'peremptory norm' defined as '... a norm accepted and recognized by the international community as a whole as a norm from which no derogation is permitted and which can be modified only by a subsequent norm of general international law having the same character' (United Nations, 2005a). Peremptory norms include prohibitions against slavery, genocide, piracy, aggressive war and torture.

States may make agreements between themselves, but the accords are not considered treaties if a peremptory norm is violated. Additional states may enter into an existing treaty between two or more states if provided for in the treaty. Termed 'accession' the additional state may or may not require (depending on the treaty) the consent of the existing signatories to the treaty to be able to enter into the treaty.

Eight steps are common in the development of multilateral treaties:

- Negotiation: A pledge to address a shared issue through a treaty, agreement or convention.
- Adoption: The conclusion of negotiation after the form and content of an agreement is reached. Adoption may be secured via meetings of parties in their capacity as participating in international organizations or an international conference convened specifically to address the shared issue.
- Signature: Parties agree to abide by the conditions of the agreement while the approval process is underway. Signature does not bind a state to a treaty but states are expected not to defeat the purpose of the treaty while it proceeds through ratification and enters into force.
- Ratification: A third party or depository holds an agreement while parties secure the necessary domestic approvals that may include domestic legislative change.
- Acceptance (approval): Binds a state to conform with the treaty while it proceeds through the ratification process.
- Confirmation: Agreement by a state to be bound by the treaty prior to it entering into force.
- Entry into force: The date when the treaty enters into force and binds all states that have ratified the treaty.
- Accession: Request by a state to become party to a treaty already negotiated.

Source: United Nations (2005a)

Certain treaties allow 'reservations', which permit states to reject or amend certain provisions of a treaty with which they disagree. However, the LOSC (discussed below) expressly prohibits reservations: it is an all-or-nothing treaty.

The UN Charter was developed after World War II to '... save succeeding generations from the scourge of war...'. The purpose of the Charter is to prohibit acts of aggression and unlawful use of force between states. Under the UN Charter, states with disputes over points of law or facts are obligated to settle disputes by peaceful means. The Charter sets out a continuum of increasingly interventionist dispute reconciliation mechanisms starting with negotiation between states/parties with disputes. If negotiation fails, the Charter directs parties to use non-binding tactics including inquiry, conciliation and mediation. This promotes dialogue, collection of the facts to an issue, and use of a third party to resolve disputes. The next step, if non-binding approaches fail and parties agree to further action, is arbitration, where an independent party evaluates proposals from the affected parties and chooses one. In arbitration, the arbitrator cannot develop an alternative option; one of the proposals from the affected parties must be selected. Lastly, parties are directed to the courts for decision. All states that are signatories to the Charter also agree to the rules of the ICJ. The ICJ is final arbiter of over 240 international treaties and the purpose of the ICJ is to settle international disputes and provide advisory opinions. States cannot be forced to settle disputes before the ICJ. Headquartered at the Hague, the ICJ is composed of 15 judges and has no powers of enforcement, which is left to the UN Security Council. Only states may appear before the ICJ and its rulings only apply to the disputed issue brought forward by the states; however, certain ICJ rulings have come to be recognized as important in determining international legal principles. The Court is not bound by the principle of *stare decisis* so that it is not obligated to consider past precedents in their deliberations. However, the ICJ carefully considers past decisions and they are regularly cited by both the Court and in argument.

The relationship between domestic and international law

Congruence between domestic and international law offers a state a number of efficiencies. These include: cost saving as a result of harmonized rules and procedures; access to international markets; standardized compliance and enforcement; and positive international reputation and optics. For many states, such as those in the European Union, the number of international laws developed in Brussels far exceeds the number of domestic statutes passed. However, states – including those in the European Union (EU) – have different practices, often based on constitutional frameworks, on the extent to which international law supersedes domestic law. As a consequence, there is little consistency in how states treat international law in their domestic legal systems. Most states accept that international treaties or conventions (any type of agreement that is legally binding) to which they are a party cannot be casually overridden by domestic legislation. However, states have different perspectives with respect to the process by which international law is accepted as part of domestic law. 'Monist' states, such as the Netherlands, are those which accept international agreements to which they are a party as having the force of domestic laws. 'Dualist' states, such as Australia and Canada, intentionally separate international and domestic laws and carefully choose which international treaties, once ratified, evolve into domestic laws. A number of states, including Germany and the US, occupy the middle ground. Here, international treaties must meet domestic legal principles (e.g. Germany reserves the right to evaluate all international treaties against German *Basic Law*) or have checks and balances to ensure that treaties represent the public interest (e.g. the US uses primarily its Senate, but also its House, in interpreting international law) (de Mestral and Fox-Decent, 2008; Law Society of New South Wales, 2010).

A murkier issue with which many states wrestle is whether customary international law can be superseded by domestic law. The United Kingdom (UK), for example, accepts customary international law as part of domestic (common) law except where it conflicts with domestic laws. Other states, including Canada and New Zealand, generally follow the UK practice. However, they use a general principle of statutory interpretation that statutes should be interpreted to the extent possible not to be inconsistent with their international legal obligations, with the obvious proviso that if such interpretation is not possible, the domestic statute prevails.

Why international law works

Often the public perception of international law is that, due to its lack of sanctions and voluntary nature, states routinely flout international conventions. This perception is fed by high-profile media coverage of trade wars, border disputes, non-compliance with UN directives, illegal fishing and specific nations (e.g. North Korea) refusing to recognize international agreements.

The reality, however, is that compliance with international law is similar to that of domestic laws. It is rarely ignored and very infrequently challenged. The reason for high compliance rates with international law is not the threat of sanctions. International compliance might be compared with national taxes: most citizens pay their taxes knowing full well that the chance of being audited is low. However, compliance with tax codes is high and the reason for this is that citizens are raised to understand the societal importance of legal behaviour. So too with international law. Rather than fearing sanctions, states recognize the benefits of compliance and cooperation.

There are many reasons for the success of international law, which are summarized from Malanczuk (1997):

- States create laws for themselves and in their interests. As such, there is little to be gained from ignoring or contravening their own laws.
- Most international law begins as customary practice and therefore, even in the absence of international law, these customs would often still be followed. As such, international law is often merely the formalization of the current practices between states.
- States are, for the most part, geographically anchored to specific territory, meaning that they must coexist with neighbouring states for long periods of time. As such, cooperation is necessary for social and economic prosperity and for avoiding groups of other states from imposing sanctions for breaking international laws.

International marine law

Unlike the world's terrestrial landmasses and islands where, with the exception of places such as Antarctica, states have established and acknowledged claims of ownership, most of the ocean continues to be free for the use of all. Consequently, the ocean has traditionally been poorly governed. The result has been little to no concern for ocean health, significant loss of human life due to poor maritime safety, and political friction among coastal and seafaring states.

Traces of international marine law date back to the Roman Empire, which declared the oceans as a commons free to all. In AD 529 Emperor Justinian declared that '...the sea is common to all, both as to ownership and as to use. It is owned by no one; it is incapable of appropriation, just as is the air. And its use is open freely to all men' (Watson, 2009). Throughout the Middle Ages coastal states began to exert their sovereignty over their adjacent territorial waters. This culminated in a 1494 Papal Bull from Pope Alexander VI resulting in the Treaty of Tordesillas where it was agreed that Spain would own all marine areas in the West Atlantic while Portugal would possess everything in the East Atlantic. Furthermore, Spain was given the Pacific and the Gulf of Mexico and Portugal the South Atlantic and Indian Ocean.

As other European nations built merchant and naval vessels to participate in colonization and international trade the Treaty of Tordesillas was rendered obsolete. The next phase of the development of international marine law consisted of the writings of Hugo Grotius, a Dutch jurist who authored the 1608 tome *Mare Liberum* (Free Sea) (Grotius, 1633). Grotius argued that since oceans cannot be 'occupied' or defended in the same manner as terrestrial environments, the sea therefore should be free to all. This doctrine resulted in the 'three mile rule', whereupon states 'occupied' (held jurisdiction) areas from the coastline out to three miles, or the maximum distance a cannon ball could be fired in the eighteenth century. In contrast, at the request of King Charles I, John Selden, an English jurist, authored *Mare Clausum* (Closed Sea) in 1635 that argued that oceans can be appropriated by states without physical occupation.

Ultimately, Grotius's proposition became the accepted practice of most states and the free seas doctrine became one of the first recognized principles of international marine law. This conviction endured until the early twentieth century when a number of states raised the need to codify international marine law. In 1930, the League of Nations convened a Codification Conference to formalize the rules of international law and identify outstanding areas of international law that had not been adequately addressed. Forty-four states attended the conference and though territorial waters was one of the three Conference themes, the only agreement that was reached was the Convention on Certain Questions Relating to the Conflict of Nationality Laws (League of Nations, 1930).

The next major development in international marine law was the 1945 decision by the US to unilaterally declare its sovereignty over its continental shelves, including the natural resources on and within the seabed in these areas. Termed the 'Truman Proclamation' this action established a precedent that was soon followed by other nations.

In 1949, the UK sued Albania in the ICJ for compensation resulting from damage to two British destroyers that had struck mines in Albanian waters in 1946. This was the first case (the *Corfu Channel Case*) of the newly formed ICJ and Albania was ordered to pay GB£843,947 in compensation. This case is significant as it provided the Court the opportunity to consider issues related to territorial seas including the coastal state's rights within these areas and the rights of other states to peaceful navigation (ICJ, 1949).

The final significant legal case prior to convening the various United Nations Law of the Sea Conferences was the 1951 *Fisheries Case* where the UK requested that the ICJ rule on the extent of Norwegian territorial waters. Norway had delineated its territorial seas using 'straight baselines' (discussed below) from headland to headland and therefore claimed control of fishing rights in these areas (ICJ, 1951). This excluded the UK from fishing in what it considered were 'high seas' areas. The ICJ ruled in favour of Norway, however, soon prompting other nations to connect headlands.

The International Law Commission (ILC) was established by the UNGA in 1948 and tasked with the development and codification of international law. In particular, the 'regime of the high seas' was identified as a priority area. Over the next decade, the ILC made a number of recommendations including freedom of the high seas, flagging for ships, the right to safe passage, acts of piracy, baselines, high seas fishing and the 12 nmi territorial sea (www.un.org/law/ilc/).

To date, over 150 international conventions with significance to the marine environment have been adopted and are summarized in Table 2.2.

Table 2.2 List of international marine conventions and non-binding agreements

Convention name	Short name	Year of adoption	Year of entry into force	Secretariat/relevant authority	Ratifying parties as of 2012	Relevant chapter
Global non-marine conventions with importance to the marine environment						
1959 Antarctic Treaty		1959	1961	Secretariat of the Antarctic Treaty	50	8
1963 Treaty Banning Nuclear Weapon Tests in the Atmosphere, in Outer Space and Under Water	Partial Nuclear Test Ban Treaty (PTBT)	1963	1963	UN: UN Office for Disarmament Affairs (UNODA)	136	
1985 Vienna Convention for the Protection of the Ozone Layer		1985	1988	UN: UN Environment Programme (UNEP)	197	
1987 Montreal Protocol on Substances that Deplete the Ozone Layer	Montreal Protocol	1987	1989	United Nations: UNEP	197	
1991 Protocol on Environmental Protection to the Antarctic Treaty	Madrid Protocol	1991	1998	Secretariat of the Antarctic Treaty	33	9
1992 Agenda 21 Preamble and Chapter 17	Agenda 21	1992	1992	United Nations: Department of Economic and Social Affairs		5
1992 Rio Declaration on Environment and Development	Rio Declaration	1992	1992	UN: UNEP		5
1992 Convention on Biological Diversity (CBD)	CBD	1992	1993	Conference of the Parties	193	5
1992 United Nations Framework Convention on Climate Change (UNFCCC)	UNFCCC	1992	1994	UNFCCC Secretariat	194	5
1997 Kyoto Protocol to the UN Framework Convention on Climate Change	Kyoto Protocol	1997	2005	UNFCCC Secretariat	190	5
1997 Montreal Amendment to Montreal Protocol		1997	1999	UN: UNEP	179	

Table 2.2 continued

Convention name	Short name	Year of adoption	Year of entry into force	Secretariat/relevant authority	Ratifying parties as of 2012	Relevant chapter
1999 Basel Protocol on Liability and Compensation for Damage Resulting from Transboundary Movements of Hazardous Wastes and Their Disposal	The Basel Protocol on Liability and Compensation	1999	Not yet in force	UN: UNEP	10	5
1999 Beijing Amendment to the Montreal Protocol on Substances that Deplete the Ozone Layer	Beijing Amendment	1999	2002	UN: UNEP	156	
2001 Stockholm Convention on Persistent Organic Pollutants	Stockholm Convention 2001	2001	2004	UN: UNEP	168	
Global marine conventions						
1884 Convention for the Protection of Submarine Telegraph Cables	1884 Cable Convention	1884	1888	France	41	
1910 International Convention for the Unification of Certain Rules of Law related to Assistance and Salvage at Sea and Protocol of Signature	Assistance and Salvage Convention 1910	1910	1913	Comité Maritime International (CMI)	86	7
1910 International Convention for the Unification of Certain Rules of Law Related to Collision between Vessels and Protocol of Signature	Collision Convention 1910	1910	1913	CMI	95	7
1914 International Convention for the Safety of Life at Sea	SOLAS 1914	1914	1919	International Maritime Organization (IMO)	162	7
1923 Convention and Statute of the International Regime of Maritime Ports	1923 Ports Convention	1923	1926	UN Secretary-General	43	

Table 2.2 continued

Convention name	Short name	Year of adoption	Year of entry into force	Secretariat/relevant authority	Ratifying parties as of 2012	Relevant chapter
1924 International Convention for the Unification of Certain Rules of Law relating to Bills of Lading and Protocol of Signature	Hague Rules 1924	1924	1931	CMI	80	
1926 International Convention for the Unification of Certain Rules concerning the Immunity of State-owned Vessels and 1934 Additional Protocol	Immunity of State-Owned Ships Convention 1926/1934	1926, 1934 (Additional Protocol)	1937	CMI	30	
1946 International Convention for the Regulation of Whaling	Whaling Convention	1946	1948	International Whaling Commission (IWC)	88	6
1948 Convention on the International Maritime Organization	IMO Convention	1948	1958	IMO	170	7
1952 International Convention for the Unification of Certain Rules relating to Civil Jurisdiction in Matters of Collision	Collision/Civil Jurisdiction Convention 1952	1952	1955	CMI	85	7
1952 International Convention for the Unification of Certain Rules relating to Penal Jurisdiction in Matters of Collision and Other Incidents of Navigation	Collision/Penal Convention 1952	1952	1955	CMI	91	7
1952 International Convention for the Unification of Certain Rules relating to the Arrest of Sea-going Ships	Arrest Convention 1952	1952	1956	CMI	94	7
1957 International Convention relating to the Limitation of the Liability of Owners of Sea-going Ships and Protocol of Signature	Limitation of Shipowners' Liability (Revised) Convention 1957	1957	1968	CMI		7

Table 2.2 continued

Convention name	Short name	Year of adoption	Year of entry into force	Secretariat/relevant authority	Ratifying parties as of 2012	Relevant chapter
1958 Convention on the Continental Shelf	United Nations Conference on the Law of the Sea (UNCLOS) 1958	1958	1964	UN Division for Ocean Affairs & Law of the Sea	58	2
1958 Convention on Fishing and Conservation of the Living Resources of the High Seas	UNCLOS 1958	1958	1966	UN	38	2
1958 Convention on the Territorial Sea and the Contiguous Zone	UNCLOS 1958	1958	1964	UN	52	2
1958 Convention on the High Seas	Geneva Convention on the High Seas	1958	1962	UN	63	2
1961 International Convention for the Unification of Certain Rules relating to Carriage of Passengers by Sea and Protocol	Carriage of Passengers Convention 1961	1961	1965	CMI	11	7
1966 International Convention on Load Lines (CLL)	CLL 66	1966	1968	IMO		7
1967 Protocol to Amend the International Convention for the Unification of Certain Rules of Law relating to Assistance and Salvage at Sea	Assistance and Salvage Protocol 1967	1967	1977	CMI	9	7
1968 Protocol to Amend the 1924 International Convention for the Unification of Certain Rules of Law Relating to Bills of Lading	Visby Rules	1968	1977	CMI	30	7

Table 2.2 continued

Convention name	Short name	Year of adoption	Year of entry into force	Secretariat/relevant authority	Ratifying parties as of 2012	Relevant chapter
1969 International Convention Relating to Intervention on the High Seas in Cases of Oil Pollution Casualties	Intervention Convention 1969	1969	1975	IMO	87	7
1969 International Convention on Civil Liability for Oil Pollution Damage (CLC)	1969 CLC/1969 Civil Liability Convention	1969	1975	IMO	36	
1969 UN General Assembly Resolution (UNGAR) 2574 D (XXIV): Question of the Reservation Exclusively for Peaceful Purposes of the Sea-Bed and the Ocean Floor, and the Subsoil Thereof, Underlying the High Seas Beyond the Limits of Present National Jurisdiction, and the Use of Their Resources in the Interests of Mankind	UNGAR 2574	1969		UN		9
1970 UN General Assembly Resolution 2749 (XXV) on Declaration of Principles Governing the Sea-Bed and the Ocean Floor, and the Subsoil Thereof, Beyond the Limits of National Jurisdiction (UNGAR 2749)	UNGAR 2749/ Declaration of Principles	1970		UN		9
1972 Convention for the Prevention of Marine Pollution by Dumping from Ships and Aircraft (OSPAR)	Oslo Convention	1972	1974	OSPAR Commission	13	7
1972 Convention on the International Regulations for Preventing Collisions at Sea (COLREG)	COLREG 1972	1972	1977	IMO	155	7
1973 Protocol Relating to Intervention on the High Seas in Cases of Pollution by Substances Other Than Oil	Intervention Protocol	1973	1983	IMO	54	7

Table 2.2 continued

Convention name	Short name	Year of adoption	Year of entry into force	Secretariat/relevant authority	Ratifying parties as of 2012	Relevant chapter
1974 United Nations Convention on a Code of Conduct for Liner Conferences	Liner Code 1974	6 April 1974, Geneva, Switzerland	6 October 1983	UN	78	7
1974 International Convention for the Safety of Life at Sea (SOLAS)	SOLAS 1974	1 November 1974, London, UK	25 May 1980	IMO	162	7
1976 Convention on Limitation of Liability for Maritime Claims (LLMC)	LLMC	19 November 1976, London, UK	1 December 1986	IMO	52	7
1977 Convention on Civil Liability for Oil Pollution Damage Resulting from Exploration for and Exploitation of Seabed Mineral Resources (CLEE)	CLEE 1977	1 May 1977, London, UK		UK	0	7
1978 Protocol Relating to the 1973 International Convention for the Prevention of Pollution from Ships (MARPOL) (including Annexes, Final Act and 1973 International Convention)	MARPOL 73/78	17 February 1978	2 October 1983	IMO	151	7
1979 Protocol Amending the International Convention for the Unification of Certain Rules relating to Bills of Lading as modified by the Amending Protocol of 23rd February 1968	Special Drawing Rights Protocol	21 December 1979, Brussels, Belgium	14 February 1984	CMI	33	7

Table 2.2 continued

Convention name	Short name	Year of adoption	Year of entry into force	Secretariat/relevant authority	Ratifying parties as of 2012	Relevant chapter
1979 International Convention on Maritime Search and Rescue (SAR)	SAR Convention	27 April 1979, Hamburg, Germany	22 June 1985	IMO	96	7
1982 Paris Memorandum of Understanding (MOU) on Port State Control	Paris MOU	26 January 1982, Paris, France	1 July 1982	The Paris Memorandum of Understanding on Port State Control	27	7
1982 United Nations Convention on the Law of the Sea	Law of the Sea Convention (LOSC or UNCLOS)	10 December 1982, Montego Bay, Jamaica	16 November 1994	UN	162	2
1988 Convention for the Suppression of Unlawful Acts against the Safety of Maritime Navigation (SUA)	SUA 1988/1988 SUA Convention/ Maritime Convention	10 March 1988, Rome, Italy	1 March 1992	IMO	156	7
1988 Protocol for the Suppression of Unlawful Acts against the Safety of Fixed Platforms Located on the Continental Shelf	1988 SUA Protocol/ Fixed Platform Protocol/SUA Protocol	10 March 1988, Rome, Italy	1 March 1992	IMO	145	7
1989 Protocol Concerning Marine Pollution Resulting from Exploration and Exploitation of the Continental Shelf	1989 Protocol	29 March 1989, Kuwait	17 February 1990	Regional Organization for the Protection of the Marine Environment (ROPME)	8	7
1989 International Convention on Salvage	Salvage Convention	28 April 1989, London, UK	14 July 1996	IMO	58	7

Table 2.2 continued

Convention name	Short name	Year of adoption	Year of entry into force	Secretariat/relevant authority	Ratifying parties as of 2012	Relevant chapter
1991 UN General Assembly Resolution 46/215 (1991): Large-Scale Pelagic Drift-Net Fishing and Its Impact on the Living Marine Resources of the World's Oceans and Seas	UNGAR 46/215	20 December 1991		UN		6
1992 Protocol to Amend the 1969 International Convention on Civil Liability for Oil Pollution Damage	1992 Protocol/CLC Protocol 1992	27 November 1992, London, UK	30 May 1996	IMO	128	7
1992 International Convention on Civil Liability for Oil Pollution Damage (Consolidated text of the 1969 Convention, incorporating the amendments of 1976, 1992 and 2000)	1992 CLC/1992 Civil Liability Convention	27 November 1992, London, UK	30 May 1996	IMO	128	7
1993 International Convention on Maritime Liens and Mortgages	Maritime Liens and Mortgages Convention	6 May 1993, Geneva, Switzerland	5 September 2004	UN	13	7
1993 Agreement to Promote Compliance with International Conservation and Management Measures by Fishing Vessels on the High Seas	Food and Agriculture Organization (FAO) Compliance Agreement	24 November 1993, Rome, Italy	24 April 2003	FAO	39	5
Part XI of the 1982 United Nations Convention on the Law of the Sea as Amended by the 1994 Agreement on the Implementation of Part XI of the 1982 United Nations Convention on the Law of the Sea	Installation/Marine Environment/ Decommission/ Decommissioning/ Mining/Seabed	28 July 1994, New York, US	28 July 1996	UN	141	2

Table 2.2 continued

Convention name	Short name	Year of adoption	Year of entry into force	Secretariat/relevant authority	Ratifying parties as of 2012	Relevant chapter
1994 Agreement Relating to the Implementation of Part XI of the United Nations Convention on the Law of the Sea of 10 December 1982		28 July 1994, New York, US	28 July 1996	UN	141	2
1995 Agreement for the Implementation of the Provisions of the United Nations Convention on the Law of the Sea of 10 December 1982 relating to the Conservation and Management of Straddling Fish Stocks and Highly Migratory Fish Stocks	Fish Stocks Agreement 1995	4 December 1995, New York, US	11 December 2001	UN	77	6
1996 Protocol to Amend the 1976 Convention on Limitation of Liability for Maritime Claims (LLMC)	LLMC Protocol 1996	2 May 1996, London, UK	13 May 2004	IMO	35	7
1996 Protocol to the 1972 Convention on the Prevention of Marine Pollution by Dumping of Wastes and Other Matter	LC Protocol 1996	7 November 1996, London, UK	24 March 2006	IMO	42	7
1997 Agreement on the Privileges and Immunities of the International Tribunal for the Law of the Sea		23 May 1997 in New York, US	30 December 2001	International Tribunal for the Law of the Sea (ITLOS)	39	2
1998 Protocol on the Privileges and Immunities of the International Seabed Authority		26 March 1998 in Kingston, Jamaica	31 May 2003	International Seabed Authority (ISA)	32	2

Table 2.2 continued

Convention name	Short name	Year of adoption	Year of entry into force	Secretariat/relevant authority	Ratifying parties as of 2012	Relevant chapter
1999 International Convention on Arrest of Ships (ARREST)	Arrest of Ships Convention/ ARREST 1999	12 March 1999, Geneva, Switzerland	Not yet in force	UN	7	7
2000 Specific Guidelines for Assessment of Platforms or Other Man-Made Structures at Sea		18–22 September 2000		IMO		
2001 International Convention on the Control of Harmful Anti-fouling Systems on Ships (HAFS)	The HAFS Convention/ AFS 2001	5 October 2001, London, UK	17 September 2008	IMO	60	5
2001 International Convention on Civil Liability for Bunker Oil Pollution Damage (BUNKERS)	Bunkers Convention/ BUNKERS 2001	23 March 2001, London, UK	21 November 2008	IMO	46	7
2004 International Convention for the Control and Management of Ships' Ballast Water and Sediments (BWM)	BWM 2004/ Ballast Water Convention	13 February 2004	Not yet in force	IMO	35	5
2005 Protocol to the 1988 Convention for the Suppression of Unlawful Acts against the Safety of Maritime Navigation (SUA)	2005 Protocol to the 1988 SUA Convention/2005 SUA Protocol	14 October 2005, London, UK	28 July 2010	IMO	17	7
2005 Protocol to the 1988 Protocol for the Suppression of Unlawful Acts against the Safety of Fixed Platforms Located on the Continental Shelf	2005 SUA Platforms Protocol/SUA PROT 2005	14 October 2005, London, UK	28 July 2010	IMO	13	7

Table 2.2 continued

Convention name	Short name	Year of adoption	Year of entry into force	Secretariat/relevant authority	Ratifying parties as of 2012	Relevant chapter
2005 Protocol for the Suppression of Unlawful Acts against the Safety of Fixed Platforms Located on the Continental Shelf	2005 SUA Platforms Protocol/SUA Protocol 2005	14 October 2005, London, UK	28 July 2010	IMO	13	7
2005 Convention for the Suppression of Unlawful Acts Against the Safety of Maritime Navigation	2005 SUA Convention	14 October 2005, London, UK	28 July 2010	IMO	17	7
2006 Maritime Labour Convention (MLC)	MLC	7 February 2006 in Geneva, Switzerland	Not yet in force	International Labour Organization (ILO)	15	7
2007 Nairobi International Convention on the Removal of Wrecks (WRC)	WRC 2007/ NAIROBI WR Convention 2007/ Nairobi WRC 2007	18 May 2007, Nairobi, Kenya	Not yet in force	IMO	5	7
2009 London Convention and Protocol/ UNEP Guidelines for the Placement of Artificial Reefs		27–31 October 2008		UN		
2009 Hong Kong International Convention for the Safe and Environmentally Sound Recycling of Ships	Hong Kong Convention	15 May 2009, Hong Kong, People's Republic of China	Not yet in force	IMO	0	7

Table 2.2 continued

Convention name	Short name	Year of adoption	Year of entry into force	Secretariat/relevant authority	Ratifying parties as of 2012	Relevant chapter
2009 Agreement on Port State Measures to Prevent, Deter and Eliminate Illegal, Unreported and Unregulated Fishing	Port State Measures Agreement	22 November 2009, Rome, Italy	Not yet in force	FAO	16	5
2010 United Nations General Assembly Resolution A/65/37A on Oceans and the Law of the Sea	UNGAR A/65/37A	7 December 2010				
Important multilateral regional marine conventions and national proclamations						
1839 Convention between France and Great Britain for Defining the Limits of Exclusive Fishing Rights		2 August 1839	17 August 1839	UN	2	
1882 Convention regulating the Police of the North Sea Fisheries	North Sea Fisheries Convention	6 May 1882	15 March 1884	Government of Netherlands	6	
1884 Convention for the Protection of Submarine Telegraph Cables	1884 Cable Convention	14 March 1884, Paris, France	1 May 1888	Government of France	41	
1936 Convention Regarding the Regime of the Straits	Montreux Convention	20 July 1936, Montreux, Switzerland	9 November 1936	Government of Turkey	9	2
1942 Treaty between Great Britain and Northern Island and Venezuela Relating to the Submarine Areas of the Gulf of Paria		26 February 1942, Caracas, Venezuela	22 September 1942	UN	3	

Table 2.2 continued

Convention name	Short name	Year of adoption	Year of entry into force	Secretariat/relevant authority	Ratifying parties as of 2012	Relevant chapter
1945 US Presidential Proclamation No. 2668, Policy of the United States with Respect to Coastal Fisheries in Certain Areas of the High Seas	Truman Proclamation	28 September 1945, Washington, US		Government of the US		2
1945 US Presidential Proclamation No. 2667, Policy of the United States with Respect to the Natural Resources of the Subsoil and Sea Bed of the Continental Shelf	Truman Proclamation	28 Sept 1945, Washington, US		Government of the US		2
1972 Convention for the Conservation of Antarctic Seals (CCAS)	CCAS	1 June 1972, London, UK	11 March 1978	British Antarctic Survey	17	8
1980 Convention on the Conservation of Antarctic Marine Living Resources (CCAMLR)	CCAMLR	20 May 1980, Canberra, Australia	7 April 1982	Commission for the Conservation of Antarctic Marine Living Resources	30	8
1980 Protocol for the Protection of the Mediterranean Sea Against Pollution from Land-Based Sources and Activities (LBS)	LBS Protocol	17 May 1980, Athens, Greece	17 June 1983	UN: UNEP	22	5
1982 Agreement Concerning Interim Arrangements Relating to Polymetallic Nodules on the Deep Sea Bed between France, the Federal Republic of Germany, the United Kingdom and the United States		2 September 1982, Washington, US	2 September 1982	US	4	9

Table 2.2 continued

Convention name	Short name	Year of adoption	Year of entry into force	Secretariat/relevant authority	Ratifying parties as of 2012	Relevant chapter
1984 Provisional Understanding Regarding Deep Seabed Matters between Belgium, France, the Federal Republic of Germany, Italy, Japan, the Netherlands, the United Kingdom and the United States	Provisional Understanding Regarding Deep Seabed Matters	3 August 1984, Geneva, Switzerland	2 September 1984	US	6	
1989 Guidelines and Standards for the Removal of Offshore Installations and Structures on the Continental Shelf and in the Exclusive Economic Zone (IMO Resolution A.672 (16))	Offshore Installations Resolution	19 October 1989		IMO		
1994 Protocol for the Protection of the Mediterranean Sea against Pollution Resulting from Exploration and Exploitation of the Continental Shelf and the Seabed and its Subsoil	Offshore Protocol	14 October 1994, Madrid, Spain	24 March 2011	UN: UNEP	6	
1995 Protocol Concerning Specially Protected Areas and Biological Diversity in the Mediterranean (SPA)	SPA and Biodiversity Protocol	10 June 1995, Barcelona, Spain	12 December 1999	UN: UNEP	18	
1996 Protocol on the Prevention of Pollution of the Mediterranean Sea by Transboundary Movements of Hazardous Wastes and Their Disposal	Hazardous Wastes Protocol	1 October 1996, Izmir, Turkey	19 December 2008	UN: UNEP	6	
1998 OSPAR Decision 98/3 on the Disposal of Disused Offshore Installations		23 July 1998, Sintra, Portugal	9 February 1999	OSPAR Commission		

Table 2.2 continued

Convention name	Short name	Year of adoption	Year of entry into force	Secretariat/relevant authority	Ratifying parties as of 2012	Relevant chapter
2001 Convention on the Protection of the Underwater Cultural Heritage	2001 Convention	2 November 2001, Paris, France	2 January 2009	UN: United Nations Educational, Scientific and Cultural Organization (UNESCO)	41	
2002 Protocol Concerning Cooperation in Preventing Pollution from Ships, and in Cases of Emergency, Combating Pollution of the Mediterranean Sea	Prevention and Emergency Protocol	25 January 2002, Valetta, Malta	17 March 2004	IMO	12	
2004 Regional Cooperation Agreement on Combating Piracy and Armed Robbery against Ships in Asia (ReCAAP)	ReCAAP	11 November 2004, Tokyo, Japan	4 September 2006	ReCAAP Information Sharing Centre	15	
2008 Protocol on Integrated Coastal Zone Management (ICZM) in the Mediterranean	ICZM Protocol	21 January 2008, Madrid, Spain	24 March 2011	UN: UNEP	7	
2008 United Nations Security Council Resolutions (UNSCRs) 1816, 1848 and 1851 on the Situation in Somalia	UNSCRs 1816, 1848 and 1851	2 June 2008 2 December 2008 16 December 2008		UN Security Council	7	
2010 United Nations Security Council Resolution 1918 on the Situation in Somalia	UNSCR 1918 2010	27 April		UN Security Council		

Table 2.2 continued

UNEP Regional Seas Programme Conventions

Convention name	Short name	Year of adoption	Year of entry into force	Secretariat/relevant authority	Ratifying parties as of 2012	Relevant chapter
Convention on the Protection of the Marine Environment of the Baltic Sea Area	Helsinki Convention	1974, amended 1992	1980, and amendments in 2000	Helsinki Commission	10	5
Convention for the Protection of the Marine Environment and the Coastal Region of the Mediterranean	Barcelona Convention	1976, amended 1995	1978, and amendments in 2004	Mediterranean Action Plan	18	5
Kuwait Regional Convention for Co-operation on the Protection of the Marine Environment from Pollution	Kuwait Convention	1978	1978	ROPME Secretariat	8	5
Abidjan Convention for Co-operation in the protection and Development of the Marine and Coastal Environment of the West and Central African Region	Abidjan Convention	1981	1984	Abidjan Convention Secretariat	33	5
Convention for the Protection of the Marine Environment and Coastal Area of the South-East Pacific	Lima Convention	12 November 1981	1986	CPPS	5	5
Regional Convention for the Conservation of the Red Sea and Gulf of Aden Environment	Jeddah Convention	1982	1985	Programme Secretariat	7	5
Convention for the Protection and Development of the Marine Environment of the Wider Caribbean Region	Cartagena Convention	1983	1986	Caribbean Environment Programme	28	5

Table 2.2 continued

Convention name	Short name	Year of adoption	Year of entry into force	Secretariat/relevant authority	Ratifying parties as of 2012	Relevant chapter
Convention for the Protection, Management and Development of the Marine and Coastal Environment of the Eastern African Region	Nairobi Convention	1985, amended 2010	1996, amendments not yet entered into force	UN: UNEP	10	5
Convention for the Protection of Natural Resources and Environment of the South Pacific Region	Noumea Convention	1986	1990	Secretariat of the Pacific Regional Environment Programme	24	5
Convention on the Protection of the Black Sea Against Pollution	Bucharest Convention	21 April 1992	15 January 1994	Black Sea Commission	5	5
Convention for the Protection of the Marine Environment of the North-East Atlantic	OSPAR Convention	1992	1998	OSPAR Commission	16	5
Convention for Cooperation in the Protection and Sustainable Development of the Marine and Coastal Environment of the North-East Pacific	Antigua Convention	1992	Not yet in force	US	8	5

Source: treaties.un.org/Pages/MSDatabase.aspx

Differences between marine law and other legal systems

As far back as Roman times marine law was differentiated from Roman (land) law. It was recognized that, regardless of nationality, culture or geographic location, maritime legal issues were predominantly international in nature and concerned with addressing safety, navigation, ship registration, contracts, liability and insurance.

Marine (or 'maritime') law originates from civil law tradition. While influenced by common law, international maritime law exhibits the key tenets of civil law, which include codification modified by aspects of customary law and past practices and traditions. There are, however, a number of key differences between marine law and other legal systems. These are discussed in Box 2.2.

Tetley (2000) defines the following three types of maritime law:

- Public international maritime law: Encompasses the legal relationships between states in respect of marine matters and consists of agreements, conventions and treaties negotiated between states or supranational organizations (e.g. EU).
- Conflict of maritime laws: The collection of rules used to resolve maritime disputes between private parties subject to the laws of different states.
- International private maritime law: Encompasses the legal maritime relationships between private parties of different states. While negotiated between states, private international maritime law primarily applies to the actions of non-state actors such as shipping companies.

Box 2.2 Differences between marine law and other legal systems

Jurisdiction: Jurisdiction is generally less important in marine law than other domestic legal systems. In most international agreements and conventions a number of jurisdictions may be involved in a marine dispute. Imagine a ship that is registered (flagged) in a state different than its ownership. Now imagine it to be involved in an incident during transit in the territory of one or more coastal states while travelling between two different port states. As such, multiple states may have some jurisdiction over a matter.

Dispute resolution: Marine law utilizes alternative dispute resolution mechanisms as a means to arbitrate disputes and an international body of jurisprudence regarding arbitration is developing. In particular, the UN Law of the Sea Convention outlines specific dispute resolution mechanisms that are mandatory for state parties to the Convention.

Self regulation: Marine law is increasingly self-regulatory and protocols are established independent of government organizations.

Financial caps: Tort (harms) law operates on the principle of *restitutio in integrum*, meaning that full compensation is due to the victim. However, from as early as the eleventh century, shipping law has operated under the principle of limitation of liability (financial caps), which caps damages at certain, agreed-upon limits. Limitation of liability applies across a number of international conventions and agreements pertaining to the transport of goods, passengers, noxious substances and oil.

Source: Cho (2010)

In addition, the collection of international marine laws that govern how states may use the ocean is sometimes referred to as 'admiralty law' or the 'law of the seas' (not to be confused with the UN Law of the Sea Convention discussed below). 'Maritime law' often refers to the 'international private marine law' discussed above that primarily applies to private parties operating in the marine environment.

UN Conventions on the Law of the Sea

By the 1950s, human uses of the oceans, driven by globalization, technological advances and human population growth changed significantly. Offshore hydrocarbon development, seabed mining, container-based shipping, high seas fishing and militarization of the oceans highlighted the limitations of the existing rules of the ocean and the relationships between nations that use the ocean.

In particular, a number of jurisdictions, led by the US, had unilaterally increased their jurisdiction beyond the 12nmi territorial seas. This precipitated friction between coastal states and seafaring nations with respect to shipping and other coastal economic activities. There was also considerable interest in seabed mining in the 1960s to 1970s, yet little agreement on its governance, whether on or off the continental shelves. States at this time were also positioning themselves to their best advantage based on geography, foreign policy and shipping to their markets. *Coastal states* blessed with significant coastal areas adopted national policies exerting control and exclusive rights over their marine areas. Also termed 'territorialists', states such as Canada and Russia considered marine resource development and conservation a national right. *Maritime states*, including the US and other geographically smaller seafaring nations (e.g. UK) advocated for the freedom of the seas. Also termed 'internationalists', maritime states were increasingly concerned that a patchwork of national regulations would complicate shipping and other resource development (McDorman, 2008). States' positions on marine sovereignty were further articulated by how they viewed uses of their marine territory by other states. *Functionalist states* believed that states should only regulate their marine areas for certain activities such as vessel-source pollution. *Unilateralist states* believed that *all* marine activities in their waters should be regulated.

The first United Nations Conference on the Law of the Sea (UNCLOS I) was held in Geneva, Switzerland in 1958. Attended by 86 states, UNCLOS I produced the following four conventions:

- Convention on the Territorial Sea and Contiguous Zone (entered into force in 1964). While this Convention acknowledged the existence of the territorial sea (including the airspace above and seabed below) and defined how the landward boundary is to be measured, the breadth of the territorial sea provoked controversy. The Convention also codified the rights of innocent passage through territorial waters and international straits. Lastly, the Convention defined the contiguous zone as 12nmi seaward of the baseline and permitted coastal states to intervene in these waters. This was hoped to prevent and punish infringement of its customs, fiscal, immigration or sanitary regulations within its territory or territorial sea (United Nations, 2005b).
- Convention on the Continental Shelf (entered into force in 1964). This Convention gave coastal states rights over the natural resources on their continental shelves. The shelves were defined as '... to the seabed and subsoil of the submarine areas adjacent to the coast but outside the area of the territorial sea, to a depth of 200 metres or, beyond

that limit, to where the depth of the superjacent waters admits of the exploitation of the natural resources of the said areas; (b) to the seabed and subsoil of similar submarine areas adjacent to the coasts of islands'. Within the continental shelf areas the Convention provided direction on permissible activities and their relation to safe navigation (United Nations, 2005c).

- Convention on the High Seas (entered into force in 1962). This Convention defined the 'high seas' as any non-territorial or non-internal waters and defined the permissible activities on the high seas along with the requirements for flagging of ships and navigational safety (United Nations, 2005d).
- Convention on Fishing and Conservation of the Living Resources of the High Seas (entered into force in 1966). This Convention allows all states to fish on the high seas and outlines measures to ensure that stocks are not overfished (United Nations, 2005e).

The second United Nations Conference on the Law of the Sea (UNCLOS II) was again held in Geneva in 1960. The Conference was intended to focus on the breadth of the territorial sea and fishery limits. However, as with so many international meetings of the era, it failed to reach agreement on any new conventions.

The third United Nations Conference on the Law of the Sea (UNCLOS III and hereafter referred to as the 'LOSC') was held in New York in 1973. While the original intent of the Conference was to regulate seabed mining, the Conference became a nine-year exercise among 160 states to rewrite the rules of the ocean. Eleven negotiating sessions were held over 585 days during this protracted period. The final Convention, adopted in 1982, established navigational rights, territorial sea limits, economic jurisdiction, rights to seabed resources, management of shipping through narrow straits and the security of the marine environment (United Nations, 1982).

The LOSC is perhaps the most complex and comprehensive international agreement ever completed. It rivals the International Declaration of Human Rights and Charter of the United Nations (both non-binding) for its scope and impact on humanity. Indeed, there is no terrestrial agreement that approaches the comprehensiveness of the LOSC.

The LOSC can be summarized in three principles (after LeGresley, 1993):

- States own and have the geographically defined rights to certain resources in their adjacent marine areas.
- Parts of the ocean are the 'common heritage of mankind'.
- States are obliged to preserve the seas and accommodate the needs of other states.

The LOSC consists of 17 parts, 400 articles and 9 annexes and currently has 162 signatories. The LOSC entered into force for those states a party to it 12 months after the signature of the 60th state, in November 1994.

The LOSC enabled the establishment of four institutions (later discussed in this chapter):

- International Tribunal for the Law of the Sea.
- Commission the Limits of the Continental Shelf.
- International Seabed Authority.
- Meeting of States Parties to the Convention.

The LOSC also led to, or informed, a number of other agreements that will be discussed later in this chapter.

The LOSC also includes a number of provisions for dispute resolution. Unlike most treaties where mechanisms for dispute resolution are contained in a separate protocol, the LOSC is unique in that dispute resolution is part of the Convention and signatories to the agreement must abide by the LOSC's dispute resolution apparatus. Unique to international law, the LOSC contains both negotiation and arbitration approaches to dispute resolution.

Under the LOSC (Article 287, Part XV), arbitration (binding) approaches include submission to:

- The International Tribunal for the Law of the Sea, which includes the Seabed Disputes Chamber.
- The ICJ.
- Ad hoc arbitration (in accordance with Annex VII).
- Special arbitration tribunals (established under Annex VIII).

For disputes related to maritime boundaries, military matters and issues under discussion by the United Nations Security Council, the LOSC allows states at any time ('optional exceptions') to opt out of compulsory adjudication. In addition, for issues regarding national sovereignty, the LOSC permits states to submit disputes to a conciliation commission, which is non-binding but uses moral suasion to pressure states into compliance with the commission's judgments.

The LOSC establishes no global forum or process where states meet to discuss the implementation of the Convention or other ocean matters. As such, addressing imminent ocean issues has fallen to the UNGA, which has passed a number of non-binding resolutions on contentious topics such as high seas driftnet fishing and piracy (McDorman, 2008). For less urgent matters, the UN Secretary-General, using authority provided for in the LOSC, has convened annual meetings of the State Parties to the Law of the Sea Convention (SPLOS). These assemblies address administrative matters such as the election of judges for ITLOS and members of the Commission on the Limits of the Continental Shelf.

Marine jurisdiction under the LOSC

In the years following Hugo Grotius' *Mare Liberum* proposal that oceans that could not be continuously occupied were free to all, many states claimed jurisdiction out to 3nmi seaward of their coasts (the 'cannon shot rule'). Despite the absence of global agreement, this distance gradually increased to 12nmi by the 1960s as many countries expanded their marine sovereignty. However, it was noted that if every state increased their jurisdiction 12nmi this would effectively bring under national jurisdiction more than 100 straits and channels many of which, including the Straits of Gibraltar (Mediterranean) and Hormuz (Persian Gulf), are important global shipping routes. This situation created tension between the major ocean-faring states and smaller coastal states with territories adjacent to straights and narrows.

While many states were taking sides on the 3 and 12nmi jurisdictional issue, other states had more ambitious claims of marine sovereignty. The 1945 Truman Proclamation unilaterally claimed jurisdiction to all natural resources on the continental shelves of the US. Argentina, Chile, Peru and Ecuador soon followed suit; but, given their very narrow continental shelves, they claimed jurisdiction out to 200nmi. Other nations, such as Canada, adopted a middle position, claiming a 100nmi jurisdiction over navigation in an effort to protect its Arctic areas from pollution.

Under the LOSC, the following jurisdictional arrangements were agreed to:

The 'baseline'

All the LOSC jurisdictional agreements originate from what is termed in the agreement as 'the baseline'. The baseline is defined as the low tide mark; all other jurisdictional zones are measured seaward from this line (Article 5). While the baseline concept works well for certain types of coastlines (i.e. ones with easily established low tide lines), other types of coastlines, including island archipelagos, estuarine/lagoon systems and Antarctic ice shelves are more difficult to ascertain. Accordingly, Article 7 permits straight lines to be drawn between points on a coastline where the coastline is convoluted or low tide marks difficult to determine.

Internal waters

All waters (marine, estuarine, freshwater) on the landward side of the baseline fall under the jurisdiction of the state. In internal waters, states may establish and enforce domestic (national) laws, regulate use and access any resource (Figure 2.1). Foreign vessels have no right of passage within internal waters unless authorized by the respective state (Article 8).

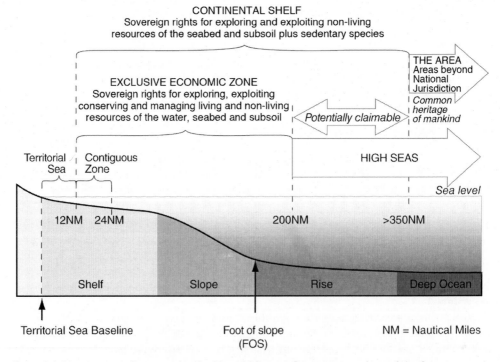

Figure 2.1 Maritime jurisdiction under the United Nations Convention on Law of the Sea

In instances, such as bays, where the width of the mouth of the bay is greater than 24nmi from headland to headland internal waters will be indented toward the baseline.

Territorial waters: the 12nmi territorial sea

From the baseline to 12nmi states are free to establish laws, regulate use and use any resource within this zone. Delineating territorial waters is optional and states are not required to exert jurisdiction to the full 12nmi. The right to designate territorial waters extends even to rocks and islets above the water line. Within this zone vessels have the right to 'innocent passage', defined as passage that is 'not prejudicial to the peace, good order or the security' of the coastal state. A vessel may not fish, pollute, discharge weapons for practice or spy on the coastal state.

Coastal states may suspend the right of innocent passage in strategic areas of their coastal seas. Warships are immune from coastal states' jurisdiction within the territorial sea and the coastal state cannot board a warship. This is even so if a crime is committed on board while in the port of the coastal state (a function of the sovereign immunity of the vessel).

For those states with strategic straits (necessary for global trade) the concept of 'transit passage' was agreed to in the LOSC. In this instance naval vessels, including submarines and ship-born associated aircraft are allowed to maintain configurations (e.g. carrying weapons) or postures (readiness) that would not be permitted under the rights of innocent passage. Ships in transit passage must adhere to international civilian regulations (e.g. navigation safety, pollution) and must traverse through the strait without stopping except in the case of emergencies. In particular, under transit passage, submarines must surface (and display their flag – although debated by the US), aircraft may not be launched from ships, and research or surveying must not be performed.

The LOSC provides consideration for archipelagic states such as Indonesia where the 12nmi zone is defined using a line joining the outermost points of the outermost islands. In the archipelagic zone ships of all states have the right of innocent passage; however, the state has the right to designate shipping lanes to ensure unobstructed passage.

Territorial waters: beyond the 12nmi territorial sea to 24nmi

Beyond the 12nmi territorial waters seaward to 24nmi lies the 'contiguous zone' where a state can enforce jurisdiction related to customs, taxation, immigration and pollution. In addition, the contiguous zone and beyond is a zone of 'hot pursuit'. In these areas states may pursue those who have violated, or are about to violate, a state's laws in the territorial sea. A vessel can be hotly pursued up to the territorial sea of another state and hot pursuit also applies to violations of domestic laws.

The Exclusive Economic Zone

The Exclusive Economic Zone (EEZ) extends from the baseline out to 200nmi. The EEZ extends outward to 200nmi even if the coastal state's continental shelf is less than 200nmi in width. Under this definition the total area of the world's EEZs is approximately 30 million square nmi (Figure 2.2). EEZs were established primarily to address fishing by foreign fleets and to extract offshore oil and gas in certain areas such as the Gulf of Mexico. EEZs are becoming increasingly important, however, as technology to extract hydrocarbons and other non-renewable resources improves.

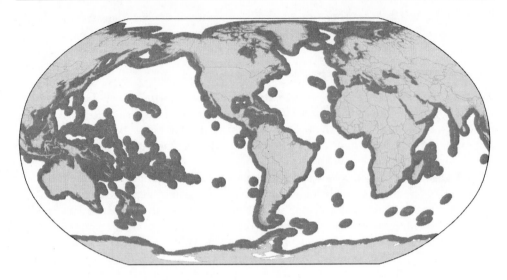

Figure 2.2 Exclusive Economic Zones of the world

The LOSC confers on the coastal state the rights to explore, exploit, conserve and manage the resources in the water column and on and under the seabed (Article 56). Included in these rights is the authority to construct (or dismantle) and regulate the operation of installations, structures and artificial islands in accordance with international regulations (Article 60). Coastal states are also required to promote the 'optimum utilization' of living resources within their EEZs. Based on best available science, states are obliged to establish catch limits for living resources and to ensure that fisheries do not negatively affect other species in their EEZs (Article 61). Any surplus catch below optimum utilization is to be offered to other states based on historical use and the needs of less developed states. In addition, the coastal state is required to publish their domestic fishing laws and regulations (Article 62). A coastal state has the right to inspect non-military ships within its EEZ and make arrests if the ship contravenes the coastal state's laws. Coastal states, however, must release arrested vessels and their crews upon posting of a reasonable bond and crews may not be imprisoned (Article 73). Coastal states may not transfer rights to resources in their EEZs to other states or their nationals when agreements are in place with landlocked developing countries, geographically disadvantaged countries or less developed countries (Article 72).

Within another state's EEZ, foreign ships or aircraft (used for spotting schools of fish or marine mammals) retain the right to fish, conduct research, construct seabed cables or pipelines, and explore and develop natural resources with the permission, and within the regulations, of the coastal state (Article 58).

Continental shelves

The EEZ and a state's continental shelf are not the same. The continental shelf is the submerged portion of a continent that extends from the baseline to the 200nmi EEZ or until the submerged continent ends, whichever is greater. The LOSC refers to the 'natural prolongation' of the land territory as the 'extended continental shelf', including the seabed and subsoil of the shelf (Article 76). The LOSC provides two constraints on the maximum extents of the extended continental shelf: the first is that the shelf may not be more than 350nmi from a state's coastal baselines. The second constraint limits the extended continental shelf 100nmi seaward from the 2500m isobath (depth contour). Within the two above constraints, continental shelves can be delineated by the following means, also termed formula lines:

- Sediment thickness: A line, termed the *Gardiner line*, is defined by the points where the thickness of the sediment is at least 1 per cent of the distance to the foot of the slope.
- Bathymetry: A formula, termed the *Hedberg formula*, defines a line by points 60nmi seaward from the foot of the continental slope where the foot of the slope is defined as the point of maximum change in the gradient at its base.

Once the Gardiner/Hedberg lines are established, then the constraint lines are applied to determine the outer limit of the continental shelf.

Approximately 85 coastal states have continental shelves beyond the 200nmi limit (e.g. Canada's east coast extends to 400nmi and beyond). In instances where states have extended continental shelves they are required to submit information to the Commission on the Limits of the Continental Shelf, which makes recommendations that the submitting states considers and may make a resubmission. Ultimately, it is the submitting state that decides where the outer limit of its shelf is and submits the coordinates and maps to the UN Secretary-General. At this time, the outer limit is final and binding on the submitting state (subject to bilateral overlapping claims of other states) (Article 76).

Coastal states have rights to the non-biological resources on and within continental shelves as well as living sedentary species (species that do not leave the seabed) that extend beyond 200nmi but not to resources (i.e. fish) within the water column in these areas (Article 77). Also, coastal states have the exclusive right to authorize and regulate drilling on the continental shelves. However, they do not have rights to the water column or airspace above the extended continental shelf; nor can coastal states' rights interfere with navigation (Article 78). States may set conditions and establish locations for the laying of seabed cables and pipelines but they may not prohibit other states from installing seabed cables and pipelines (Article 79). The rights of coastal states to the extended continental shelf are not conditional on occupation, proclamation or use (Article 77).

Coastal states are required to make payments to the International Seabed Authority (see below), based on revenue produced from the extended continental shelf, which in turn will redistribute to other states. This, in turn, is based on equitable sharing taking account of the needs and interests of less developed states. Payments are to be made annually after the first five years of production on a sliding royalty scale starting at 1 per cent of the value of production and increasing by 1 per cent annually until year 12, after which it remains at 7 per cent (Article 82).

High seas

The high seas are best understood as the water column beyond the 200nmi EEZ where domestic law, outside of those states with accepted extended continental shelves, no longer applies. The LOSC affirms the reservation of the high seas for peaceful purposes (Article 88), prohibits high seas sovereignty claims (Article 89) and affirms the right of all states to operate ships under their national flags (Article 90). The LOSC further declares that the high seas are open to all states, including those that are landlocked and/or developing (Article 87). High seas rights conferred on all states include freedom of: navigation, fishing, scientific research, overflight, laying submarine cables or pipelines, and the construction of artificial islands or structures (Article 87).

The LOSC provides the administrative framework for shipping on the high seas (Articles 90–99). Specific provisions include direction on the duties of flag states (Articles 90–94) including the requirement for states to set the conditions (construction, training, seaworthiness, communications) under which ships may sail under their flags (Articles 91 and 94). Ships are therefore bound to the national laws of their flag (Article 92) while flag states are required to maintain a register of ships flying their flags (Article 94). Flag states are also obliged to investigate any reports by other states of non-compliance with national or international laws by flag state ships (Article 94).

The LOSC provides immunity to warships and ships (e.g. survey ships) on non-commercial government business (Article 96). For those ships involved in collisions or other navigational incidents, the arrest and detention of a ship may only be undertaken by the flag state. Furthermore, disciplinary proceedings against masters and crew may be undertaken only by the flag state or state(s) of which the master and crew are nationals. Lastly, only the state that issued the certificates or licenses of a disciplined master or crew member may revoke these privileges (Article 97).

All ships must render assistance to persons and vessels in distress and coastal states must establish and maintain effective search and rescue services (Article 98). No ship may transport slaves (Article 99) and states shall cooperate to intercept ships carrying illegal drugs (Article 108).

Articles 100–107 address piracy on the high seas. The LOSC binds all states to cooperate in the elimination of piracy (Article 100) and defines piracy as any illegal act of violence, detention or depredation against a ship or aircraft and/or its crew or cargo. Piracy can transpire in areas beyond national jurisdiction (ABNJ) (Article 101), and it includes the mutiny of a warship, government ship or aircraft by its crew (Article 102). The LOSC defines a pirate ship as any ship or aircraft controlled by persons with the intent as outlined in Article 101 and allows any state to seize or arrest a pirate ship (or aircraft), using government ships or aircraft, and prosecute pirates under the state's national laws (Articles 105 and 107).

The LOSC also deals with broadcasting. With the exception of distress calls, no ship may broadcast to the general public from the high seas. Persons engaged in unauthorized broadcasting may be prosecuted by the flag state, coastal states (receiving the transmissions), state of which the broadcasters are nationals, or states where the broadcasting equipment was installed (Article 109).

Warships may not board ships of another flag state unless the ship is involved in piracy, slave trade, unauthorized broadcasting or is suspected to be without nationality (Article 110). To discourage this practice, ship owners may receive compensation for any loss or damage from states whose warships or aircraft wrongly board ships believed to be contravening Articles 100–109.

Foreign ships believed to be contravening national or international laws in territorial waters may be pursued by domestic warships or aircraft into the high seas. This right of 'hot pursuit' begins once the offending ship has been warned by visual and auditory means to stop. It ends when the ship that is being pursued enters the territorial sea of its flag state (Article 111). Ship owners may receive compensation for any loss or damage from states whose warships or aircraft incorrectly seize or arrest ships believed to be contravening national laws.

As mentioned previously, all states have the right to lay pipelines and submarine cables in ABNJ (Article 112). However, states have several obligations regarding pipelines and cables. For instance, damaging or interfering with them is necessarily a punishable offence. They are also required to pass domestic laws to indemnify ship owners for damages that are caused by attempts to avoid harm to pipelines (Article 113).

The LOSC affirms the rights of nationals from all states to fish on the high seas (see Chapter 6). This is subject to regional fisheries agreements and treaties, as well as the rights of coastal states regarding straddling stocks and highly migratory species (Article 116). All states, however, must cooperate on the conservation of living resources (fish, marine mammals, sea birds) on the high seas and, where one or more states use the same resources, negotiate agreements (or establish regional fisheries management organizations) to ensure optimum sustained yield (Articles 117–119). States are also required to base decisions on the best available science. This science is to be shared between states through sub-regional, regional and international organizations (Article 119).

Islands

Under the LOSC islands have the same jurisdictional regime (territorial sea, contiguous zone, EEZ and continental shelf) as other land masses. The exception is what the LOSC terms 'rocks', which are emergent lands that are unable to support human habitation or economic activity (Article 121).

Enclosed or semi-enclosed seas

The LOSC defines enclosed or semi-enclosed seas as a '…gulf, basin or sea surrounded by two or more States and connected to another sea or the ocean by a narrow outlet or consisting entirely or primarily of the territorial seas and exclusive economic zones of two or more coastal States'. Examples of enclosed or semi-enclosed seas include some of the world's most important water bodies and include the Black, Baltic, Caribbean and Mediterranean seas and the Persian Gulf, Gulf of Mexico and Gulf of Thailand. The management of enclosed or semi-enclosed seas is convoluted by a number of problematic facts: some states have not ratified the LOSC (e.g. Turkey, the US); other states have not delineated their marine boundaries (e.g. Caribbean states); some states have strategically important seaways (e.g. South China Sea); and some states have interests in seabed petroleum resources (e.g. North Sea). Moreover, some areas with these kinds of seas are not even recognized as states (e.g. Taiwan, Northern Cyprus) (CEPAL, 2004).

The LOSC dictates that states bordering on enclosed or semi-enclosed seas cooperate on: the management of living marine resources; protection and preservation of the marine environment; scientific research; and working with other states and international organizations (Article 123).

The Area

The 'Area' is formally defined in the LOSC as those non-biological resources on or under the seabed outside of the continental shelf areas. An example of such a resource came from the HM. Challenger expeditions from 1872–1876 when nodular polymetallic structures in the deep seabed that contained high amounts of manganese where discovered (Bezpalko, 2004). By the 1960s the commercial potential of these manganese nodules was recognized. As the resource was 3600–5500m below the surface, the deep seabed outside of the continental shelves became a region of interest during the LOSC negotiations. The LOSC defines these non-biological resources as 'minerals' (Article 133) regardless of whether or not they are metallic in nature. The Area does not apply to the water column or airspace above the seabed outside of continental shelves (Article 135).

Under the LOSC, the Area is subject to the '…common heritage of mankind' and to be used exclusively for peaceful, beneficial purposes. Consequently states should cooperate to achieve these ends (Articles 136 and 138). The LOSC provides for the economic development of the Area under the conditions that development contributes to the world economy and growth in international trade (Article 150). Specifically, all states, regardless of geographic location, degree of economic development or whether landlocked or not have equal rights to participation and access to the Area (Article 141). According to the LOSC the economic windfall derived from the Area should be shared among all states (Article 139).

No state is permitted to claim any territory or the natural resources within the Area. However, mineral resources obtained from the Area may be appropriated (owned) (Article 137). In addition, archaeological artefacts or discoveries of historical significance must be managed for the benefit of humanity all the while recognizing the importance of the culture of origin (Article 149). States bear the responsibility for actions of their nationals (individuals, corporations or government) and are liable for their actions (Article 139). Activities in the Area are overseen by the International Seabed Authority (the 'Authority' – see below), which reports to the UNGA (Articles 140, 143–146 – see below).

Rights of passage through straits under the LOSC

Straits are marine corridors frequently used by ships to transit between, and occasionally within, states. Straits are generally narrow bodies of water geographically constrained by land (or ice) and may connect EEZs, high seas, national waters or any combination of these areas. Given the importance of global trade, straits frequently have significant economic or military importance to both the coastal and maritime states (Figure 2.3).

Examples of well-known straits include:

- The Strait of Hormuz bordered by Iran, the United Arab Emirates and the Omani enclave of Musandam (14.3km at its narrowest point). Twenty per cent of the world's petroleum is transported through the Strait of Homuz making the Strait a strategically important site for ensuring the uninterrupted flow of petroleum products.
- The Strait of Gibraltar, bordered by Morrocco and Gibraltar (13km at its narrowest point) connects the Mediterranean Sea to the Atlantic Ocean.
- The Strait of Malacca, bordered by Malaysia and Indonesia (2.7km at its narrowest point) is traversed by more than 50,000 ships annually carrying approximately 25 per cent of the world's traded goods and is the shortest route between Asia and the Middle East.

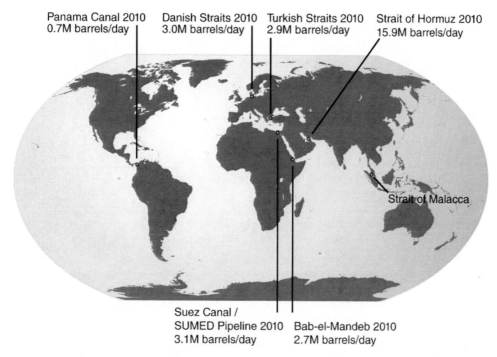

Panama Canal 2010
0.7M barrels/day

Danish Straits 2010
3.0M barrels/day

Turkish Straits 2010
2.9M barrels/day

Strait of Hormuz 2010
15.9M barrels/day

Strait of Malacca

Suez Canal /
SUMED Pipeline 2010
3.1M barrels/day

Bab-el-Mandeb 2010
2.7M barrels/day

Figure 2.3 Globally important shipping straits and volume of crude oil and petroleum products shipped in 2010 for comparison purposes

Source: www.eia.gov

As seaborne trade increased in the 1960s and 1970s, states realized the risk of moving goods through straits and therefore encouraged the codification of the rights and duties of coastal and maritime states operating in straits. The LOSC articulates the rights of passage through straits in Part III of the Convention.

First and foremost, the LOSC mandated that all ships and aircraft enjoy the right of transit passage between one part of the high seas or EEZ and another part of the high seas or EEZ (Article 38). However, if an alternative high seas route exists to the strait there is no right of passage (Article 36). No rights of passage were conferred in internal waters unless the establishment of strait baselines encompassed waters historically used for transit (Article 35). Rights of passage in no way affect marine jurisdiction of coastal states bordering straits as outlined in Part II of the Convention (Article 34). No non-transit related activity, including marine scientific research and hydrographic studies is prohibited without the permission of the coastal state(s). Where these activities are permitted by the coastal state they are subject to the rules of the Convention (Articles 38, 40). Ships during transit passage are required to proceed without delay, avoid any non-transit related activities, refrain from any threat or use of force against the bordering coastal state(s) (Article 39) and comply with international navigation and pollution regulations. Bordering states may establish sea lanes and traffic separation schemes so long as they conform to international regulations and are communicated to seafarers (Article 41). States bordering straits may establish regulations governing the safety of navigation, protection from pollution, fishing and commerce in the strait so long as these regulations do not impede the right of innocent passage (Article 42). Bordering states may communicate known dangers to passage within straits but may not suspend transit passage (Article 44).

Rights of archipelagic states under the LOSC

Part IV of the LOSC governs the unique nature of archipelagic states, which are defined in the LOSC as '...states constituted wholly by one or more archipelagos and may include other islands' (Article 46). The LOSC further defines archipelagos as a group of islands that are socially, politically and geographically interconnected. Many states, including Indonesia (17,500 islands), Philippines (6800 islands) and Fiji (300 islands) are entirely archipelagic in nature and, when EEZs are mapped, the territorial waters of archipelagic states are extensive and complex. This had led to overlapping jurisdictions, difficulties in interpreting internal and territorial waters and creating political uncertainties for shipping.

The rights and responsibilities of archipelagic states were raised as far back as 1888. It was then that it was suggested that islands constituting a state should have a separate method of determining territorial waters (IDI, 1891). At the first two UNCLOS Conventions in 1958 and 1960, Indonesia and the Philippines championed the notion of recognizing archipelagic states as differentiated from other coastal (e.g. Canada) and marine states (e.g. Germany) (Coquia, 1983).

Nevertheless, it was not until the LOSC finished that the rights of archipelagic states were codified into international law. Archipelagic states may identify archipelagic waters (mostly equivalent to internal waters) by drawing strait baselines that join the outmost points of the outermost islands (and reefs exposed at low tide) (Article 47) so long as:

- The main islands included in this area and the ratio of the area of water to land (including atolls) does not exceed 9:1.
- Baselines do not exceed 100nmi; however, up to 3 per cent of the baselines may exceed 100nmi to a distance of 125nmi.
- Baselines will generally configure to the geographic configuration of the archipelago.
- Baselines do not exclude access to the high seas of another state.
- Baselines do not infringe on or remove access to another state's territorial waters.
- Baselines must be identified on charts or their geographic locations provided to the Secretary-General of the United Nations.

Territorial seas, contiguous zones and EEZs will be measured from the baselines drawn as per Article 47 (Article 48). Archipelagic states have authority over the airspace over their archipelagic waters, as well as the water column, seabed and subsoil resources (Article 49). However, unlike internal waters, the LOSC provides for the right of innocent passage in archipelagic waters (Article 50). This right may only be suspended for protection of the state's security and only after proper notification to other states. Furthermore, archipelagic states must recognize traditional fishing rights and existing agreements with bordering states. As in the case with other coastal states, they too must respect the rights of undersea cables transiting archipelagic waters (Article 52). Archipelagic states may designate sea lanes and air routes through their archipelagic waters consistent with international regulations (Article 53).

No non-transit related activity, including marine scientific research and hydrographic studies, is permitted without the permission of the archipelagic state. Where these activities are permitted by the archipelagic state they are subject to the rules of the Convention (Articles 38, 40). Ships during transit passage are required to: proceed without delay, avoid any non-transit related activities and refrain from any threat or use of force against the archipelagic state (Article 39). Archipelagic states may communicate known dangers to passage within straits but may not suspend transit passage (Article 44).

Rights of landlocked, geographically disadvantaged and developing states under the LOSC

The LOSC is unique among international legal instruments in its approach to states without the means or access to capitalize on oceanic opportunities whether these opportunities are through trade, fishing or the development of non-renewable ocean resources (Figures 2.4 and 2.5 and Table 2.3). The LOSC affirms the rights of these states to participate in the ocean economy as well as mandating the means (e.g. wealth redistribution for economic activity generated in the Area, sharing of surplus fisheries capacity) and mechanisms (e.g. International Seabed Authority to oversee the redistribution of wealth). Specifically, the LOSC addresses the rights of landlocked, geographically disadvantaged and developing states. It also addresses the obligations of other nations to support these rights, which are discussed below (Vasciannie, 1990).

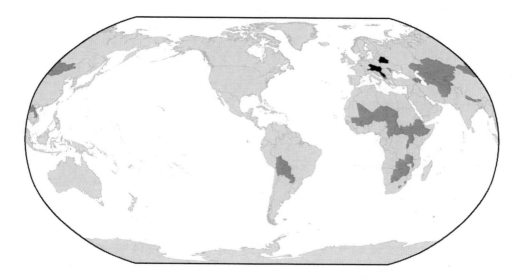

Figure 2.4 Landlocked developed states (darker shading) and landlocked developing countries (lighter shading)

Source: www.un.org/special-rep/ohrlls/lldc/list.htm

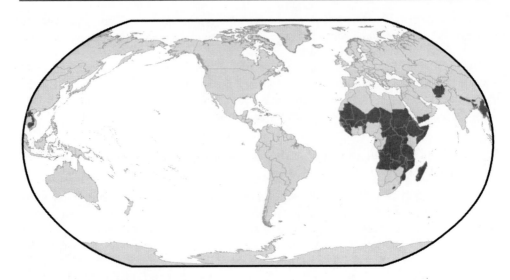

Figure 2.5 Least developed countries
Source: www.unohrlls.org/en/ldc/25/

Rights of landlocked states under the LOSC

Landlocked states are those states without any coastline fronting a sea or having ready access to the sea. There are currently 48 landlocked states, mostly composed of less developed states in Africa and Asia (termed 'landlocked developing states' or LDSs by the UN) and 22 of these states are party to the LOSC (UN, 2012) (Figure 2.4). With the exception of European nations such as Luxembourg and Switzerland, most landlocked states are economically disadvantaged by cumbersome transit costs, protracted transit times and the need to negotiate with coastal states for port access (Bowen, 1986). Landlocked nations, and particularly the 31 landlocked developing countries (LLDCs), seek to participate in ocean development, particularly with respect to transportation, fisheries and seabed mining.

In response to these desires, the LOSC confers certain rights on landlocked states. Landlocked states have the right to participate in the exploitation of the oceans, including activities in the Area (Article 140). Coastal states are directed to allocate a portion of the surplus fish catch within EEZs (based on availability, environmental concerns and nutritional needs) to LLDCs. This is negotiated through bilateral and multilateral agreements with coastal states (Article 69). Article 87 affirms that the high seas are open to all states and that no special rights or privileges are conferred upon coastal or maritime states. Article 125 asserts the right of landlocked states to access the sea using all manner of transportation; however, these rights must be secured through agreements with coastal states. Articles 148 and 160 also commit coastal and developed states to assist LLDCs in participating in activities in the Area, including participation in marine science and research (Articles 143 and 256). Article 82 affirms that wealth produced from the development of states' extended continental shelves will be shared taking account of the needs and interests of less developed states. Articles 266–278 commit more developed states to cooperate with landlocked states to develop and transfer marine technology.

For more developed landlocked states (Andorra, the Czech Republic, Hungary, Luxembourg, Serbia, Switzerland) the LOSC permits their participation in the exploitation of marine resources only within the EEZs of more developed states in the same geographic region (Article 69).

Rights of geographically disadvantaged states under the LOSC

Geographically disadvantaged states (GDSs) are those coastal states – including states bordering enclosed or semi-enclosed seas – that require access to the EEZs of other adjacent states to satisfy their nutritional needs (Article 70).

Using bilateral or multilateral agreements with coastal states, the LOSC affirms the rights of GDSs to utilize the surplus catch of other coastal nations in the same geographical region or sub-region. The LOSC also promotes the participation of GDSs in the Area, recognizing the unique challenges of these states to participate in the economic benefits of the oceans' vast expanse. In addition, the LOSC commits to technology transfer to enable developing states to participate in activities in the Area (Article 150). Furthermore, GDSs have a right of notification of, as well as participation in, marine research undertaken in an adjacent coastal state (Article 254).

Rights of less developed states under the LOSC

Less developed states (termed 'least developed countries' by the UN) are not defined under the LOSC; but are defined by the UN as those states with a three-year gross national income (GNI) under US$750 that also fall below thresholds as defined by the Human Assets Index (HAI) and Economic Vulnerability Index (EVI) (United Nations, 2004) (Figure 2.5). In 2012, 48 states met the UN test of least developed countries, of which 16 states were landlocked and 11 states consisted of small island states (United Nations, 2012). Collectively, these 48 states comprised a total population of 832.3 million people with an average age under 25 years old (United Nations, 2012).

In addition, while not included in the LOSC, a companion set of states to the least developed countries, termed small island developing states (SIDSs) is a particularly useful concept in applying the LOSC (Table 2.3). SIDSs were first recognized as a distinct group of states at the United Nations Conference on Environment and Development in June 1992 (Chapter 5) based on the shared challenges of: small resource bases; undiversified economies; difficulties accessing international markets; susceptibility to national disasters; high population growth; and high costs of energy, infrastructure and communications (United Nations, 2010). There are currently 41 SIDSs identified by the UN Office of the High Representative for the Least Developed Countries, Landlocked Developing Countries and the Small Island Developing States (United Nations, 2010).

The LOSC affirms the same rights to developing states as geographically disadvantaged states. These rights include: right to utilize the surplus catch of other coastal states (Article 62); right of navigation and passage through straights (Article 38) and in the high seas (Article 87); and right to the economic benefits to the Area, including technology transfer to participate in the development of the Area (Article 148).

In addition, more developed states are obligated to assist less developed states in the development of scientific and research capacity related to the protection of the marine environment (including ensuring developing states receive priority in funding for these

Table 2.3 List of small island developing states

UN Members

Antigua and Barbuda	Federated States of Micronesia
Bahamas	Mauritius
Bahrain	Nauru
Barbados	Palau
Belize	Papua New Guinea
Cape Verde *	Samoa *
Comoros *	São Tomé and Principe *
Cuba	Singapore
Dominica	St. Kitts and Nevis
Dominican Republic	St. Lucia
Fiji	St. Vincent and the Grenadines
Grenada	Seychelles
Guinea-Bissau *	Solomon Islands *
Guyana	Suriname
Haiti *	Timor-Lesté *
Jamaica	Tonga
Kiribati *	Trinidad and Tobago
Maldives *	Tuvalu *
Marshall Islands	Vanuatu *

Non-UN Members

American Samoa	Guam
Anguilla	Montserrat
Aruba	Netherlands Antilles
British Virgin Islands	New Caledonia
Commonwealth of Northern Marianas	Niue
Cook Islands	Puerto Rico
French Polynesia	US Virgin Islands

Note: *Also less developed countries
Source: www.unohrlls.org/en/sids/46/

programmes). They are also obligated to make less developed states aware of the effects of their activities on marine environmental quality (Articles 202–203). Articles 266–278 commit developing states to cooperate with landlocked states to develop and transfer marine technology.

Marine scientific research

Part XIII of the LOSC articulates how marine research activities are to be undertaken both inside and outside of territorial waters. The Convention first affirms that all states have the right to undertake scientific research, including least developed countries, LLDCs and GDSs (Article 238), that research is to be for peaceful purposes (Article 240) and that marine research cannot be used as a means to assert territorial claims (Article 241). States are obliged to cooperate on marine research, including assisting least developed countries (Articles 242–243), and to disseminate the results of this work (Article 244). All states may undertake research in the Area or in the water column beyond the EEZ (Article 256).

Coastal states have complete control over research activities in their territorial seas, EEZs and continental shelves (Articles 246–247). Other states must secure consent of the coastal state before undertaking research in these areas; however, coastal states are expected to allow marine research to occur in their EEZs and continental shelves so long as the activities benefit mankind and conform to national procedures. Marine research is prohibited also from creating structures, drilling into the continental shelf or exploiting living and non-living natural resources (Article 246).

Successes and challenges of the LOSC

Overall, the LOSC is viewed as one of the most successful international conventions and the term 'Constitution for the Oceans' adequately represents the magnitude of the Convention. Given the scope and scale of issues, there were no expectations from states that were involved in drafting the Convention that they would 'get it right' the first time. However, to date, most states respect and adhere to the Convention and the processes that were outlined in the LOSC have for the most part progressed in an orderly fashion (Freestone, 2013).

Nonetheless, with the hindsight of 30 years, one can see that there are a number of issues that were neglected during drafting of the Convention, as well as problems that surfaced after the completion of the LOSC. Summarized by Freestone (2012), they include:

- Underwater cultural heritage resources were not addressed but have since been managed under the UNESCO Convention on Underwater Cultural Heritage.
- The concept of managing to maximum sustainable yield (MSY) is a central tenant of the LOSC (Chapter 6) yet has been repeatedly discredited as a fishery management tool.
- The Convention fails to provide an adequate definition of marine scientific research resulting in disagreement and conflict between states who conduct research-related activities (e.g. hydrographic surveys, ocean fertilization) within other states EEZs.
- The Convention is concerned with the relations between states rather than consideration of individual-related issues such as human rights.
- Climate change (Chapter 5) was never envisioned as a future issue when the Convention was drafted.
- While the LOSC addressed the high seas and deep seabed areas, the Convention was not a comprehensive treatise on ABNJ addressing topics such as bio-prospecting and marine protected areas (Chapter 5).

In addition, Churchill (2012) determines that at least one third of the 160+ state parties to the LOSC are in breach of one or more of the Convention's provisions in the following ways:

- A number of states have established baselines, territorial seas and EEZs contrary to the LOSC direction.
- Not all states are in compliance with Article 94 of the Convention related to seaworthiness of ships flying their flag.
- Many states are in breach of their obligations under Article 61 (2) to maintain living resources within their EEZs as well as Article 117–119 to conserve living resources of the high seas.

- A number of states are failing to preserve or protect rare or fragile marine ecosystems (Article 194 (5)).

International organizations that establish and oversee international marine law

Permanent Court of Arbitration

The Permanent Court of Arbitration (PCA) was established by the Convention for the Pacific Settlement of International Disputes in 1899. The PCA's original mandate was to facilitate dispute resolution between states and has evolved to provide dispute resolution services to states, state entities, intergovernmental organizations and private parties. There are currently over 100 member states and the PCA is based in The Hague, Netherlands.

The first marine arbitration undertaken by the PCA was the 1910 North Atlantic Coast Fisheries dispute between the US and Great Britain. A fundamental part of this dispute was how the 3nmi boundary for territorial waters was measured for bays. Great Britain (representing what is now Canada) contended that 'bays' were defined from headland to headland regardless of size, thus excluding American fishers from most Canadian coastal nearshore waters. The US retorted that only smaller 'bays' less than 6nmi from the entrance should be included as territorial waters. The PCA ruled that, for bays larger than 3nmi from headland to headland at the entrance, territorial waters should be defined by the 3nmi boundary following the geography of the coastline.

The PCA plays an important role in dispute resolution under the LOSC. Part XV of the Convention sets out the options for dispute resolutions. However, should a state not express a preference for a particular means of dispute resolution or states disagree on what means to use, an arbitral tribunal (constituted under Annex VII of the LOSC) becomes the default means of dispute resolution. The majority of dispute resolutions to date have involved disputes over maritime boundaries and since the LOSC came into force in 1994, five of six of the Annex VII dispute resolutions have been arbitrated under the PCA. They are as follows (www.pca-cpa.org):

- Bangladesh versus India (marine boundary delineation), instituted in October 2009 and is still pending.
- Ireland versus UK ('MOX Plant Case' concerning radioactive contamination of the Irish Sea by the UK), instituted in November 2001 and terminated through a tribunal order issued on 6 June 2008.
- Guyana versus Suriname (marine boundary delineation), instituted in February 2004 and decided by a final award rendered on 17 September 2007.
- Barbados versus Trinidad and Tobago (marine boundary delineation), instituted in February 2004 and decided by a final award rendered on 11 April 2006.
- Malaysia versus Singapore (land reclamation activities by Singapore in the Straits of Johor), instituted in July 2003 and terminated by an award on agreed terms rendered on 1 September 2005.

International Court of Justice

The ICJ was established in 1945 by the Charter of the United Nations and began operating in 1946. The ICJ is one of the six principle organs of the UN and, like the PCA, is headquartered in The Hague. The court is composed of 15 judges elected for nine-year terms by the UNGA and UN Security Council. The Court is responsible for settling legal disputes submitted to it by states (termed 'Contentious cases') and providing advisory opinions on legal questions referred to it by authorized UN organs and specialized agencies (termed 'Advisory proceedings'). The ICJ's first case was the *Corfu Channel Case* and since then the ICJ has adjudicated a number of marine disputes (discussed in Box 2.3).

Box 2.3 International Court of Justice marine-related cases

- Territorial and Maritime Dispute (Nicaragua versus Colombia). Nicaragua claims ownership of the maritime features (cays) off her Caribbean coast and instituted proceedings against Columbia in 2001. Honduras and Costa Rica submitted applications for permission to intervene in the dispute but permission was not granted by the ICJ.
- Maritime Dispute (Peru versus Chile). Peru instituted proceedings against Chile in 2008 to delineate marine boundaries between the two nations.
- Whaling in the Antarctic (Australia versus Japan). Australia initiated proceedings against Japan in 2010, contending that Japan continues to ignore the global moratorium on commercial whaling under the International Convention for the Regulation of Whaling.
- Maritime Delimitation in the Black Sea (Romania versus Ukraine). Romania initiated proceedings against the Ukraine in 2004 to delineate marine boundaries between the two nations.
- Territorial and Maritime Dispute between Nicaragua and Honduras in the Caribbean Sea (Nicaragua versus Honduras). Nicaragua initiated proceedings against the Honduras in 1999 to delineate marine boundaries between the two nations.
- Fisheries Jurisdiction (Spain versus Canada). Spain initiated proceedings against Canada in 1995 following the seizure of a Spanish vessel fishing outside of Canada's EEZ, but on Canada's continental shelf. The Kingdom of Spain contended that no international law permits the application of domestic law and the seizure of vessels on the high seas. The ICJ ruled that it lacked jurisdiction to adjudicate the dispute in 1998.
- Oil Platforms (Islamic Republic of Iran versus US). Iran initiated proceedings against the US for violating the 1957 'Treaty of Amity' and for destroying offshore oil and gas platforms owned by the National Iranian Oil Company in 1987 and 1988. The ICJ rejected the claims of both states in 2003.
- Maritime Delimitation and Territorial Questions between Qatar and Bahrain (Qatar versus Bahrain). Qatar initiated proceedings against Bahrain in 1991 to resolve sovereignty over the Hawar islands, sovereign rights over the shoals of Dibal and Qit'at Jaradah, and the delimitation of the maritime areas of the two states. The ICJ ruled that Qatar has sovereignty over Zubarah and Janan Island and that the low-tide elevation of Fasht ad Dibal falls under the sovereignty of Qatar. It also found that Bahrain has sovereignty over the Hawar Islands and the island of

Qit'at Jaradah and the court drew a single maritime boundary between the two states.

- Maritime Delimitation between Guinea-Bissau and Senegal (Guinea-Bissau versus Senegal). Guinea-Bissau initiated proceedings against Senegal in 1991 to delineate marine boundaries between the two states. The ICJ removed the case from its docket when the two states reached agreement on the management of the disputed area.

- Maritime Delimitation in the Area between Greenland and Jan Mayen (Denmark versus Norway). Denmark initiated proceedings against Norway in 1988 to delineate marine boundaries between the two states. The ICJ established a delimitation line for both the continental shelf and the fishery zones in the area between Greenland and Jan Mayen.

- Delimitation of the Maritime Boundary in the Gulf of Maine Area (Canada/US). In 1979 Canada and the US requested the ICJ to delineate a single line to delimit both the continental shelf and the 200-mile exclusive fishery zone in the Gulf of Maine. The ICJ issued its ruling in 1984.

- Continental Shelf (Tunisia/Libyan Arab Jamahiriya). In 1977 Tunisia and Libya requested the ICJ determine the principles and rules of international law as applied to the ownership of their respective continental shelves. The ICJ issued its ruling in 1982 using the principles of an 'equitable solution' to establish a segmented boundary delineating continental shelves.

- Aegean Sea Continental Shelf (Greece versus Turkey). Greece initiated proceedings against Turkey in 1976 to delineate marine boundaries between the two states. The ICJ determined it was without jurisdiction to entertain the application.

- Fisheries Jurisdiction (Federal Republic of Germany versus Iceland). Germany initiated proceedings against Iceland in 1972 after Iceland extended its jurisdiction over fisheries from 12nmi to 50nmi. The ICJ ruled that Iceland could not unilaterally exclude other states from its fisheries beyond 12nmi; however, Iceland was entitled to preferential rights to the fish in its coastal waters due to the dependence of Iceland on fishing.

- Fisheries Jurisdiction (United Kingdom of Great Britain and Northern Ireland versus Iceland). The UK initiated proceedings against Iceland in 1972 after Iceland extended its jurisdiction over fisheries from 12nmi to 50nmi. The ICJ ruled that Iceland could not unilaterally exclude other states from its fisheries beyond 12nmi; however, Iceland was entitled to preferential rights to the fish in its coastal waters due to the dependence of Iceland on fishing.

- North Sea Continental Shelf (Federal Republic of Germany/Netherlands and Federal Republic of Germany/Denmark). Germany and the Netherlands, as well as Germany and Denmark, requested the ICJ to state the principles and rules of international law applicable to resolving delimiting boundaries of their respective continental shelf in 1967. In 1969 the ICJ simultaneously ruled in both cases that there was no basis in customary law for delineating continental shelf areas based on equidistance from land. They directed the three states to carry out the delimitations on the basis of equitable principles.

- Constitution of the Maritime Safety Committee of the Inter-Governmental Maritime Consultative Organization. In 1959, The Assembly of the Inter-Governmental Maritime Consultative Organization requested the ICJ to provide an opinion on whether the Maritime Safety Committee of the Inter-Governmental Maritime Consultative Organization was constituted in accordance with the Convention for the Establishment of the Organization. In 1961 the ICJ ruled that that the Maritime Safety Committee was not properly constituted since the Convention required the Assembly to elect the eight states with the largest amount of registered shipping tonnage and this had not been done.

- Antarctica (UK versus Chile and UK versus Argentina). The UK initiated proceedings against Chile and Argentina in 1955 concerning disputed territory in the Antarctic. Both Chile and Argentina refused to submit to the jurisdiction of the ICJ and the case was removed from the ICJ docket in 1956.
- Fisheries (UK versus Norway). The UK initiated proceedings against Norway for claiming exclusive fishing rights beyond a 10nmi baseline in 1949. The ICJ ruled that the 10nmi baseline was not an established custom in international law and that Norway's claims were not inconsistent with international law.
- Corfu Channel (United Kingdom of Great Britain and Northern Ireland versus Albania). The UK initiated proceedings against Albania for compensation resulting from damage to two British destroyers that had struck mines in Albanian waters. The ICJ ordered Albania to pay £843,947 in 1949.

Source: www.icj-cij.org/docket/index.php?p1=3&p2=3

Organizations under the LOSC that establish and oversee international marine law

International Seabed Authority

The Authority (or ISA) was established by Part XI of the LOSC in combination with the 1994 Agreement relating to the Implementation of Part XI of the United Nations Convention on the Law of the Sea of 10 December 1982. This Agreement and Part XI of the Convention are to be interpreted and applied together as a single instrument.

The Authority was established 16 November 1994 when the LOSC came into force. Headquartered in Kingston, Jamaica it became operational in June 1996 and operates as an autonomous international organization within the UN system (www.isa.org). The powers of the Authority are multifarious: within its dominion is the coordination of marine scientific research (Article 143), the transfer of technology (Article 144) and the protection of the marine environment and human life (Articles 145–146).

The Council is responsible for implementing the seabed provision in the LOSC, including assigning rights to explore and develop geographic areas, approving work plans and setting environmental standards. The Council established the Legal and Technical Commission to advise it on applications, technical reviews and environmental assessments. The Commission has developed *Regulations on Prospecting and Exploration for Polymetallic Nodules in the Area* and the *Regulations on Prospecting and Exploration for Polymetallic Sulphides and Cobalt-Rich Ferromanganese Crusts in the Area* that summarize the customary law related to the deep seabed (www.isa.org).

To date, 10 organizations representing 13 states have received approval for the exclusive exploration of areas, each up to 150,000 square km. Seven of the areas are in the southern parts of the North Pacific Ocean and one area is in the Indian Ocean. Each organization signs a 15-year contract (work plan) with the Authority in which half of the tenure area is relinquished to the Authority after eight years.

Currently, ambiguity exists over whether the mineral resources of the Southern Ocean are regulated by the Authority under Part XI of the LOSC and the 1994 Implementation Agreement for Part XI, or the Antarctic Treaty Consultative Parties under Article 7 of the Protocol on Environmental Protection to the Antarctic Treaty (Vidas, 2000). With the exception of scientific research, the Protocol prohibits any activity related to mineral resources and therefore conflicts with the 1994 Implementation Agreement. To date, however, the incongruity remains moot since deep sea mineral resources have not been discovered in the Southern Ocean.

International Tribunal for the Law of the Sea

The ITLOS was established under Annex VI of the LOSC to adjudicate disputes arising from the Convention. The panel is composed of 21 judges elected by member states and is headquartered in Hamburg, Germany. However, the Tribunal may sit and exercise its functions wherever necessary. Each state is limited to one member (judge) on the Tribunal and at least three members are required from each geographic group as defined by the UN. Members are elected for nine years and can be re-elected. States party to a dispute who are not represented on the Tribunal are permitted to nominate an ad hoc judge with full powers of participation for the duration of the dispute they are selected to participate in. Parts XI and XV of the LOSC govern how disputes are referred to the Tribunal.

While tribunal decisions are final they are only binding on parties to the dispute. Key Tribunal rulings are discussed in Box 2.4.

Box 2.4 International Tribunal for the Law of the Sea rulings

- The M/V 'SAIGA' Case (Saint Vincent and the Grenadines versus Guinea – case 1). On 28 October 1997 Guinea seized the oil tanker *SAIGA*, registered in the Saint Vincent and the Grenadines, on the basis the ship was involved in smuggling activities. The ship was arrested in international waters after a pursuit from Guinean waters. Saint Vincent and the Grenadines submitted their application on this matter 13 November 1997. Guinea contended that the Tribunal lacked jurisdiction in the matter. The Tribunal (divided 12 to 9) ordered the prompt release of the vessel and its crew from detention in Conakry, Guinea.
- The M/V 'SAIGA' (No. 2) Case (Saint Vincent and the Grenadines versus Guinea – case 2). Following the Tribunal order to release the *SAIGA* on 4 December 1997 Guinea prosecuted the master of the vessel resulting in a fine of $15 million, confiscation of the vessel and six months imprisonment for the master (all convictions suspended). The Tribunal ordered Guinea to refrain from carrying out its national court decision.
- Southern Bluefin Tuna Cases (New Zealand versus Japan; Australia versus Japan – cases 3 and 4). On 30 July 1999 Australia and New Zealand submitted requests to the Tribunal for provisional measures (an injunction) against Japan to immediately halt an experimental fishery for southern bluefin tuna. On 9 August 1999 Japan

filed a request to the Tribunal to deny any provisional measures for the fishery. Australia and New Zealand requested an arbitration procedure under Annex VII of the LOSC. On 27 August 1999 the Tribunal ordered the three parties to resume negotiations with the goal of reaching an agreement on measures to conserve and manage southern bluefin tuna and that the parties should restrict their catches. The Tribunal also ordered a halt to the experimental fishery unless the experimental fishery catch was included in the total allowable catch.

- The 'Camouco' Case (Panama versus France – case 5). The Panamanian-registered vessel *Camouco* was arrested by French authorities for fishing for Patagonian tooth-fish in the EEZ of the French Southern and Antarctic Territories on 28 September 1999 and escorted to Reunion. On 17 January 2000 Panama filed a request to the Tribunal for the prompt release of the vessel and its master. On 7 February 2000 the Tribunal order the release of the vessel and its crew upon payment of a $1.2 million bond.
- The 'Monte Confurco' Case (Seychelles versus France – case 6). The Republic of Seychelles-registered *Monte Confurco* was arrested by a French frigate on 8 November 2000 while fishing for toothfish in the EEZ of the Kerguelen Islands in the Southern Indian Ocean. The local district court ruled that the vessel could be released upon posting a bond of 56.4 million French francs (FF). The Seychelles applied to the Tribunal to have the vessel and Master released and the bond reduced to FF9.8 million. On 18 December 2000 the Tribunal ordered the prompt release of the vessel and its master by France upon posting a bond of FF18 million.
- Case concerning the Conservation and Sustainable Exploitation of Swordfish Stocks in the South-Eastern Pacific Ocean (Chile/EU – case 7). The Tribunal was requested on 20 December 2000 by Chile and the EU to evaluate whether the European member states were meeting their LOSC obligations with respect to swordfish conservation in the South East Pacific and whether the LOSC permits Chile's domestic swordfish conservation measures to be applied on the high seas. On 17 December 2009 the Tribunal discontinued the case at the request of the parties as they had resolved the dispute between them.
- The 'Grand Prince' Case (Belize versus France – case 8). The Belize registered fishing vessel *Grand Prince* was arrested by a French frigate 26 December 2000 while fishing for toothfish in the EEZ of the Kerguelen Islands in the Southern Indian Ocean. On 21 March 2001 Belize applied to the Tribunal to have the vessel and master released and the bond reduced to EU€206,149. On 20 April 2001 the Tribunal ruled that it had no jurisdiction in the matter as it could not be proven that Belize was the flag state of the vessel when the application was made.
- The 'Chaisiri Reefer 2' Case (Panama versus Yemen – case 9). On 3 May 2001 Yemen arrested the fishing vessel *Chaisiri Reefer 2* for allegedly violating Yemeni fisheries laws. Panama requested the Tribunal to order the immediate release of the vessel; however, on 3 July 2001 the case was removed from Tribunal's list of cases as Panama and Yemen reached agreement on the release of the ship and its crew.
- The MOX Plant Case (Ireland versus UK – case 10). On 25 October 2001 Ireland requested the Tribunal to adjudicate a dispute between Ireland and the UK over the opening of a mixed oxide fuel (MOX – reprocessed spent nuclear fuel) plant in Sellafield on the Irish Sea. On 3 December 2001 the Tribunal ruled that: a) it had jurisdiction over the dispute; b) provisional measures (an injunction) were not

required as the plant would not be operational until 2002; and c) directed both parties to cooperate on exchanging information and monitoring the environmental effects of the MOX plant.

- The 'Volga' Case (Russian Federation versus Australia – case 11). On 7 February 2002 the Australian military arrested the fishing vessel *Volga* on the high seas in the Southern Ocean for allegedly illegally fishing for Patagonian toothfish within the Australian EEZ. Russia requested the Tribunal ordered the release of the Volga under a AU$500,000 bond. On 23 December 2002 the Tribunal order the release of the *Volga* pending provision of an AU$1.92 million bond.

- Case concerning Land Reclamation by Singapore in and around the Straits of Johor (Malaysia versus Singapore – case 12). On 5 September 2003 Malaysia requested the Tribunal enact provisional measures against Singapore for land reclamation activities alleged to impinge on Malaysian territory. On 8 October 2003 the Tribunal ruled that provisional measures were not required as Malaysia failed to demonstrate the activities would cause irreversible damage and ordered both parties to cooperate through joint studies and information sharing and ordered Singapore not to conduct its reclamation activities in a manner that might harm the marine environment or affect Malaysian rights.

- The 'Juno Trader' Case (Saint Vincent and the Grenadines versus Guinea-Bissau – case 13). On 26 September 2004 the fishing vessel *Juno Trader* was arrested by Guinea-Bissau for allegedly illegally fishing in its national waters. Saint Vincent and the Grenadines requested that the Tribunal order the prompt release of the vessel and her crew. On 18 December 2004 the Tribunal ordered the release of the vessel under a €300,000 bond.

- The 'Hoshinmaru' Case (Japan versus Russian Federation – case 14). On 1 June 2007 the fishing vessel *Hoshinmaru* was arrested for alleged Russian fisheries violations in Russia's EEZ. Japan requested that the Tribunal order the release of the vessel under terms and conditions the Tribunal considered reasonable. On 7 August 2007 the Tribunal ordered the release of the *Hoshinmaru* under a 10 million roubles bond.

- The 'Tomimaru' Case (Japan versus Russian Federation – case 15). On 31 October 2006 the fishing vessel *Tomimaru* was arrested for alleged Russian fisheries violations in Russia's EEZ. Japan requested that the Tribunal order the release of the vessel under terms and conditions the Tribunal considered reasonable. On 7 August 2007 the Tribunal ruled that Japan's application was without object (a claim that has become pointless) due to a ruling by Russian courts to confiscate the vessel. The Tribunal noted that confiscation may upset the balance between coastal and flag states.

- Dispute concerning delimitation of the maritime boundary between Bangladesh and Myanmar in the Bay of Bengal (Bangladesh/Myanmar – case 16). The first states to submit a marine boundary dispute to ITLOS. ITLOS issued its judgement in 2012 and ruled that the LOSC Article 15 governed the approach to boundary delineation in this dispute and that equidistance modified by relevant circumstances was the guiding principle for delimiting the territorial sea.

- Responsibilities and obligations of states sponsoring persons and entities with respect to activities in the Area (Request for Advisory Opinion submitted to the Seabed Disputes Chamber – case 17).

- The M/V 'Louisa' Case (Saint Vincent and the Grenadines versus Kingdom of Spain – case 18). On 1 February 2006 Spain arrested the Saint Vincent and the Grenadines registered vessel *Louisa* for allegedly contravening Spanish national marine environmental laws while conducting sonar and cesium magnetic surveys of the sea floor of the Bay of Cadiz to support oil and gas exploration activities. Saint Vincent and the Grenadines requested the Tribunal order the release of the vessel under provisional measures. On 23 December 2010 the Tribunal ruled that there was no real and imminent risk to the rights of the parties and received assurances from Spain that the vessel was being monitored for environmental impacts while being held in Spanish custody.
- The M/V 'Virginia G' Case (Panama/Guinea-Bissau – case 19). On 21 August 2009 Guinean authorities arrested the oil tanker *Virginia* G while refuelling fishing vessels within Guinea-Bissau's EEZ. The tanker was released after 14 months in Guinean custody, however, Panama claimed that the vessel was damaged during this time and applied to the Tribunal for reparations.

Source: www.un.org/depts/los/convention_agreements/texts/unclos/annex6.htm

In addition, ITLOS is responsible for a number of other chambers discussed below.

Seabed Disputes Chamber

The Seabed Disputes Chamber is established under Annex VI, Section 4, Article 35 of the LOSC and governed by Part XI (s. 5) and Part XV of the LOSC. The Seabed Disputes Chamber has jurisdiction within the International Seabed Area (the Area) and is composed of 11 judges selected by the Tribunal every three years. Any mineral disputes in the Area must be settled by the Chamber and parties are not permitted to choose another type of court or tribunal. The Chamber has jurisdiction over disputes between (the LOSC Part XI, section 5, Article 187):

- States over the interpretation of Part XI of the LOSC and associated Annexes.
- States and the Authority concerning the function and actions of the Authority.
- Any type of party with respect to interpretations of contracts or conflicts between parties operating in the Area.
- A state sponsored contractor and the Authority.
- A party and the Authority for any damage arising out of wrongful acts in the exercise of the Authority's powers and function.

With respect to non-contractual matters, state parties in a dispute may submit cases to a three member ad hoc chamber convened by the Seabed Disputes Chamber or a special chamber of the ITLOS. With respect to contractual matters, including disputes over work plans, transfer of technology and financial matters, parties will be submitted to binding commercial arbitration.

To date the Chamber has ruled on one significant case. In 2008, the Authority received applications submitted by Nauru Ocean Resources Inc. (sponsored by the Republic of Nauru) and Tonga Offshore Mining Ltd (sponsored by the Kingdom of Tonga) for approval of a plan of work for deep seabed mineral exploration. In 2010, Nauru requested that the Authority seek an advisory opinion from the Chamber regarding the extent of the liabilities of a state sponsoring 'seafloor mining in international waters'. Nauru and Tonga were both concerned that the potential liabilities in undertaking seafloor mining could amount to more than the states' gross domestic products (GDPs), which effectively excluded developing states from participating in seafloor mining activities.

In 2011 the Chamber produced an Advisory Opinion (case 17 of ITLOS) titled 'Responsibilities and Obligations of States Sponsoring Persons and Entities With Respect To Activities in the Area'. The Opinion determined that states are not liable for damage caused by state-sponsored contractors if the states have domestic laws, regulations and administrative measures to secure compliance by persons under its jurisdiction (Plakokefalos, 2012).

Chamber of Summary Procedure

The Chamber of Summary Procedure is constituted annually and consists of five members and two alternates. The purpose of this chamber is to quickly resolve issues without extensive formalities or procedures.

Chamber for Fisheries Disputes

The Chamber for Fisheries Disputes, established in 1997, consists of nine members and is established under the Special Chambers provision in Article 15, paragraph of the ITLOS Statute. The Chamber for Fisheries Disputes was established to address disputes concerning the conservation and management of marine living resources.

Chamber for Marine Environment Disputes

The Chamber for Marine Environment Disputes, established in 1997, consists of seven members and is established under the Special Chambers provision in Article 15, paragraph of the ITLOS Statute. The Chamber for Marine Environment Disputes was established to address disputes concerning the protection and preservation of the marine environment.

Chamber for Maritime Delimitation Disputes

The Chamber for Maritime Delimitation Disputes, established in 2007, consists of 11 members and is established under the Special Chambers provision in Article 15, paragraph of the ITLOS Statute.

Special chambers

Article 15, paragraph 2 of the ITLOS Statute enables the establishment of special chambers. The first application of a Special Chamber was in 2000 to address the Case concerning the Conservation and Sustainable Exploitation of Swordfish Stocks in the South-Eastern Pacific Ocean (Chile/European Community). In 2007, a Special Chamber was established to address maritime boundary disputes.

Commission on the Limits of the Continental Shelf

In accordance with Article 76 of the LOSC the Commission on the Limits of the Continental Shelf (the 'Commission') is established by Annex II of the LOSC. The purpose of the Commission is to make recommendations on the seaward boundaries of the continental shelves for those states whose continental shelves extend beyond their EEZs and to assist states in preparing applications for review by the Commission (Article 3). The Commission is comprised of 21 members with backgrounds in marine mapping/geology elected by state parties to the LOSC for terms of five years with eligibility of re-election (Article 2). States have ten years after ratifying the LOSC to submit evidence on the extent of their continental shelves (Article 4).

Chapter summary

- For most of human history, oceans have been managed as *res nullis*, meaning 'nobody's property'.
- International law is the framework for how states conduct themselves at the national level.
- The objective of international law is to establish acceptable practices or behavioural norms between states to regularize and make predictable international interactions.
- International law is established through agreements between states, customs, general principles of law and judicial rulings.
- The central differences between domestic (national) and international law are: who makes it (legislatures versus states); to whom it applies (persons versus states); who interprets it (judges/bureaucracies versus states); the near absence of sanctions in the international system; and courts (largely irrelevant in international law).
- The United Nations Convention on the Law of the Sea is the primary legal agreement governing the conduct between states.
- The Convention: defines the rights of states to waters adjacent to their territory, defines parts of the ocean as 'common to mankind', and obliges states to preserve the ocean and accommodate the needs of other states.
- The Convention establishes the International Seabed Authority and International Tribunal for the Law of the Sea

References

Anand, R. P. (1983) *Origin and Development of the Law of the Sea*, Martinus Nijhoff, The Hague.

Bezpalko, I. (2004) 'The deep seabed: Customary law codified', *Natural Resources Journal*, 44, 867–887.

Bowen, R. E. (1986) 'The land-locked and geographically disadvantaged states and the law of the sea', *Political Geography Quarterly*, 5 (1), 63–69.

Boyle, A. (1999) 'Some reflections on the relationship of treaties and soft law', *International and Comparative Law Quarterly*, 48, 901–913.

Brown, E. D. (1994) *The International Law of the Sea*, 1–2, Dartmouth, Aldershot.

CEPAL (United Nations Economic Commission for Latin America and the Caribbean) (2004) *Major Issues in the Management of Enclosed or Semi-enclosed Seas, with Particular Reference to the Caribbean Sea*, LC/CAR/L.24, www.eclac.cl/publicaciones/xml/1/20811/L0024.pdf, accessed 13 February 2013.

Cho, D. (2010) 'Limitations of 1992 CLC/FC and enactment of the special law on M/V Hebei Spirit incident in Korea', *Marine Policy*, 34 (3), 447–452.

Churchill, R. (2012) 'The persisting problem of non-compliance with the Law of the Sea Convention: Disorder in the oceans', *The International Journal of Marine and Coastal Law*, 27, 813–820.

Coquia, J. R. (1983) 'Development of the archipelagic doctrine as a recognized principle of international law', *Philippine Law Journal*, 58, 143–171.

De Mestral, A. and Fox-Decent, E. (2008) 'Rethinking the relationship between international and domestic law', *McGill Law Journal*, 53, 573–648.

EIA (Energy Information Adminstration) (2012) *World Oil Transit Chokepoints*, US Energy Information Administration, www.eia.gov/countries/regions-topics.cfm?fips=WOTC#hormuz, accessed 14 January 2013.

Freestone, D. (2012) 'The law of the sea convention at 30: Successes, challenges and new agendas', *The International Journal of Marine and Coastal Law*, 27, 675–682.

Freestone, D. (ed.) (2013) *The 1982 Law of the Sea Convention at 30: Successes, Challenges and New Agendas*, Martinus Nijhoff, The Hague.

Grotius, H. (1633) *Freedom of the Seas or the Right which Belongs to the Dutch to Take Part in the East Indian Trade*, Oxford University Press, New York.

Hattendorf, J. B. (2007) *The Oxford Encyclopedia of Maritime History*, Oxford University Press, Oxford.

Higgins, R (2005) *Problems and Process: International Law and How We Use It*, Oxford University Press, Oxford.

ICJ (International Court of Justice) (1949) *Corfu Channel Case (United Kingdom of Great Britain and Northern Ireland v. Albania)*, ICJ Reports, www.icj-cij.org/docket/index.php?p1=3&p2=3&k=cd&case=1&code=cc&p3=4, accessed 13 February 2013.

ICJ (1951) *Fisheries Case (United Kingdom v Norway)*, ICJ Reports, www.icj-cij.org/docket/index.php?p1=3&p2=3&code=ukn&case=5&k=a6&p3=0, accessed 13 February 2013.

IDI (Institut de Droit International) (1891) 'Règlement international des prises maritimes', ANNUAIRE Hamburg Session, 1891, 11, 37, www.idi-iil.org/idiF/resolutionsF/1882_turin_01_fr.pdf, accessed 10 February 2013.

International Tribunal for the Law of the Sea (2011) 'Case no. 17', www.itlos.org/fileadmin/itlos/documents/cases/case_no_17/adv_op_010211.pdf.

Juda, L. (2002) 'Rio plus ten: The evolution of international marine fisheries governance', *Ocean Development & International Law*, 33 (2, 109–144.

Law Society of New South Wales (2010) *The Practitioners Guide to International Law*, www.lawsociety.com.au/idc/groups/public/documents/internetyounglawyers/065481.pdf, accessed 19 February 2013.

League of Nations (1930) Convention on Certain Questions Relating to the Conflict of Nationality Law, Treaty Series, 179 (4137), 89.

LeGresley, E. (1993) *The Law of the Sea Convention*, publications.gc.ca/Collection-R/LoPBdP/BP/bp322-e.htm, accessed 14 February 2013.

Malanczuk, P (1997) *Akehurst's Modern Introduction to International Law*, Routledge, Oxford.

Marean, C. W., Bar-Matthews, M., Bernatchez, J., Fisher, E., Goldberg, P., Herries, A. I., Jacobs, Z., Jerardino, A., Karkanas, P., Minichillo, T., Nilssen, P. J., Thompson, E., Watts, I. and Williams, H. M. (2007) 'Early human use of marine resources and pigment in South Africa during the Middle Pleistocene', *Nature*, 449 (7164), 793–794.

McDorman, T. L. (2008) *Salt Water Neighbors: International Ocean Law Relations Between the United States and Canada*, Oxford University Press, New York.

Plakokefalos, I. (2012) 'Seabed disputes chamber of the international tribunal for the law of the sea: Responsibilities and obligations of states sponsoring persons and entities with respect to *Journal of Environmental Law*, 24 (1), 133–143.

Rothwell, D. R. and Stephens, T. (2010) *The International Law of the Sea*, Hart Publishing, Oxford, Oregon.

Selden, J. (1635) *Mare Clausum, Seu, de Dominio Maris Libri Duo: Primo, Mare, Ex Iure Naturae, Seu Gentium, Omnium Hominum Non Esse Commune, sed Dominii*, Proquest, Ann Arbor, MI.

Tanaka, Y. (2012) *The International Law of the Sea*, Cambridge University Press, Cambridge.

Tetley, W. (2000) 'Uniformity of international private maritime law – the pros, cons and alternatives to international conventions – how to adopt an international convention', *Tulane Maritime Law Journal*, 24, 775–856.

United Nations (1982) *United Nations Convention on the Law of the Sea of 10 December 1982*, http://www.un.org/depts/los/convention_agreements/texts/unclos/UNCLOS-TOC.htm, accessed 14 February 2013.

United Nations (2004) *United Nations, Committee for Development Policy Report on the sixth session (29 March-2 April 2004)*, Economic and Social Council Official Records, Supplement No. 13, www.un.org/special-rep/ohrlls/ldc/E-2004-33.pdf, accessed 10 February 2013.

United Nations (2005a) *Vienna Convention on the Law of Treaties 1969*, United Nations, Treaty Series, 1155, 331, untreaty.un.org/ilc/texts/instruments/english/conventions/1_1_1969.pdf, accessed 14 February 2013.

United Nations (2005b) *Convention on the Territorial Sea and the Contiguous Zone*, Treaty Series, 516, 205, untreaty.un.org/ilc/texts/instruments/english/conventions/8_1_1958_territorial_sea.pdf, accessed 13 February 2013.

United Nations (2005c) *Convention on the Continental Shelf*, Treaty Series, 499, 311, untreaty.un.org/ilc/texts/instruments/english/conventions/8_1_1958_continental_shelf.pdf, accessed 13 February 2013.

United Nations (2005d) *Convention on the High Seas*, Treaty Series, 450, 11 and 82, untreaty.un.org/ilc/texts/instruments/english/conventions/8_1_1958_high_seas.pdf, accessed 13 February 2013.

United Nations (2005e) *Convention on Fishing and Conservation of the Living Resources of the High Seas*, Treaty Series, 559, 285, untreaty.un.org/ilc/texts/instruments/english/conventions/8_1_1958_fishing.pdf, accessed 13 February 2013.

United Nations (2010) *Official Development Statistics for Small Island Developing States*, www.unohrlls.org/UserFiles/File/Elle%20Wang%20Uploads/ODA%20for%20Sids%2006-08.pdf, accessed 11 February 2013.

United Nations (2012) *Least Developed Countries (LDCs) Fact sheet*, www.unohrlls.org/UserFiles/File/UN_LDC_Factsheet_053112.pdf, accessed 10 February 2013.

Vasciannie S. C. (1990) *Land-Locked and Geographically Disadvantaged States in the International Law of the Sea*, Oxford University Press, New York.

Vidas, D. (2000) 'Emerging law of the sea issues in the Antarctic maritime area: A heritage for the new century?' *Ocean Development and International Law*, 31 (1–2), 197–222.

Watson, A. (2009) *The Digest of Justinian*, Volume 1, University of Pennsylvania Press, Philadelphia, Pennsylvania.

An introduction to policy

Instruments for marine law and policy

Introduction

'Law', in the domestic or national context, is broadly understood to be the standards, principles and procedures that must be followed in society and administered through court systems. The definition of 'international law' is less concrete but broadly understood to be the means by which acceptable practices or behavioural norms between states are codified.

Currently, there is no universally agreed upon definition of 'policy'. Most broadly, however, 'policy' can be defined as plan or course of action to influence and determine decisions, actions and other matters. Deriving from the Greek word *polis*, or 'city', policy is akin to 'citizenship' or belonging to a group in some way. In early Greece, *nomos* was analogous to the term 'law'. However, the concept was broader than law as currently defined and included the customs, principles and need for the *polity* (citizens and their government) to understand *nomos* in order for the laws to be understood. Plato was an early progenitor of the concept of policy and advanced the idea that citizens would better comprehend laws if they understood the motivation underlying their creation.

Policies may be enacted in any situation in which one or more individuals are involved in a common activity or are part of a distinct group or culture. Schools, churches, sports clubs, industry, volunteer groups and governments all establish policies.

Countless decisions are made every day by organizations, be they government or non-government. Most of these decisions (e.g. whether to issue a building permit) are not policy decisions but business decisions informed by policies. Policy, therefore, is a way to set strategic direction, intent and objectives to address problems that are often complex and controversial. Patton and Sawicki (1993) noted that issues requiring the development of policies have the following characteristics:

- They are poorly defined.
- Their solutions cannot generally be proven to be correct before application.
- Solutions may not achieve the intended results.
- The best solutions are seldom the least costly and appealing.
- The adequacy of the solution is often difficult to measure against the 'public good'.
- The fairness of solutions is impossible to measure objectively.

Policy measures are effective in a socio-economic system by modifying the behaviour of economic agents. These agents may be producers, consumers or suppliers of services. In general terms, public policies aim to modify the course of socio-economic and natural events by (adapted from FAO, 2009):

- Directly supplying goods, services or purchasing power (e.g. the supply of public or merit goods such as ferry services, managing sediment and shorelines, financing the development of maritime industries).
- Promoting/encouraging/supporting (e.g. stimulate the adoption of new technologies for fishing gear modification, promotion of marine export products, supporting sustainable fisheries).
- Imposing/enforcing (e.g. vessel pollution, vessel safety).
- Saving/preserving (e.g. restoration of coastal habitat, preserving coral reefs).
- Preventing (e.g. the spread of invasive species).
- Discouraging (e.g. reducing marine debris).

Policy specific to the actions of government is termed *public policy*. Public policy is made or initiated by government and states the intent of government. It is expressed as laws, regulations, decisions and government actions. Public policy is interpreted and implemented by public and private actors (Birkland, 2011).

There is some debate on the scope of public policy. Certain authors subscribe to the belief that public policy constitutes the sum of all government activities and others suggest that public policy represents the actions of government and the intentions that determine those actions. Recent discussions on public policy trend towards the later definition, in which public policies generally refer to changes that are envisioned or in the early stages of implementation rather than changes that are already in place. In this respect, public policy as a 'statement of intent' can be thought to be different than protocols or procedures, which are operational in nature.

Public policies have been further subdivided into different types. *Regulatory policies* prescribe specific behaviours or limit types of actions to achieve certain objectives. Regulatory policies are frequently associated with law enforcement, business regulation (e.g. business practices), controlling access to public goods (e.g. air, water) and health and safety (e.g. workplace safety, drug approval). Regulatory policies often have the force of law and provide little to no benefits to those that are regulated.

Distributive or redistributive policies apportion a good or a benefit to a certain segment of the population in order to address one or more public problems. The costs of these benefits are shared across all members of the jurisdiction to which the policy applies. Distributive policies are widely used, especially by politicians in democratic societies. Thus, distributive policy is a tool that can improve a politician's chances for re-election ('pork barrel' politics) as well as push governments into deficit to fund new policy initiatives (Stein and Bickers, 1995). Distributive policies may include funding increases for education, subsidies for agriculture, support for research, collection and dissemination of information or the award of infrastructure projects.

Redistributive policies tax one segment of society to provide benefits to another. Redistributive policy is most often used to transfer resources from the 'haves' to the 'have-nots', and often in the form of social programmes. Common examples of these programmes include income stabilization (e.g. unemployment and retirement benefits), welfare, health care, housing and income distribution. However, certain governments use redistributive policies to support policies that aim to reduce the amount of wealth redirected from the richer to poorer segments of a society (e.g. trickle-down economics). The EU has several international redistribution programmes to address the effects of globalization on European nationals including the European Social Fund, European Agricultural Guidance and

Guarantee Fund and the European Regional Development Funds. These programmes currently utilize more than 30 per cent of the EU budget (Burgoon, 2010).

Redistributive policies are 'revenue neutral' in that the overall expenditures from general revenue are unchanged. This engenders a dynamic of winners and losers and, as such, redistributive policies can be intensely political and divisive.

Constituent policies either benefit the public at large or serve the government. Constituent policies may establish government structure, document the rules of government conduct (including financial and personnel management) or solidify roles and responsibilities of government organizations. The establishment of new functions of government (e.g. environmental regulatory agencies in the 1970s) or changes in government practices (e.g. adoption of new accounting practices) are good examples of constituent policies.

Symbolic policies (sometimes termed pseudo policies) confer neither economic benefits nor costs on any particular group but instead appeal to collective values, beliefs and perspectives. Examples of symbolic policies include banning armed conflict, prohibitions on burning of national flags and even signing international conventions that pressure reluctant states to reduce greenhouse gases (Böhringer and Vogt, 2003).

Relationship between law and public policy

The relationship between law and public policy can range from nearly non-existent (e.g. criminal law and crime prevention programmes may both address crime but otherwise have no bearing on each other) to inextricably interconnected (e.g. financial laws inform financial policies and vice versa). If public policy is defined as the actions of government, then policies may enable the formulation of laws and, conversely, laws may enable the development and implementation of policies.

However, there are clear boundaries between law and policy: while both laws and policies are enacted by governments, legal issues are stated and analysed within the processes and procedures of the legal system. Depending on the type of legal system, legal issues are evaluated solely on the basis of written law, past precedent or both. Law is concerned with the legality of an action and – except in the punishment phase – is blind to extenuating circumstances, ethical and moral consideration and the broader social good. Policy, by contrast, is concerned with the efficacy of an action and is evaluated using the principles of administration and management. Policy issues are informed by citizen and stakeholder input, science, social good, value systems and acceptability to citizens.

Other key differences between law and public policy are that public policy is designed to be understood and easily communicated to those whom it affects and those without a legal background. It also can be developed and implemented much faster than most laws. In a society where technology can rapidly result in unforeseen dilemmas such as bioethics, personal privacy and infectious diseases, policy approaches allow governments to quickly address emerging issues that would otherwise go unattended due to the complexity of creating new law. Lastly, policy is a method of testing new ideas to see whether they should at some point become enshrined in laws.

The need for marine policy

The unique structure and function of oceans means that their management is necessarily different from terrestrial environments (see Chapter 1). To more effectively discuss the efficacy of the various legal and policy approaches to marine management, these unique characteristics must be taken into account.

Common property characteristics

As discussed in Chapter 2, much of the marine environment is a common property resource (open access) and, as such, is freely available to all (Hardin, 1968). Past experiences with common property resources (e.g. air and water pollution) has repeatedly demonstrated that resources are poorly managed (degraded, depleted) when owned by no one (an absence of consumption rights). Unfettered access to marine resources leads to a rapid depletion of resources due to 'gold rush' behaviour as well as low profits stemming from overcapitalization and excessive costs. In turn, these low profits occasion dangerous working conditions and poor product quality (Deacon, 2009).

Most discussion around the perils of open access relate to the rapid depletion of fisheries. This depletion is a result of fishers having few economic incentives to consider the long-term sustainability of fish stocks. However, other marine sectors – including transportation, energy development and mining – also lack incentives to minimize environmental harm and maximize long-term opportunities.

Unpriced positive externalities

The marine environment provides a range of social and environmental benefits that are enjoyed by humanity. These public benefits, termed *ecological goods and services*, are widely valued and necessary to support social and economic activities. Yet they remain unpriced (meaning they have no direct economic value). Examples of positive externalities include biodiversity and ecological values (e.g. nutrient cycling), aesthetics and recreational amenities, health, cultural attachments and water and air quality.

Marine environments provide many important services that are not normally considered to have economic value: the oceans assimilate human wastes from terrestrial and freshwater environments, and, through fixing atmospheric CO_2, buffer the effects of climate change. In addition, the oceans contain innumerable raw materials that could provide new sources of food, fibre, medicines and technologies. Maintaining functioning marine environments, then, can be viewed as 'insurance' against current and future human negative impacts on terrestrial and marine environments. Meanwhile, forward-thinking marine policies can convert unpriced externalities into social and environmental benefits. For example, the strategic disposal of retired ships and offshore platforms can create artificial reefs that benefit fish populations and produce opportunities in fishing or tourism.

Pollution and other negative externalities

The value and productivity of marine resources can also be threatened by pollution, ocean warming and acidification (as a result of climate change), habitat loss, invasive species and other factors that impose costs that are not reflected in market processes. Just as a market framework can undervalue a resource when positive externalities are present (because the full set of beneficiaries are unrepresented), the presence of negative externalities can lead the market to overestimate the value of (and hence overproduce) a good or activity (Davis and Gartside, 2001).

Characteristics of marine policy

The relationship between public policy and law within a state's coastal waters (territorial waters and EEZs) is similar to the domestic relationship between law and policy within a state's sovereign lands. The primary difference is that international law intrudes into a state's coastal waters to a greater extent than a state's terrestrial areas. Specific examples include the right of innocent passage and the requirement to accommodate the needs of other states (e.g. right to fish, undertake research).

The relationship between public policy and law in ABNJ is very different. Through international conventions and agreements, the high seas and deep seabeds are managed as a commons. As such, international law in ABNJ is largely voluntary, operates outside the court system (with exceptions in the case of disputes) and is not punitive (see Chapter 2). The law of the high seas therefore has many similarities to what would normally be accepted as domestic policy.

ABNJ are governed by an amalgam of symbolic and regulatory (voluntary) policies. While the voluntary nature of the regulatory policies in ABNJ would suggest that they may be less effective than equivalent domestic regulation, this has not been the case. The voluntary nature of ABNJ agreements has resulted in some of the most innovative and cooperative conventions conceived. While many conventions, including those on pollution and dumping, have rates of compliance exceeding similar domestic laws, other conventions, including those on environmental protection, are aspirational in nature and, as such, are symbolic policies.

Given that international organizations have limited budgets and few to no powers of taxation, distributive and redistributive policies are limited to appeals to states that are largely symbolic in the hope of ensuring rights of passage, access to developing nations to fish and limits on pollution. As ocean governance evolves, constituent policy is frequently used in the establishment and improvement of governance mechanisms.

Introduction to marine policy instruments

At present, no single authority, akin to the nation state, sets the overall rules on activities on, in and under the world's oceans. As discussed in earlier chapters, the legal powers of governments over marine resources vary from sovereign rights in nearshore areas to 'economic' rights in offshore areas. While perhaps an oversimplification, the primary purpose of marine management is to sustain economic, cultural and environmental productivity over time. This is the essence of sustainability. It is achieved through policies that correct for inefficiencies in consumption and negative externalities. Depending on the circumstances, a number of types of policies – termed 'policy instruments' – are applied to this end.

Policy instruments are often broadly classified as *regulatory* (legal), *economic* (market-based) or *informational* (persuasive). Some authors include *cooperative* (voluntary) instruments as a fourth type. Commonly categorized as 'carrots, sticks or sermons', policy instruments have evolved over time. 'First generation' regulatory instruments addressed immediate concerns (e.g. social failure, tragedy of the commons) that required government intervention. 'Second generation' instruments use markets to correct for human behaviour, limit externalities or in instances where government regulation has failed. 'Third generation' instruments are 'suasive', meaning that the regulated enact self-regulation in their own best interests (Table 3.1).

Table 3.1 Types of policy instruments

Regulatory instruments	Economic instruments	Cooperative instruments	Informational instruments
Command and control	Environmental taxes/charges	Voluntary agreements information	Environmental
Performance-based	Tradeable permits	Roundtables	Education
Process-based	Subsidies	Mediation	Awareness (e.g. eco-labels)
Co-regulation	Financial funding	Certification	

◄──►

High   Low

Degree of state intervention

Source: Adapted from Böcher (2012)

These policy instruments are not mutually exclusive. The differences between standards and taxes, for example, may not be readily apparent to those affected by the policies (Bemelmans-Videc *et al.*, 1998). The World Bank (1997, Table 3.2) suggests four categories of environmental policies: using markets, creating markets, regulations or informational instruments (including public engagement). The approaches presented here emphasize the importance of individual conditions. Some examples of policy instrument applications for the marine environment can be found in Table 3.3.

Regulatory instruments

The objectives of regulations are manifold. Such instruments exist to minimize damage to natural resources, ensure equitable access to valuable resources and monitor the use of these resources. They also encourage or restrict certain behaviours, ensure competition and establish positive social outcomes. In other words, regulation balances rational individual behaviour with the broader societal good to maintain both the economic and environmental integrity of the 'commons'.

While regulation broadly refers to the process by which activities are administered, regulation as a legal construct establishes legislated obligations that require permits or prohibit

Table 3.2 Taxonomy of policy instruments

Regulations	Using markets	Creating markets	Informational instruments
Standards	Liability	Property rights	Information supply
Bans	Subsidy reduction	Tradable permits	Research
Permits and quotas	Environmental charges	and rights	Education and public
Zoning/plans	and taxes	International offset	outreach
	User charges	systems	Transparency mechanisms
	Deposit-refund systems		Product certification
	Targeted subsidies		Performance measures
	Insurance		Collaborative governance
	Grants and loans		Direct government

Source: Adapted from World Bank (1997)

Table 3.3 Sample applications of policy instruments for marine management

Policy instrument	Example marine applications
Direct provision	Provision of marine parks Wastewater treatment Establishing shipping lanes
Detailed regulation	Spatial zoning and planning Regulation of fishing (by dates, equipment, location) Bans on trade to protect biodiversity (e.g. seahorses) Bans on dumping Low sulphur fuel standards
Flexible regulation	Water quality discharge standards Ballast water treatment
Tradable quotas or rights	Individually tradable fishing quotas Emissions permits
Taxes, fees or charges	Fishing licenses Park fees Industrial pollution fees
Subsidies and subsidy reduction	Fisheries
Deposit-refund schemes	Hazardous waste management
Creation of property rights	Private marine parks Exclusive fishery rights
Common property resources	Community-based management
Legal mechanisms, liability	Liability bonds for oil and gas transport and mining, hazardous waste management
Voluntary agreements	Standards for marine ecotourism operators Sharing of transboundary fish stocks
Information provision, labelling	Marine Stewardship Council certification for sustainable fisheries
International treaties	International treaty on Law of the Sea

Source: Adapted from Sterner and Coria (2012)

particular activities. Regulations control risks and reduce harms. For example, domestic law may allow for the establishment of pollution limits. Regulation then sets specific limits and may mandate how the limits are to be met (e.g. type of technology). It is therefore a legal tool. When considered in the domestic context, regulatory tools are a traditional form of policy. States (or sub-national governments) act as a trustee, controlling access and use of environmental resources. Regulations can include standards, bans, permits or quotas. They can also limit the temporal or spatial extent of an activity (through zoning).

Regulation is also used in international law. International conventions on pollutants, greenhouse gases, ocean dumping and endangered species are based on regulatory approaches. These approaches are popular because they promise predictability and effectiveness. In addition, both the regulator and the regulated understand what is required of them. Regulation can be particularly effective in waters in ABNJ and states' EEZs (see Box 3.1) since point-source discharges, including ships, aircraft and platforms, are known and can be monitored. In contrast, regulation is less effective in territorial waters. These areas are subject to considerable non-point sources of pollution (i.e. from multiple sources including land sources), making it difficult to ascribe discharges to specific culprits.

Box 3.1 Regulating fisheries discards

The unintentional catch (bycatch) of non-target species that are thrown back to the sea are known as discards. Currently, approximately 8 per cent of the total global fisheries catch, amounting to approximately 7 million tons, is discarded, with shrimp and demersal (bottom) trawl fisheries responsible for about half of these discards. While most discards are returned to the sea as dead or non-viable, certain types of discards (e.g. marine mammals, turtles) are returned to the sea with the expectation that they are viable. Discards contribute to both economic inefficiencies and ecosystem impacts. As such, they are recognized as a major fisheries management problem that can be improved through better regulatory policies. A number of factors contribute to discards including:

- biological (spatially and temporally coincident fish populations, similarities in size and habitat use);
- technological (fishing gear unable to distinguish between species);
- behavioural (customs or practices that contribute to bycatch);
- economic (inability or unwillingness to adopt selective fishing gear);
- institutional (lack of regulations prohibiting bycatch).

The primary approaches to regulating fisheries discards are via the following types of regulatory instruments:

- Input controls:
  - Mandating selective fishing gear to avoid bycatch;
  - Spatial and temporal fishery closures;
  - Limiting the number and size of vessels;
  - Bycatch rules.
- Output controls:
  - Catch limits to cap the proportion and total amount of bycatch;
  - Mandatory landings to improve information on bycatch and incent fishers and processors to utilize bycatch.
- Monitoring and reporting:
  - Catch log recording and control;
  - Landing reports and control;
  - Inspections at sea;
  - Monitoring of fishing grounds.

Non-regulatory tools may include economic incentives to adopt certain gear types, transferable quotas to concentrate efforts on fewer but better equipped boats, and education to change fishers' and consumers' behaviour.

Source: modified from Johnsen and Eliasen (2011)

While all regulatory approaches are coercive in nature they can take a number of forms:

- *Coercive* or *command and control regulation* sets a specified, measurable level of activity (e.g. pollution maximum) and the 'control' dictates how the level of activity is to be achieved (e.g. technology). Command and control approaches evolved from engineering policies and were first widely applied in the US in the early 1970s through the Clean Air Act and Clean Water Act.
- *Performance-based regulation* specifies the outcomes ('output standards') to be achieved but not the means by which these outcomes are to be attained. Performance-based regulation can benefit the regulator since regulations framed as objectives/outcomes are relatively easy to draft. A central benefit to the regulated includes the ability to choose the most efficient processes and technologies to meet the regulation. On the downside, difficulties with performance-based regulation include additional costs for the regulator to verify compliance. In addition, performance-based regulations may be difficult to develop if the desired outcomes are not apparent. Most states that use performance-based regulation have adopted 'safe harbour' ('deemed to comply') provisions that limit liability and incent the regulated to develop innovative practices (OECD, 2002).
- *Process-based regulation* is an incentive-based approach that requires the regulated to develop processes that minimize risks. The premise behind this approach is that businesses have an economic incentive to avoid regulation and, consequently, will develop lower cost solutions to mitigating risks than would be achieved through command and control or performance-based regulation. The most frequent marine use of process-based regulation is managing seafood safety. Australia, Canada and the US allow seafood producers/processors to develop management plans that identify and address risks at key points of the production cycle.
- *Co-regulation* is a partnership between the regulator and regulated that moves the burden of identifying and mitigating risks onto the regulated. Under this model, codes of practice are jointly developed that are enforced by industry or industry associations rather than states or international organizations. The benefit of co-regulation to the regulator is reduced costs (e.g. monitoring, enforcement), voluntary participation and greater compliance. Benefits to the regulated include lower costs than traditional regulation, industry-wide compliance and demonstration of responsibility to consumers. A challenge with co-regulation, however, is that industry self-regulation may result in anti-competitive practices such as exclusion of new entrants into an industry (OECD, 2002).

Regulations are enforced through legal tools that include penalties, fines, liability rules and performance bonds. They focus on the punitive consequences attached to violations of conditional access or misuse. Regulations or other market-based instruments must be backed up with some threat: in the case of non-compliance this is typically financial. When a law is broken, fines are imposed to remove the advantages gained from non-compliance. While penalties for conviction of an offence range from fines to imprisonment, in practice, legal maximum penalties are rarely imposed. As a result, violators often assess the risks of being caught and calculate the potential fines as part of the cost of doing business.

However, when the offense is serious (e.g. affecting public safety), fines or other sanctions are not considered sufficient and punishments under the criminal law system may be applied to offending individuals. Liability for damages and risks related to individuals seeking compensation also influences environmental responsibility. Liability can be partial, if the cause of the damage is due to negligence or lack of acceptable precautions; or it can be full, where injured parties have the right to compensation regardless of any precautions taken. Although both forms of liability should induce necessary environmental precautions, the costs for litigation often prevent affected individuals from pursuing compensation. Partial liability can also reduce a company's precautionary measures. Since reporting on environmental performance acknowledges that products or practises may be harmful, the defence of ignorance is removed.

Strict or full liability makes sense where, as with oil spills, precautions can best be taken by one of the parties (and where large numbers of third-party victims make bargains between the parties difficult or impossible) (Box 3.2). Strict liability gives more rights to claimants, but has been criticized for escalating demand on the court system and preventing economic activity that is exposed to such risks. However, even strict liability may not be a sufficiently forceful instrument: in some cases there may be a long chain of causality, considerable distance in space or time between negligent action and realized environmental cost. Also, a confusing array of parties may be involved, some of which might no longer exist. To be effective in such cases liability must be joint and retroactive, so that all parties involved face the consequences (Sterner and Coria, 2012).

Regulatory standards

Prescriptive *regulatory standards* are the most commonly used instruments in natural resource management, likely due to their simplicity in principle (e.g. polluter pays) and typically short timelines for application. Regulatory standards can take one of three forms: *ambient standards* place limits on the total concentration of a pollutant (e.g. contaminants, noise) in a geographic area; *emission standards* place limits on individual sources (e.g. ships, platforms); and *design standards* specify a particular type of pollution control technology or production process (e.g. oil-spill mitigation requirements for commercial vessels).

Design standards are, however, widely criticized for their lack of incentives to develop alternative solutions that arrive at the same desired result (e.g. lifecycle or 'cradle to grave' approaches). Yet, depending on the situation, there are a number of reasons to consider design standards, including situations where (after Sterner and Coria, 2012):

- Technical and ecological information is complex.
- Only a few technologies exist, and one is superior.
- Monitoring costs are high, or monitoring pollution is difficult.

An example of a specific mandated technology is the regulation for the prevention of pollution by oil addressed in Annex I of the International Convention for the Prevention of Pollution from Ships (MARPOL) (Lagring et al., 2012, Chapter 7). The Convention not only includes requirements for technical standards in construction (e.g. double hulls for oil and chemical tankers) and equipment (e.g. oil/water separators), but also inspection (including enhanced surveys on existing tankers), record keeping (e.g. oil record books, cargo record books) and in-port procedures (e.g. receipts for the discharge of wastes to shore, reception facilities, tank cleaning in port) (Faure and Wang, 2006).

Regulatory approaches may restrict the location or timing of an activity (e.g. marine protected areas and no-take zones) or ban a particular product or process (e.g. fishing gear

---

Box 3.2 Liability and marine oil spills in the US

In the case of marine oil spills in the US, there is a wide range of laws with complex implications for liability. The Oil Pollution Act of 1990 (OPA 90) was enacted largely in response to the major spill in Prince William Sound, Alaska in 1989 that had resulted from the grounding of the *Exxon Valdez*. The OPA 90 makes parties responsible for oil spills strictly liable for damages caused. It 'channels' liability for oil spills by specifying exactly who is to be treated as the responsible party. In the case of spills from vessels, the owner/operator of the vessel is the responsible party. For offshore facilities such as the Deepwater Horizon (responsible for the world's second largest marine oil spill in 2010 – Chapter 7), the holder of the drilling permit (e.g. BP) is the responsible party.

   The OPA 90 also includes liability caps, traditional in maritime law, that include some exceptions (e.g. for gross negligence) and vary depending on the type of spill and damage caused. Prior to OPA 90, liability was limited in most cases to the value of the vessel. However, the OPA 90 did not preempt criminal or state law so the availability of alternative legal regimes (civil or criminal) and jurisdictions (state and federal) add significant complexity and uncertainty to the damages phase of litigation. As a result of the *Exxon Valdez* disaster, the federal government sought recovery of damages to Prince William Sound by filing criminal charges under the Migratory Bird Treaty Act (MBTA) for causing the death of protected birds and the Refuse Act, for dumping of 'refuse'. Since the MBTA is a strict liability statute, the prosecution does not have to show that the defendants intended to harm wildlife or prove that the defendants knew their actions would lead to an oil spill to find liability. Violations of these laws carried penalties and required restitution in the form of compensation for damages to the natural environment. As a result of civil and criminal charges, Exxon pled guilty and reached a settlement involving significant payments to the federal government for cleanup and damages. Claimants in a separate class action lawsuit were initially awarded US$5 billion, but Exxon spent 20 years appealing the award. Finally, in 2008 the Supreme Court reduced assessed damages to $500 million. It remains to be seen whether private suits for economic damages due to the Deepwater Horizon spill will be subject to the OPA 90 limits, currently capped at $75 million.

Source: Faure and Wang (2006, 2008)

---

restrictions) (Pauly, 1997; Johnstone, 1999; Agardy, 2010). Marine examples include bans on dynamite fishing and the use of cyanide fishing for the live fish trade, as well as proposed bans on fishing gear like the 'hulbot-hulbot' in the Philippines (a form of seining that involves throwing a large rock tied to a net into the sea and dragging it underwater) (White *et al.*, 2002). Because of the destructive impact to coral reefs – a critical habitat and spawning ground of the fishery resource – these bans are considered reasonable.

Success of regulatory approaches in the marine environment

A major criticism of regulatory policy is that to be effective, monitoring and enforcement are required indefinitely. Situations where monitoring is problematic – including non-point source pollution where many emissions are small, dispersed and mobile – may necessitate regulating preventative design standards. In the case of hazardous wastes, it may not be the use of the product but its disposal that is the chief concern.

In addition, regulatory approaches are often challenging to operate in developing economies. Institutions may be ineffectual, underfunded or corrupt; monitoring and enforcing regulations can be near impossible. The immediate pressure of building the economy or feeding the population can override investment in guaranteeing the sustainability of even the most valuable resources (Sterner and Coria, 2012).

The following examples illustrate the variety of marine activities that have been regulated both in territorial waters and ABNJ.

A review of the regulation of sea cucumber fisheries in the territorial waters of 62 states concluded that depleted and overexploited fisheries were on average managed by 2.6 regulatory measures (Purcell et al., 2013). In contrast, better-managed fisheries were regulated by an average of 4.7 regulatory measures. The most successful regulatory measures included fleet (vessel) controls, limited entry controls and harvest closures (see Chapter 6). However, enforcement capacity and poverty were also strongly correlated to the sustainability of sea cucumber fisheries world-wide.

Albert et al. (2013) review domestic (including sub-national) and international regulatory approaches to the management of ballast water, a source of biological invasive species (see Chapter 7). They conclude that some jurisdictions (e.g. IMO) establish regulatory standards that emphasize international consensus, practicability and cost of treatment in exchange for a higher risk of invasions. Meanwhile, other jurisdictions (e.g. US) have questioned whether existing ballast water standards sufficiently protect against invasive species. Troublingly, the study concludes that without a better scientific understanding of what standards are required to prevent or minimize invasions, regulations are, by and large, merely guesswork.

Schinas and Stefanakos (2012) assess the economic impacts on international shipping as a result of the IMO MARPOL Annex VI (see Chapter 7) requirements to use low sulphur fuels (to reduce SOx emissions from ship propulsion systems). Presently, regulatory standards for areas outside and inside the sulphur emission control areas (SECAs – currently the Baltic and North Seas and North American coastal areas) are 3.5 per cent and 1.5 per cent sulphur content by mass respectively. These standards are scheduled to decline to 0.5 per cent (in 2020) and 0.1 per cent (in 2015) respectively. Ships, then, may find themselves transiting seas with different fuel standards, thus requiring them either to switch fuels or to use more expensive fuels throughout their entire journey. As a result of these regulatory changes, increases in fuel costs are estimated to be approximately 20 per cent and 90 per cent inside and outside of SECAs respectively. These increases may affect the competitiveness of shipping over shorter distances because shippers often have rail or trucking options available to them.

Tributyltin (TBT) has been a popular anti-fouling compound used in marine paints since the 1960s (Gipperth, 2009). TBT compounds are hazardous to marine life and may enter the human food chain. The domestic and international response to regulating organotin compounds (including TBT) is twofold: first, many states have established ambient standards for maximum concentrations in marine waters and sediments; and second, states began to ban or regulate organotin compounds in the 1980s. These domestic efforts were followed by the 2001 IMO anti-fouling convention that entered into force in 2008 (see Chapter 7). Parties to the convention are required to prohibit and/or restrict the use of harmful anti-fouling systems on ships flying their flag. So too are ships not entitled to fly their flag but which operate under their authority and all ships that enter a port, shipyard or offshore terminal of a party.

While banning organotin compounds and establishing ambient standards are well intentioned, they have created both anticipated and unanticipated consequences. The anti-fouling convention does not regulate alternatives to organotin compounds. So a number of new chemicals have appeared on the market that appear to have similar deleterious effects as TBT. Shippers, in an effort to control costs, may register their vessels in a flag state that is not a party to the convention or that lacks domestic organotin regulations (the 'race to the bottom'). Moreover, states with stores of organotin compounds that they can no longer use may dispose of them in an unsafe manner. Perhaps more ironically, bans on these substances may inadvertently increase species invasions (Gipperth, 2009).

Economic instruments

Critics of regulatory approaches have proposed economic instruments as a complementary or alternative approach to achieving public policy objectives. Rather than trying to prescribe behaviour through regulation, market-based approaches rely on consumers responding to market driven price signals to change behaviours and achieve policy goals. Expanding on the 'polluter pays' principle, economic instruments incorporate the 'user/beneficiary pays' principle along with the principle of full cost recovery. If a policy changes the cost of a good, service, activity, input or output, then it is a *market-based instrument* (MBI).

There are many definitions for MBIs depending on the context in which they are applied. Most commonly, MBIs are policies that encourage specific behaviours through market signals and incentives rather than through explicit directives regarding mechanisms, levels or methods. While regulations can also affect the market, they allow little flexibility in the means by which goals are achieved. Furthermore, regulations do not impact cost or price directly. Thus, MBIs may be able to achieve policy goals more cost effectively than regulatory mechanisms. Nevertheless, MBIs and regulatory instruments are not mutually exclusive, as any policy still requires appropriate legislative or regulatory backing.

MBIs can also be employed to provide diverse social benefits. These may include payments to indigenous peoples for protecting goods and services, developing alternative income sources in rural/remote areas, or fostering income diversification opportunities (Greiner, 2013). Usually, however, MBIs are normally associated with addressing environmental or public health issues including pollution (e.g. waste management, greenhouse gases), resource use (e.g. fisheries) or protecting environmental goods and services (e.g. pollination, waste assimilation).

The earliest applications of MBIs were effluent charges (polluter pays) instituted in France (1968), Netherlands (1969) and Germany (1981). One of the most well-known applications of MBIs is fuel taxes that, through increasing the cost of driving, result in more energy efficient cars. In Europe, for example, 40–60 per cent of the cost of fuel is taxes. As a result, European cars produce two to three times less CO_2 emissions than equivalent vehicles in the US.

MBIs are generally classified into the following categories:

- *Price-based instruments*: Taxes, charges or subsidies that establish (or help to establish) a specific regulated price for resources or outputs that are deemed to have a social or environmental cost associated with them. Advantageously, they can work as a positive or negative incentive. For example, a price may be set for the creation of a unit of pollution.

- *Quantity-based instruments*: Also termed 'rights-based' instruments, these establish quantitative limits on resource use (or environmental goods and services). Advantages are that rights and obligations are often tradable. For example, permit trading systems and offset schemes create a 'right' to pollute that can be transferred or sold.
- *Market enhancement instruments*: Also known as 'market-friction' approaches, these include a range of policy actions and approaches, from subsidies to establishing property rights. Their purpose is to correct existing market imperfections or 'frictions' that lead to poor resource use. Eco-labelling programmes, for example, make it easier for consumers to select goods and services that align with their environmental preferences.

Ultimately, MBIs are designed to ensure that resources are used more efficiently. However, they also offer an opportunity to generate revenue. Making the markets work more efficiently involves establishing prices that address market failures, be they externalized costs or underpriced resources. The proceeds of taxes or levies can provide further environmental subsidies, reduce broader income taxes or be channelled into constructive programmes such as monitoring and enforcement.

Use of MBIs requires the following conditions to be met (after Stavins, 2003):

- A pre-existing market or capacity to create or award property rights.
- A measurable target resource that is influenced by cost and profit considerations.
- A target resource that, if MBIs are applied, can be used to attain public policy goals.
- A homogenous metric (indicator) that can be used as the focus of the instrument. The metric should correspond to and drive behavioural changes that are consistent with the resource management objectives.

Importantly, the effectiveness of MBIs will generally be a function of how well each of the above conditions is met. The focus on quantitative metrics means that MBIs typically have already identified measurable goals, such as catch quotas, kilos of discharge or production dollars. MBIs with potential application for marine resource management include charges and taxes, subsidies and tax concessions, property rights and market support, and tradable permits and quotas (Greiner *et al.*, 2000). A summary of of MBIs with application to marine management issues can be found in Table 3.4.

Price-based instruments: taxes and charges

A charge or tax can be envisioned as a 'price' to be paid on the use of the environment. This 'price' can be levied on producers or consumers and is a very effective means of changing behaviour. Taxes that are levied to internalize social or environmental costs – such as pollution – on private third parties causing the externality are known as *Pigouvian taxes*. The objective of Pigouvian taxes is to drive down pollution output by ensuring the economic benefits of further production exceed the environmental costs associated with it. User charges are a related method, but rather than taxing activity levels or particular behaviours, they involve a fixed charge for the use of environmental services. These charges are typically assessed as an access or license fee. Product charges or taxes could be applied to goods

Table 3.4 Summary of advantages and disadvantages of market-based instruments application in the marine environment

Instrument type	Advantages	Disadvantages	Relevance
Emission and effluent charges/taxes	Low transaction and compliance costs for firms or individuals Promote technological innovation Create long-term incentives Raise revenue Create flexibility for polluters Useful when damage per unit of pollution varies little with quantity of pollution	Setting charge/tax at the right level Require monitoring	Discharge from point sources When monitoring is viable at a reasonable cost When pollution abatement is feasible
Product/user charges	Reduce use of harmful products (for product charges) Raise revenue Create flexibility for users Simple to administer	Setting charge at the right level Require monitoring Often weak link to pollution	Products used in large quantities Where not feasible to monitor pollution from individual sources For products whose demand or output is sensitive to price changes Effective when sources are numerous and damage per unit of pollution varies little with pollution quantity
Deposit-refund systems	Reduces volume of waste/pollution Encourage safe disposal, reuse and recycling Create flexibility for users	Transaction costs may be high Markets for recycled products may not be well developed	Most effective if applied to products with an existing distribution system and where reuse and recycling is technically and economically feasible Where problems are related to waste disposal
Tradable permits	Allocation of resources to highest valued use Reduced information needs for regulators Certainty regarding pollution or resource use levels Compliance cost can be reduced Flexibility for polluters Create long-term incentives	Establishing an efficient market Setting overall level and initial allocation of permits Transaction costs involved in trading	Where environmental impact is independent of pollution source When environmental impact does not correlate with time of production Where there are enough sources to establish a market Effective when damage per unit of pollution varies with amount of pollution

Table 3.4 continued

Instrument type	Advantages	Disadvantages	Relevance
Subsidies and tax concessions	Encourage actions to overcome environmental problems	Externalities are not internalized by polluter May reward poor environmental performers May pay those who would undertake activity even without subsidy Cost to budget May stimulate too much activity	Where other instruments do not work or are too expensive
Property rights/market support	Enable goods/services to be identified, which can then have value attached, which in turn allows trading to occur Improves efficiency of market	May not be feasible where environmental benefits exist, for which payment cannot be extracted Jurisdictional issues can arise – e.g. international 'commons' such as the ocean or atmosphere offer limited opportunity to enforce property rights on the full set of users	Where ownership of 'environment' is uncertain or non-existent Where significant information asymmetry is present Where environmental or social values are unpriced

Source: adapted from UNEP (2004)

that have an environmental impact anywhere along the line from production to consumption. They can be assessed on a product, such as the use of petroleum, or on a product characteristic, such as the carbon content in petroleum (Sterner and Coria, 2012).

Taxes and charges are suited to situations where the individuals responsible for the problem can be readily identified. However, the costs of monitoring and enforcement may still be high because of the need to ensure that all individuals responsible for the problem are subjected to the tax. Where a tax or charge is issued, the party responsible for the damage to the environment bears all of the costs of changing resource use or management practices.

More complex price-based instruments include integrated strategies, such as ecological fiscal reform (EFR) or ecological tax reform (ETR). These approaches adjust existing taxes to phase out products and to discourage certain practices. Alternatively, new ecological taxes provide incentives to reduce environmental impacts and potentially 'recycle' the revenue from the new taxes. Revenue can be recycled in numerous ways including: reducing existing tax rates; providing new credits or subsidy programmes; providing refunds to taxpayers to encourage positive behaviours; or support the development and upgrades to green technologies. While determining the level of taxes necessary to drive behavioural change can be hard to predict, complementary tax reduction enhances political acceptability. It does so by ensuring there is no increase in the overall tax burden (Bernow et al., 1998).

Sterner and Coria (2012) suggest that taxes and charges can be appropriate in situations where:

- The risk of 'excessive economic costs' dominates concerns about potential costs associated with insufficient environmental action.
- The action or commodity to be taxed is the direct cause of the negative externality being targeted, or is closely connected to it.
- The reasons for the tax/charge are understood by affected parties and there is an understanding of how revenues will be used.
- There is clarity over the rate of tax/charge necessary to change behaviour(s) such that externalities are reduced or eliminated.
- Unfair competitive advantages are avoided.
- There is a desire for improved cost recovery or expanded revenue.

Quantity-based instruments: tradable permits/rights

Tradable permits allocate property rights over a resource that was previously common property. This property could be in the form of air or marine fisheries. The purpose of permits is to mitigate risks and impacts. Permit systems broadly work in two ways: first, 'cap and trade' systems establish an allowable overall level of pollution or extraction, and create a limited number of permits that allow an activity. Markets then establish economic values for the permits. Users that stay below their allotted level may sell their surplus permits to others or apply them to offset other practices in their business. The second system is 'baseline and credit'. These systems do not limit the absolute amount of pollution or extraction. Instead, a baseline allowance is established for each user (e.g. facility, landowner) and users may sell unused credits to others. Both systems operate on the premise that rewarding those who can reduce their externalities at the lowest costs will maximize the economic value extracted from all permits. These two systems can also include 'banking' and 'borrowing' (intertemporal trading) schemes. This incents polluters to voluntarily reduce emissions through retrofitting or closing down less efficient sources. The credits from these actions can be applied to future projects, which, in turn, accelerate emissions reductions. In addition, banking can dampen price fluctuations.

Tradable permits evolved from regulatory permitting approaches that were deemed to be more efficient if permits were allowed to be bought and sold. Tradable permits are normally used once regulations have fully allocated either the maximum acceptable pollution or the resources to be exploited. Perhaps the most successful permit scheme is the US cap and trade system for sulfur dioxide (SO_2). In 2001, a mere six years after its initiation, the system reduced SO_2 emissions by 40 per cent and reduced compliance costs by 43–55 per cent (Burtraw and Palmer, 2003).

Numerous fisheries are managed today using tradable permits (individual transferrable quotas – see Chapters 6 and 10) and the IMO is currently investigating whether tradable permits should be applied to shipping in order to reduce CO_2 emissions. The growth in permit trading is fostered by technological advances that make it economically feasible to monitor use and changes in environmental quality. This information informs decisions on economic trade-offs and alternative resource use options (Johnstone, 1999; Salzman and Ruhl, 2000).

Stavins (2003) suggests that trading systems are likely to be feasible and appropriate for situations in which:

- There is a willingness to grant private property or exclusive access rights to a common property resource.
- The traded commodity is relatively homogeneous (e.g. CO_2, a fish stock).
- Resource or environmental conservation is a principal objective.

Tradable permits have been shown to be effective in theory (through economic modelling) and in practice. Among the benefits of this system are: certainty that environmental or resource use thresholds will not be exceeded, economic growth will continue without compromising environmental quality or resource yields, and finding the most cost-effective methods to reduce production of externalities. Disadvantages to permit trading schemes are centred on both the costs of establishing initial allocations and operational transactions. Transaction costs include obtaining market-relevant information, negotiating trades and compliance and verification costs. A summary of considerations in the design of permit trading schemes is provided in Box 3.3.

Market enhancement instruments

Substantial gains can be made in environmental protection simply by reducing existing frictions (inefficiencies) in market activity. As discussed previously, market enhancement instruments are the means by which existing market imperfections or 'frictions' are corrected. Most instruments operate in two ways. First, they improve the information available to consumers, investors or producers. This enables them to make more informed (efficient) decisions about the environmental or social consequences of their purchases, their investments or their production. Second, they reduce transaction costs that lead to improved environmental and social outcomes. Some examples of market friction reductions are discussed below.

Subsidies and tax concessions

A subsidy is a payment for certain activities that the government wishes to promote. A tax concession reduces tax assessments on those engaged in such activities. The objective of subsidy and tax reduction measures is to provide incentives to reduce costs and expand activities that address environmental and social problems. There are many similarities between subsidies and tax concessions and, from an economic perspective, they are generally considered as 'negative taxes' in contrast to taxes and charges.

There are several types of subsidies. *Production subsidies* are government payments to producers either to increase the production of a good or service (e.g. fishing, offshore wind energy) or decrease some externality (abatement) of production (e.g. sewage treatment). *Export subsidies* are government payments to producers to gain international market access. *Employment subsidies* are government payments to reduce unemployment, train workers or encourage the growth of certain fields. *In-kind* subsidies are non-monetary support provided by governments. The provision of infrastructure, services to assist exporters, preferential access to natural resources, or limitations on liability exemplify in-kind subsidies. *Procurement subsidies* are government purchases (e.g. shipbuilding) to support domestic producers or certain sectors. *Market price support* is when governments establish minimum prices or enact tariffs and trade barriers to protect certain producers.

All of these subsidies may be time limited to encourage growth in a certain part of the

Box 3.3 Considerations in establishing permit trading systems

Defining what commodity is to be traded: Most permit trading systems specify physical quantities such as tonnes of a pollutant (e.g. SO_2, CO_2) or harvested resources, such as fish. If these quantities are difficult to measure and enforce, however, then it may be more feasible to measure transactions such as fuel purchases instead of emissions.

Defining what sectors trading will be applied to: Permit trading (even for a single pollutant) can be applied to certain sectors (e.g. coal power generation) or all sectors (e.g. all industrial facilities that emit SO_2). Schemes that cover all sectors are believed to be more economically efficient as abatement costs are spread over a much larger part of the economy.

Defining where a commodity is to be traded: In the case of pollutants, their impacts and abatement costs (mitigating actions) may vary depending on geographic location. As such, allocations and permit prices may need to be adjusted to account for these differences.

Determining what other measures may be required to achieve objectives: Permit trading alone may be insufficient to achieve anticipated reductions in pollution or extractive use. Other complementary measures such as regulatory schemes, subsidies and tax credits (to foster innovation and technological change) may be required for permitting systems to be successful.

Determining who is allowed to hold a permit: Permits may be assigned directly to 'downstream' users (e.g. polluters, fishers), 'upstream' users (e.g. producers or processors of fuels) or both ('hybrid system').

Determining how permits are allocated: Permits can be granted either to existing users (grandfathering) or through a competitive process (auctions). Grandfathering is straightforward as existing resource users are granted permits that may be adjusted for the output of each user. Furthermore, when users exit the market their allocations can be reallocated to new entrants. The disadvantages of grandfathering are that the practice creates instant windfalls for users that receive free allocations and that the opportunity to raise revenue (through auctions) is lost. In contrast, auctions set a true market price for emissions/resource yields and do not limit new entrants into the market. Auctions also provide an opportunity to raise substantial revenue and can be structured to give local/indigenous participants an opportunity to receive allocations.

Setting the rules for participation: Rules can be established on who is allowed to participate (e.g. third-party brokers, small users), whether participation is voluntary or mandatory, the minimum size of the trades, or whether government approval of trades is required.

Establishing compliance and penalty regimes: Compliance regimes set dates by which users must report on their actual allocations, something often not straightforward. Intense trading often occurs prior to these dates as users in deficit attempt to purchase credits from users in surplus. Penalty regimes set the rules (e.g. financial penalties) for non-compliant users.

Source: modified from OECD (2004, 2008)

economy, or they may become a permanent part of the economy. Subsidies may be appropriate in cases where a socially or environmentally beneficial activity can be reliably targeted. Producing a benefit is generally not equivalent to ceasing to generate a cost. Offshore wind farms, for example, may receive subsidies or tax concessions since 'green energy' is necessary to meet greenhouse gas reduction targets. In this instance, the subsidy is either a one-time payment to help launch the operation or a deficiency payment to compensate for the difference between the target price and the actual market price of the product or service. However, subsidies frequently have unintended effects and often become 'perverse subsidies'. For example, many energy-exporting states heavily subsidize domestic energy use (which nullifies conservation efforts) and many agricultural states subsidize agricultural industries (which can drive consumer prices up and result in overproduction).

Subsidies have a long history in marine fishing. Currently, the world's fishing fleet receives annual subsidies of up to $19 billion. The consequence has been extreme excess fishing capacity of 40–60 per cent, or, stated differently, up to 2.6 million unnecessary fishing vessels (Sumaila *et al.*, 2012). These subsidies lower retail prices (and thus increase consumer demand) and create a 'race to fish'. Unsustainable catches and poor safety practices for fishers are the sad result.

Tax concessions are similar to subsidies in that both approaches attempt to grow targeted aspects of the economy. Concessions can be conceived as a subsidy to a producer's tax burden. Concessions can take diverse forms: *tax exemptions*, where no tax is paid; *tax credits*, where reduced tax is paid, or *tax deferrals*, where tax payments are delayed. Tax concessions are criticized because they are often less transparent than subsidies and, once enacted, are difficult to repeal.

Property rights and market support

The peril of open access (advocated by Grotius, Chapter 2) is a frequent theme found throughout this text. Individuals, corporations and states have few incentives to produce, regulate, manage or monitor the production of marine-derived goods and services efficiently in ocean areas that are shared with others. Expressed by Hardin (1968) as the 'tragedy of the commons', the 'everybody's property is nobody's property' problem of open marine access often leads to rapid depletion of resources due to 'gold rush' behaviour. Low profits due to overcapitalization and excessive costs, dangerous working conditions due to low profit margins and poor product quality further illustrate this peril (Deacon, 2009). While most discussion around the perils of open access relates to the rapid depletion of fisheries, other sectors, including transportation, energy development and mining, also lack incentives to minimize environmental harm and maximize long-term opportunity.

From an international perspective, the global ocean commons is gradually being enclosed. The proliferation of declarations of EEZs over the past several decades has resulted in 95 per cent of the world's fisheries and nearly all of the world's offshore hydrocarbon resources (only found on continental shelves) vesting within state governments. The establishment of EEZs has been a significant step towards economic rationality and enclosing the commons. Nonetheless, many of the most important global fisheries, including tuna, occur primarily outside of EEZs.

Marine property rights, therefore, vest primarily with the state, which may then confer ownership onto sub-regional governments, indigenous groups or private parties. However, states may allow open access to certain resources within their jurisdictions or they may lack the ability to control and regulate access (e.g. failed states). Yet, there are some situations where it may be more beneficial for resources to remain in the commons. Situations where state control is weak and citizens are poor may benefit from retaining a common property system (Sterner and Coria, 2012).

An emerging trend in the creation and trade of property rights is assigning rights and value to environmental services such as carbon sequestration and biodiversity protection. The development of markets for these types of services in the marine environment is challenging for two reasons. First, the diverse nature of the ecosystem services produced in different geographic areas gives rise to measurement and accounting difficulties. Second, the relationship between actions and environmental outcomes is poorly understood. This understanding is complicated by the time lags involved, and the unpredictability of variables such as ocean currents and temperature. Almost inevitably, problems of who should be legally responsible for underwriting the environmental improvements are generated (Davis and Gartside, 2001).

Informational instruments

Information significantly affects the policy-making environment, so much so that it can also be considered a policy instrument in and of itself. Information provides the foundation for all instruments, as generally every policy requires data on technology and ecology. There are a broad range of explicit educational and information-based policy tools. These can act as stand-alone instruments, or supplement and reinforce the effectiveness of other mechanisms. Within the context of the marine policy environment the main informational categories include supply, research, education and public outreach campaigns, transparency mechanisms and product certification.

Information supply

Information supply is fundamentally important in changing attitudes and behaviours. Studies have shown that environmentally inappropriate behaviour often arises not from selfish motivations but rather from ignorance. There is a significant relationship between engagement in environmentally responsible behaviours and a knowledge of environmental issues and conservation activities. This is consistent with surveys and community consultations with resource users, which identify research-based advice as a major need for improving management. Recognizing the central role of information supply can assist the establishment of effective environmental policies.

Where informed resource users have self-interest in protecting biodiversity, appropriate information programmes can occasion substantial benefits. For example, enlightened self-interest may be sufficient to encourage traditional fishers to voluntarily discontinue the use of a net type that damages fish stocks and that would ultimately impact their livelihood. In this instance, the fishers would be informed that there are exclusionary devices that reduce bycatch and the effort and time spent sorting their catch would be minimized (Clucas, 1998; Kelleher, 2005). Although the precautionary principle has come a long way in influencing policy decisions made with little baseline scientific or economic understanding, setting serious limitations before any scientific connections have been established is usually seen as overly restrictive. Lack of information, however, is a common source of failure in a range of policies. Providing information is a powerful means by which governments can facilitate market operation and improve decision-making by their participants.

Research

When it comes to providing information on the impacts and effects of different environmental activities, research is required for an evidence-based understanding. Funding scientific and economic research aimed at understanding the dynamics and interrelationships in ecological and economic systems can assist in determining values and risks for natural resource management purposes, and reflect these in public policy. Some of the benefits of research are private or, at least, can be recovered efficiently through the sale of goods and services, while other benefits are of a 'public good' nature. Research is usually most cost-effective when it is conducted by scientists or specialist organizations in association with resource users who have local expertise, on behalf of policy-makers (Greiner *et al.*, 2000).

Education and public outreach

Advertising and granting awards to those who adopt best management practices potentially provide powerful incentives to shift public perception concerning environmental problems (Box 3.4). Well-defined advertising campaigns can simultaneously provide information and influence behaviour. There is also evidence that prizes and awards can play an important role in raising awareness of environmental issues and changing attitudes. In cost–benefit terms, prizes and awards are usually very inexpensive to create and maintain, yet they yield meaningful results. If they are developed at multiple levels, awards can engage the interest of substantial sections of the community, often securing free publicity and raising consciousness and influence motivation at minimal cost. However, it must be acknowledged that such campaigns have not always been successful; much is contingent on the tactics employed (McKenzie-Mohr and Smith, 1999).

Environmental education and awareness campaigns, like voluntary initiatives are ultimately considered a soft policy option. Focusing on encouraging change rather than requiring action or providing broad financial support to do so, they typically function best as part of a larger strategic approach. Educational initiatives play an important role in motivating participation and sustaining behavioural changes in individuals, households and industry. Ambitious education and awareness campaigns have also been proven to effectively reduce the need for economic incentives and regulatory enforcement.

Transparency mechanisms: corporate and state of the environment reporting

Governments seeking to advance sustainable development are increasingly employing strategies that encourage, support, mandate or directly demonstrate more environmentally sound business practices. Disclosure is designed to support market-based incentives (such as product differentiation), public awareness and pressures and government enforcement. Transparency mechanisms can be advanced on at least two different levels: disclosure of government inspection and factory audit reports, or disclosure of corporate environmental responsibility (CER) practices, including both the positive and negative impacts of business operations on a range of environmental issues (Box 3.5). Requiring that industry track and transparently disseminate information on the quality of their environmental practices can encourage action through accountability (Wahba, 2008). Australia, China, Denmark, the EU, France, India, Germany, Norway, Spain, Sweden and the US are among the countries

---

Box 3.4 World Wildlife Fund's International Smart Gear Competition

One example of a marine conservation award is the World Wildlife Fund's (WWF) International Smart Gear Competition. First held in 2005, the competition brings together the fishing industry, research institutes, universities and government to inspire and reward practical, innovative fishing gear designs that reduce bycatch – the accidental catch and related deaths of sea turtles, birds, marine mammals, cetaceans and non-target fish species. WWF believes the Smart Gear Competition will help catalyze a wide-ranging, multidisciplinary response by encouraging creative thinkers everywhere to share their ideas for modified fishing gears and procedures that increase selectivity for target fish species and reduce bycatch. The competition is open to eligible entrants from any background, and entrants have included gear technologists, fishermen, engineers, chemists and inventors.

WWF offers more than $50,000 in prize money to attract innovative ideas that may prove to be a valuable solution to some of the most pressing bycatch problems in fisheries around the globe. Financial support for the competition is provided by a number of government departments including National Oceanic and Atmospheric Administration (NOAA), Canadian Department of Fisheries and Oceans (DFO), as well as support from a number of foundations and corporations.

An international panel made up of gear technologists, fisheries experts, representatives of the seafood industry, fishermen, scientists, researchers and conservationists judge the entries.

Winning both the Grand Prize ($50,000) and the Special Tuna Prize ($7500) in 2011 was an entry from the captain of a Japanese tuna vessel. The design, called the 'Yamazaki Double-Weight Branchline', works by increasing the sinking rate of fishing gear, making it more difficult for seabirds to chase baited hooks. Used in conjunction with other devices, it can reduce seabird mortality by almost 90 per cent. Runners up, each receiving $10,000, include the 'SeaQualizer', designed to reduce fish mortality in the recreational fishing industry, and the 'Turtle Lights for Gillnets', aimed at reducing the capture of marine turtles in gillnets. This device reduces turtle interactions by up to 60 per cent without affecting target catch rates or catch value. Following the award, WWF works with the inventors to develop their ideas into usable products that enable smarter fishing.

Source: www.smartgear.org

---

that have developed governmental policy initiatives to promote sustainability reporting or environmental, social and governance (ESG) disclosure. Stock exchanges in Brazil, China, Malaysia, Pakistan, Singapore and South Africa are also playing a pivotal role in requiring or recommending listed companies to disclose sustainability performance information. Recent research from the auditing firm, KPMG, shows that in 2011 95 per cent of the Fortune Global 250 companies voluntarily provide information on their sustainability policies and performance – an increase from 50 per cent in 2008. Studies have also found that there are some significant correlations between corporate management's perceived importance of environmental factors and environmental reporting practices, suggesting environmental disclosure influences corporate practices (Marshall, 2003). While the impact

---

Box 3.5 Biodiversity audits

Building on the success of applying transparency mechanisms to industry, an emerging motivational strategy is the biodiversity audit. This innovative approach involves a small audit team containing scientific and other expertise as well as including respected members of the local community. Their role is to assess the biodiversity and significance of a specific area. Audits are conducted periodically, and results published locally to raise awareness of biodiversity issues and to enable longitudinal comparisons of the biodiversity characteristics. If audits are coupled with a system of prizes and awards, their capacity to raise consciousness and influence attitudes could be further advanced in a manner that is cost-effective and efficacious.

Source: Hill (2005)

---

of transparency is hard to separate from the effects of other regulatory policies, some facilities have responded to rankings or public reporting obligations. Ultimately, increasing community and facility awareness of emissions – be it through internet access to pollutant registries, inventories, right-to-know publications or reports – has made it easier to evaluate sources of environmental contaminants and analyse policy-instrument effectiveness.

Product certification

Also known as eco-labels, environmental product certification comprises a number of labelling systems for food and consumer products. The roots of eco-labelling can be found in the growing global concern for environmental protection on the part of the public, businesses and governments. A certified eco-label identifies a product that meets specified environmental performance criteria (standards). Eco-labels are often voluntary, but disclosing the environmental impact of some products can also be mandated by law. Directed at consumers, the intent of certification and labelling is to make it easy to take environmental concerns into account when shopping. Some labels quantify pollution or energy consumption by way of index scores or units of measurement; others simply assert compliance with an established standard or certification (Salzhauer, 1991).

The International Organization for Standardization (ISO) has developed a series of standards and guidelines for use in various programmes, including eco-labelling. Type I environmental labelling, or the ISO 14024 standard, relates to a wide range of environmental issues associated with a product or service. The certification criteria are developed by a third party and are based on lifecycle considerations (i.e. assessment of product from raw material extraction to end of life). Type I certification is limited to environmental leaders in a given product or service category.

Currently, the Ecolabel Index (www.ecolabelindex.com) tracks 432 eco-labels in 217 states that span 25 product sectors. Marine eco-labelling can be separated into four broad types:

* Seafood product labels: These have independent certification based on agreed upon standards, third-party certification, chain of custody and certification of individual stocks. In most instances the products are labelled to inform consumer choice.

Examples include Marine Stewardship Council (discussed in Box 3.6), Naturland and Friend of the Sea.

- Consumer awareness initiatives: These labels recommend to consumers what types of fisheries products to buy based on overviews of current information on overall stock status and fishing methods. Individual products are not labelled, but restaurants can partner to display the relevant labels. Examples include SeaChoice, Seafood Watch and FishWise.
- Eco-labelled shipping: These labels are relatively new and currently the German Blue Angel programme is the most renowned. The label's launch coincides with publication by the European Commission of a new EU strategy on air pollution from shipping that aims to reduce SO_2 and particulate emissions in the short term and NOx and other problems in the long term (ED 21/11/02).
- Tourism eco-labels: These initiatives are aimed at strengthening sustainability practices through tourism services. With Canada's Clean Marine Green Leaf Eco-rating Program, operators follow environmentally sound marina practices. Galapagos Quality is a seal of quality given to tourism businesses in Galapagos that have voluntarily committed to meeting environmental standards and requirements.

Box 3.6 Product certification example: Marine Stewardship Council

The Marine Stewardship Council's fishery certification programme and seafood eco-label recognize and reward sustainable fishing. The Council is a global organization working with fisheries, seafood companies, scientists, conservation groups and the public to promote the best environmental choice in seafood. Council certification assures consumers world-wide that seafood products bearing the Marine Stewardship Council eco-label come from fisheries that have passed rigorous, scientific, independent assessment for environmental and management practices. In order to become certified, fisheries must meet the Council's environmental standard for sustainable fishing.

The Council standard has three overarching principles that every fishery must prove that it meets:

- **Principle 1: Sustainable fish stocks**. The fishing activity must be at a level that is sustainable for the fish population. Any certified fishery must operate so that fishing can continue indefinitely and is not overexploiting resources.
- **Principle 2: Minimizing environmental impact**. Fishing operations should be managed to maintain the structure, productivity, function and diversity of the ecosystem on which the fishery depends.
- **Principle 3: Effective management**. The fishery must meet all local, national and international laws and must have a management system in place to respond to changing circumstances and to maintain sustainability.

What does this mean in practice?

Under the Marine Stewardship Council programme every fishery is measured against these principles. The actions that different fisheries take to show they meet the three principles vary in each case, taking into account the unique circumstances of the fishery. For example, a fishery could be asked to show that:

- **Principle 1**: It has reliable data on the age and gender patterns of fish populations to prevent too many young fish being caught and that other factors that affect the health of the stock – such as illegal fishing – have been considered.
- **Principle 2**: Measures are in place to limit bycatch (living creatures caught unintentionally). This could mean changing how fish trimmings are discarded so that seabirds are not drawn towards hazardous fishing gear.
- **Principle 3**: Vessel owners have signed a code of conduct, shared GPS (global positioning system) data or undertaken research to ensure their fishery is well managed. Effective management also ensures that all vessels will, for example, change their fishing gear or, when required, respect closed zones.

Eligibility for certification is assessed and awarded by third-party certifiers. When a fishery meets the Council standard for sustainable fishing its certificate is valid for five years. During this period the performance of the fishery will be reviewed at least once a year to ensure that it continues to meet the standard. After five years, the fishery must be reassessed in full if it wishes certification. Once a fishery has been certified, all companies in the supply chain – from boat to plate – that want to sell seafood from the fishery with the blue Council eco-label must be certified as meeting the Council chain of custody standard.

Despite their intentions, Marine Stewardship Council programmes have been criticized for their failure to demonstrate positive environmental impacts in many fisheries as well as certifying fisheries that that may not be sustainable (Jacquet et al., 2010).

Performance measures

Performance measures can be part of an environmental reporting or certification process, or an independent mechanism that assesses whether the goals and objectives of existing policies have been reached. Measures identify strategic outcomes, track progress and determine where adjustments should be made. They are valuable policy tools for improving accountability, informing decision-makers, demonstrating programme effectiveness and increasing political support and financial resources. Performance measures can be qualitative or quantitative assessments of outputs, outcomes or processes. However, identifying the right suite of measures and appropriate targets can be challenging, and managing complex outcomes over time requires expertise and financial support. In addition, achieving targets can be influenced by factors outside the policy process, making high-level outcomes difficult to assign and account for internally (Mazur, 2010; Shaw et al., 2010).

Voluntary instruments

One of the most effective ways to build motivation is to pursue structures that build community and industry ownership. Voluntary initiatives are an alternative option to more expensive legislative policy tools or programmes. Voluntary instruments and mechanisms rely neither on coercion nor on the continuation of substantial financial incentives; rather, they rely on voluntarism and self-regulation. Among the mechanisms are self-regulation by industry, projects supported by non-government organizations, community groups or stakeholders, and direct contracts with stakeholders (e.g. covenants on resource use). Voluntary instruments may also involve challenging business, industry or government to improve their environmental standard or practices: in return they garner public acknowledgement or avoid future regulation. While the costs of development and administration for voluntary initiatives are lower, the lack of regulatory consequences has made their effectiveness highly controversial.

Voluntary initiatives are particularly vulnerable to 'free riders', where those who choose not to participate avoid the costs of improvements incurred by competitors. In addition, not all resource users will make rational choices. Policy must also take account of the minority who may be irrational, incompetent or inflexible, underlining the need for a regulatory safety net. The lack of reporting or monitoring also makes it difficult to assess whether and where progress has been made. Proponents of voluntary programmes suggest they promote an ethic of resource custodianship and a cooperative model for achieving environmental goals. However, there are concerns that working closely with industry lowers the bar and prevents setting more ambitious targets. As a result, government proposals for voluntary initiatives in response to environmental concerns are often viewed as falling short of necessary or serious action.

Whether voluntarism and self-regulation are dependable in delivering targeted environmental outcomes depends on the extent of the gap between the public interest in environmental protection and the private interests of resource users. Where society's interest in protecting resources and the resource user's interest substantially coincide, self-regulation may be a cost-effective and appropriately non-interventionist strategy. Generally, the smaller the gap between interests, the more dependable voluntary mechanisms will be.

Unfortunately, in many circumstances, there is a considerable gulf between the public interest in resource conservation and the private interests of individual consumers. For example, whereas many fishermen may identify with the conservation objectives of voluntary protection of nursery grounds for juvenile fish, their self-interest in protecting resources is not so apparent in the short term. More commonly, there is a perceived tension between maximizing the use of resources and protecting juvenile stocks. While the preservation of nursery grounds may arguably provide long-term benefits to users (increasing the survival of juvenile fish, reducing physical disturbance to fish habitat), these benefits are less immediate and tangible than the increase in short-term catch yields that fishing the nursery grounds promises. For economically marginalized resource users, evidence suggests that short-term production payoffs are often perceived to outweigh the longer term benefits of conservation, presenting a major limitation of self-regulation and other motivational-based approaches to resource protection.

The application of voluntary instruments may be most effective at a small scale or in situations where parties already participate in self-regulation and more coercive policy is not necessary. Because of the risks and lack of dependability, relying on voluntary instruments as an exclusive means for environmental protection can work only if resource users appreciate the value of protection and comprehend how their own self-interest is served in protecting it. Informational and educational incentives are extremely important in this respect. However, in virtually all cases voluntary programmes need to be supported by other mechanisms such as price, property rights and regulation. Nevertheless, voluntarism still has an important role, particularly where environmental threats require active community participation. The challenge is to build a stewardship ethic, normalize the necessary conservation practices and harness altruistic behaviour.

Policy-makers also need to be cognizant of the traps of using financial incentives to jump-start voluntary programmes. If payments for conservation are subsequently withdrawn, then the benefits of those conservation efforts may be lost. Moreover, removal of the financial support may convert a carefully cultivated custodial ethic into one that is antagonistic towards the costs of conservation. One way to address the risk of reactive behaviour is to emphasize that any subsidies are only transitional in nature.

Although it is important to recognize and design policy that acknowledges self-interest, it is not a full explanation for an individual's behaviour. In contrast, altruism and respect for broad conservation objectives can also drive behaviour, and mechanisms that seek to support and harness such motivations can be integrated into a policy approach. If, for example, limited resources were put into encouraging and supporting resource users willing in principle to enter voluntary agreements, this might provide a cost-effective mechanism, even though it is only applicable to a small minority. If it were coupled with appropriate tax concessions, then many more users might be encouraged to participate. Successful programmes may also promote awareness of resource conservation, thereby contributing to motivational change – a factor acknowledged in the CBD.

To summarize, except in those specific circumstances where the public interest in protecting biodiversity conservation and private interest substantially coincide, and where biodiversity loss is reversible, voluntary mechanisms are most successful when employed as a supplement to other instruments to avert irreversible losses. These mechanisms, in most cases, need to be supported by policies that ensure dependability and recognize the need for precautionary measures (Boardman, 2009).

Chapter summary

- At present, no single authority, akin to the nation state, sets the overall rules on activities on, in and under the world's oceans.
- 'Policy' can be defined as plan or course of action to influence and determine decisions, actions and other matters.
- Policy instruments are often broadly classified as regulatory (legal), economic (market-based) or informational (persuasive). Some authors include cooperative (voluntary) instruments as a fourth type.

References

Agardy, T. (2010) *Ocean Zoning: Making Marine Management More Effective*, Earthscan, London.

Albert, R. J., Lishman, J. M. and Saxena, J. (2013) 'Ballast water regulations and the move towards concentration-based numeric discharge limits', *Ecological Applications*, 23 (2), 289–300.

Bemelmans-Videc, M., Rist, R. and Vedung, E. (1998) *Carrots, Sticks & Sermons: Policy Instruments & Their Evaluation*, Transaction Publishers, New Brunswick, NJ.

Bernow, S., Costanza, R., Daly, H., DeGennaro, R., Erlandson, D., Ferris, D., Hawken, P., Horner, J. A., Lancelot, J., Marx, T., Norland, D., Peters, I., Roodman, D., Schneider, C., Shyamsundar, P. and Woodwell, J. (1998) 'Ecological tax reform', *BioScience*, 48, 193–196.

Birkland, T (2011) *An Introduction to the Policy Process: Theories, Concepts, and Models of Public Policy Making*, M.E. Sharpe, Armonk, NY.

Boardman, R. (2009) *Canadian Environmental Policy and Politics: Prospects for Leadership and Innovation*, Oxford University Press, Don Mills Ontario.

Böcher, M. (2012) 'A theoretical framework for explaining the choice of instruments in environmental policy', *Forest Policy and Economics*, 16, 14–22.

Böhringer, C. and Vogt, C. (2003) 'Economic and environmental impacts of the Kyoto Protocol', *Canadian Journal of Economics*, 36 (2), 475–494.

Burgoon, B. (2010) 'Betwixt and between?: The European Union's redistributive management of globalization', *Journal of European Public Policy*, 17 (3), 433–448.

Burtraw, D. And Palmer, K. (2003) *The Paparazzi Take a Look at a Living Legend: The SO2 Cap-and-Trade Program for Power Plants in the United States*, Resources for the Future, Washington, DC.

Clucas, I. (1998) 'A study of the options for utilization of bycatch and discards from marine capture fisheries', *FAO Fisheries Circular 928*, FAO, Rome.

Davis, D., and Gartside, D. F. (2001) 'Challenges for economic policy in sustainable management of marine natural resources', *Ecological Economics*, 36 (2), 223–236.

Deacon, R. T. (2009) *Creating Marine Assets: Property Rights in Ocean Fisheries*, PERC Policy Series, 43, Bozeman, Montana.

FAO (Food and Agriculture Organization) (2009) *Quantitative Socio-Economic Policy Impact Analysis: A Methodological Introduction*, FAO, Rome.

Faure, M. and Wang, H. (2006) 'Economic analysis of compensation for oil pollution damage', *Journal of Maritime Law and Commerce*, 37, 179–217.

Faure, M., and Wang, H. (2008) 'Financial caps for oil pollution damage: A historical mistake?', *Marine Policy*, 32 (4), 592–606.

Gipperth, L. (2009) 'The legal design of the international and European Union ban on tributyltin antifouling paint: Direct and indirect effects', *Journal of Environmental Management*, 90, 86–95.

Greiner, R. (2013) 'Social dimensions of market-based instruments: Introduction', *Land Use Policy*, 31, 1–3.

Greiner, R., Young, M. D., McDonald, A. D. and Brooks, M. (2000) 'Incentive instruments for the sustainable use of marine resources', *Ocean Coastal Management*, 43 (1), 29–50.

Hardin, G. (1968) 'The tragedy of the commons', *Science*, 162 (3859), 1243–1248.

Hill, D. A. (2005) *Handbook of Biodiversity Methods: Survey, Evaluation and Monitoring*, Cambridge University Press, Cambridge.

Jacquet, J., Pauly, D., Ainley, D., Dayton, P., Holt, S. and Jackson, J. (2010) 'Seafood stewardship in crisis', *Nature*, 467, 28–29.

Johnsen, P. and Eliasen, S. Q. (2011) 'Solving complex fisheries management problems: What the EU can learn from the Nordic experiences of reduction of discards', *Marine Policy*, 35 (2), 130–139.

Johnstone, N. (1999) 'The compatibility of tradable permits with other environmental policy instruments', in OECD (ed.) *Implementing Domestic Tradable Permits for Environmental Protection*, OECD, Paris.

Kelleher, K. (2005) 'Discards in the world's marine fisheries: An update', *FAO Fisheries Technical Paper 470*, FAO, Rome.

Lagring, R., Degraer, S., de Montpellier, G., Jacques, T., Van Roy, W. and Schallie, R. (2012) 'Twenty years of Belgian North Sea aerial surveillance: A quantitative analysis of results confirms effectiveness of international oil pollution legislation', *Marine Pollution Bulletin*, 64 (3), 644–652.

Marshall, R. S. (2003) 'Corporate environmental reporting: What's in a metric?', *Business Strategy and the Environment*, 12 (2), 87–106.

Mazur, E. (2010) 'Outcome performance measures of environmental compliance assurance', *Environment Working Papers*, 18 (18), OECD, Paris.

McKenzie-Mohr, D. and Smith, W. (1999) *Fostering Sustainable Behaviour: An Introduction to Community-Based Social Marketing*, New Society Publishers, Gabriola Island, British Columbia.

OECD (Organization for Economic Co-operation and Development) (2002) *Regulatory Policies in OECD Countries – From Interventionism to Regulatory Governance*, Annex 2, OECD, Paris.

OECD (2004) 'Political economy of tradable permits – competitiveness, co-operation and market power'. in F. J. Convery, L. Dunne, L. Redmond and L. B. Ryan (eds) *Greenhouse Gas Emissions Trading and Project Based Mechanisms*, OECD, Paris.

OECD (2008) *Environmentally Related Taxes and Tradeable Permit Systems in Practice*, OECD, Paris.

Patton, C. V. and Sawicki, D. S. (1993) *Basic Methods of Policy Analysis and Planning*, Prentice Hall, Englewood Cliffs, New Jersey.

Pauly, D. (1997) 'Points of view: Putting fisheries management back in places', *Reviews in Fish Biology and Fisheries*, 7 (1), 125–127.

Purcell, S. W., Mercier, A., Conand, C., Hamel, J.-F., Toral-Granda, M. V., Lovatelli, A. and Uthicke, S. (2013) 'Sea cucumber fisheries: Global analysis of stocks, management measures and drivers of overfishing', *Fish and Fisheries*, 14, 34–59.

Salzhauer, A. L. (1991) 'Obstacles and opportunities for a consumer ecolabel', *Environment*, 33, 10–37.

Salzman, J. and Ruhl, J. B. (2000) 'Currencies and the commodification of environmental law', *Stanford Law Review*, 53 (3), 607–694.

Schinas, O. and Stefanakos, Ch. N. (2012) 'Cost assessment of environmental regulation and options for marine operators', *Transportation Research Part C*, 25, 81–99.

Shaw, S., Grant, D. B. and Mangan, J. (2010) 'Developing environmental supply chain performance measures', *Benchmarking: An International Journal*, 17 (3), 320–339.

Stavins, R. N. (2003) 'Experience with market-based environmental policy instruments', in K. Mäler and J. R. Vincent (eds) *Handbook of Environmental Economics*, Elsevier, Amsterdam.

Stein, R. M. and Bickers, K. N (1995) *Perpetuating the Pork Barrel: Policy Subsystems and American Democracy*, Cambridge University Press, Cambridge.

Sterner, T. and Coria, J. (2012) *Policy Instruments for Environmental and Natural Resource Management*, RFF Press, New York.

Sumaila, U. R., Cheung, W., Dyck, A., Gueue, K., Ling, H., Lam, V., Pauly, D., Srinivasan, T., Swartz, W., Watson, R. and Zeller, D. (2012) 'Benefits of rebuilding global marine fisheries outweigh costs', *PLoS ONE*, 7 (7), 1–12.

UNEP (United Nations Environment Programme) (2004) *Economic Instruments in Biodiversity-Related Multilateral Environmental Agreements*, UNEP, Nairobi, Kenya.

Wahba, H. (2008) 'Does the market value corporate environmental responsibility? An empirical examination', *Corporate Social Responsibility and Environmental Management*, 15 (2), 89–99.

White, A. T., Courtney, C. A. and Salamanca, A. (2002) 'Experience with marine protected area planning and management in the Philippines', *Coastal Management*, 30 (1), 1–26.

World Bank (1997) *World Development Report 1997: The State in a Changing World*, Oxford University Press, Oxford.

Chapter 4

Developing and analysing policies

An introduction to policy analysis

Introduction

Chapter 3 introduced the various types of policy instruments and their potential application in marine environments. This chapter explores the development and analysis of policies with a focus on the characteristics of policy development, considerations in selecting policies and methods for selecting and analysing policies.

Policy analysis is a very extensive yet ill-defined topic. Broadly, it encompasses all aspects of how problems are identified, decisions are made, policies are implemented and outcomes are measured. Simply stated, policy analysis is the business of informing decision-making. Those that undertake policy analysis (often policy analysts) try to realize a number of expectations. They are expected to understand the intricacies, complexity and political context of their business areas. While developing viable alternatives to the status quo, they need to understand and be able to communicate the implications of their solutions. Additionally, they must provide cost–benefit analyses to their alternatives and then recommend a feasible course of action to decision-makers. As a core function of all governments, policy analysis also applies to any organization that establishes programmes intended to resolve social, economic or environmental problems and can be undertaken by any member of the organization, whether internal or external to it.

What separates policy analysis from planning and management is that policy analysis is a strategic tool to set strategic direction, intent and objectives to address a particular problem. In contrast, planning and management are used to implement a range of policies. For example, a local government may decide that residential population densities are to be increased in order to preserve green space and to demonstrate the principles of smart growth. Planners and managers will subsequently be tasked to implement the policy and may establish processes to change zoning bylaws, establish bike lanes or determine additional infrastructure necessary to support high densities.

Participants in, or those affected by the policy process, are often termed 'actors' or 'agents'. They are generally represented in one of the following three general types:

- Citizen (public) actors: Represent interests in the environment, health, recreation, gender, aboriginal etc. May include scientists and researchers, media and non-governmental organizations.
- Market actors: Represent the interests tied to the production of public goods and services. May include industry, lobbyists, think tanks, organized labour etc.
- State actors: Represent elected and appointed government officials, executive,

legislative, judicial system administrators. May include government scientists and researchers or public servants whose job it is to represent citizen or market interests.

Approaches to policy analysis

Policy analysis has been characterized in myriad ways and it is outside of the scope of this text to provide a thorough treatment of the academic perspectives on these delineations. However, we should consider two types of policy analysis that are in wide use:

1   *Prospective policy analysis*, also termed *ex ante, a priori* and *pre hoc* policy analysis is used prior to or during the development of policies to inform policy development, selection and recommendation to decision-makers. Prospective policy analysis is prescriptive in that it recommends potential solutions. It is also predictive in that it attempts to forecast the likely outcomes for various policy options and uses existing data in both quantitative and qualitative analyses to inform decision-making.
2   Retrospective policy analysis, also termed *ex post, a posteriori* and *post hoc* policy analysis is applied once a policy as been adopted and implemented. Retrospective policy analysis is evaluative and used to assess whether the policy is meeting its goals and objectives. It often requires data collection and research to describe or interpret whether the policy is meeting specific outputs and outcomes. In certain cases, retrospective analysis will evaluate the present condition of a target (e.g. fish population) of the policy and try to recreate a past historical context to determine whether the policy is working. Most complex policies are difficult to analyse using retrospective analysis since ascribing causal relationships between specific policies and observed outcomes can be exceedingly complicated.

The policy cycle

A number of analysts and governments use what is loosely termed the 'policy cycle' to guide policy development and evaluation. The notion that an ordered approach to decision-making could improve the quality, durability and transparency of government decision was first proposed by Lasswell (1951) and later by May and Wildavsky (1978). The policy cycle is an academic construct designed to be both descriptive and normative. This entails that the policy cycle captures how prudent decisions are made as well as providing a framework to guide government decision-making. It attempts to deliver an archetype that is sensitive to political realities: this realpolitik includes external influences, the need to make decisions quickly and the need to adopt policies palatable to the electorate. In decades past, policies were formulated through a closed, elitist process where only a few individuals had access to decision-makers. In recent decades, the electorate, stakeholders and non-government organizations have had much more involvement in the development of policy. Lately, open government initiatives, participatory democracy and access to information have involved citizens – at least in democratic states – to an extent not contemplated by the early progenitors of the policy cycle.

There is currently no single, agreed-upon model of policy development. However, many interpretations of the policy cycle have common elements that are converging towards a shared understanding of the development and application of policies (Table 4.1, Figure 4.1). All descriptions of the cycle entertain a problem-solving process involving discrete states

Table 4.1 Examples of different interpretations/definitions of the stages of policy development (the 'policy cycle')

	Patton and Sawicki (1986)	Howlett and Ramesh (1995)	Bridgeman and Davis (2000)	Munger (2000)	Bardach (2011)
Stage in policy cycle					
Problem recognition	Verify, define and detail the problem	Agenda setting	Identify issues	Problem formulation	Define the problem
Proposal of a solution	Establish evaluation criteria	Policy formulation	Policy analysis	Selection of criteria	Assemble evidence
Choice of a solution	Identify alternative policies	Decision-making	Policy instruments	Comparison of alternatives	Identify options
Putting the solution into effect	Assess alternative policies	Policy implementation	Consultation	Political and organizational constraints	Select evaluation criteria
Monitoring the results	Display and distinguish among alternatives	Policy evaluation	Coordination	Implementation and evaluation	Project outcomes
	Implement, monitor, and evaluate the policy		Decision		Confront trade-offs
			Implementation		Decide
			Evaluation		Tell your story

with multiple actors and institutions. Also, all models also suggest a feedback mechanism where policies are improved or evolve into new policies. A summary of the various stages of the policy cycle is provided below.

The policy cycle: problem definition

The policy cycle begins with *problem definition* where an issue comes to the attention of government(s). At this point the problem is not well defined, the various policy actors (as defined above) may have divergent views on the problem, and government may have, or be presented with, a number of incomplete proposals to solve the problem. In this stage, the magnitude and extent of the problem is defined, comparisons to other similar problems are made to see if solutions already exist, data and information are used to further define the problem and estimates of resources required to address the problem are assembled. The problem definition stage is complete when the problem is meaningfully stated, affected parties are identified and the level of importance of the problem is understood. In defining a problem, key considerations include whether the matter:

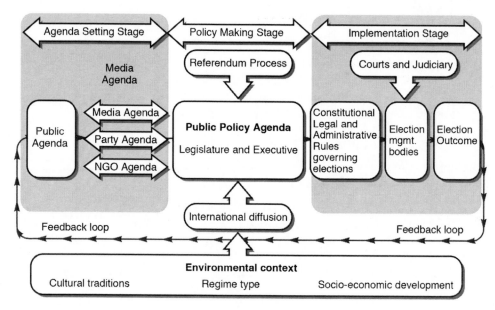

Figure 4.1 The policy cycle and associated actors and institutions
Source: reproduced with permission from Norris (2011)

- is wholly or partly within the scope/mandate of the organization?
- can be solved when historical contexts, cultural milieus and current values are considered?
- originates as a result of past policies (or lack thereof) or is of an intractable nature (e.g. climate change) in the short and medium term?
- originates as a result of a deficiency of data or information?
- requires incremental or transformational change to affect desired outcomes?

The policy cycle: proposal of a solution

The second stage of the policy cycle (recognizing that there are differences in opinion – see Table 4.1) is *proposal of a solution/policy formulation* where policies are articulated. This is usually undertaken by governments, or in the case of marine environments, by intergovernmental organizations (e.g. the IMO). In this stage, multiple solutions are developed with associated risks, opportunities, benefits and costs. Draft solutions may be developed 'in house' by government(s) or developed through stakeholder consultations or public processes. Solutions normally consider a wide range of options/alternative policies, including the status quo and 'stretch' options. For each option, an assessment of expected effects and impacts as well as the best and worse case scenarios should be undertaken in addition to evaluating the likelihood that the option will solve the problem. Many complex policy solutions must consider historical contexts and cultural practices (including indigenous use), contrasting value systems and past policy decisions. Furthermore, they must predict changes to the social (e.g. demographics, technology) and natural (e.g. climate change) environments.

Key considerations in developing proposed solutions include:

- Administrative ease (e.g. changing laws).
- Costs and benefits (e.g. determining who pays and who benefits).
- Effectiveness (e.g. likelihood of success).
- Equity (e.g. whether one or more groups are disadvantaged relative to other groups)
- Legality (e.g. whether a solution is likely to be challenged in court).
- Political acceptability (e.g. whether citizens will favourably react to the policy).
- Measurability (e.g. can output and outcomes be measured).

The policy cycle: decision-making

The third stage of the policy cycle is *decision-making* where government(s) adopt a particular solution. In the global context, decision-making often occurs during international conventions and therefore the decision becomes synonymous with and referred to as 'convention' or 'agreement' (Chapter 2). Of significant import, decision-makers may not accept a recommended solution and may merge options, change options or ask staff to undertake further analyses and return with revised solutions. This is a normal situation in the formulation of both domestic and international policy. Most solutions are rejected by government decision-makers because they – in reality or perception – are:

- too expensive;
- too difficult to implement;
- punitive on certain sectors (e.g. fishing) or states (e.g. less developed states);
- too much of a divergence from established business practices (i.e. setting a new precedent);
- unlikely to be understood by the public;
- unlikely to solve the problem;
- likely to result in unintended consequences;
- politically risky.

The policy cycle: policy implementation

The fourth stage of the policy cycle can best be described as *policy implementation*, often the most challenging aspect of policy development. Implementing complex policies may require wholesale changes in business processes and practices, human behaviour and infrastructure and technology. Fees and taxes, resource reallocation between users, rules or regulations and the structure/organization of government(s) can also be significantly altered. Many, if not most decision-makers underestimate the difficulties in implementing policy and expect new policies to roll out swiftly and without problems. Yet, marine policies may take many years to fully implement, especially if the policy mandates change in infrastructure (e.g. double hulled tankers) or technology (e.g. turtle excluders for fishing gear).

Key considerations in policy implementation include:

- Identifying appropriate pilot projects.
- Development and presentation to decision-makers of an implementation plan and communications plan.

- Development of a monitoring plan that evaluates both implementation and effectiveness of the policy.
- Setting future dates when the policy will be reviewed.

The policy cycle: policy evaluation

Lastly, *policy evaluation* is the final stage in the policy cycle. In this stage the monitoring plan is implemented and the policy is reviewed at periodic intervals to determine its progress. It should be noted, however, that policy evaluation can be undertaken at any point in the policy cycle and there is much in common between the problem definition and evaluation stages. Theodoulou and Kofinis (2004) define policy evaluation as:

- the assessment of whether a set of activities implemented under a specific policy has achieved a given set of objectives;
- the effort that renders a judgement about programme quality;
- information gathering for the purposes of making decisions about the future of the programme;
- the use of scientific methods to gauge the success of its implementation and its outcomes.

Several types of policy evaluations have been proposed and there is no consensus concerning methods and standards. Evaluation at the various policy cycle stages, however, can be one of two types: *formative evaluations* are undertaken when policies are either in development or aspects of the policy are already in place. Formative evaluations allow for evaluation and continuous improvement in the first four stages of the policy cycle. In contrast, *summative evaluations* are used for mature policies to assess their effectiveness and to determine if changes in the policy are required (Smith and Larimer, 2009).

With respect to the different aspects of policy that could be evaluated, Theodoulou and Kofinis (2004) propose the following four types of policy evaluation. *Process evaluations* determine how well a policy (or programme the policy established) is being administered and focus less on whether the policy is achieving its desired outcome. Process evaluations are generally undertaken by staff directly involved in the programme in order to find administrative efficiencies in the delivery of the programme. *Outcome evaluations* strive to measure outputs (e.g. number of ships compliant with mandated technologies) to demonstrate the degree of compliance with the policy. In other words, outcome evaluation is used to measure tangible and available results of the policy, often before the policy matures, to provide an indication of whether the policy is working. Outcome evaluation is therefore not used to determine whether the policy is effective in meeting its stated goals; rather it measures tangible progress towards these goals. *Impact evaluation* evaluates whether the policy is achieving its stated objectives as the decision-makers intended. Impact evaluation is what most consider when thinking about policy appraisal and evaluates whether the policy is working and/or has produced unintended effects. *Cost–benefit analysis* assesses the financial costs of the policy (or programme the policy creates) against the societal benefits. In particular, applying cost–benefit analysis to mature programmes enables them to be modified. Also, they can be assessed against the economic analysis in the programme formulation stage to determine how closely predictive cost–benefit models align with reality.

Regardless of the type of policy evaluation used there are four drivers for policy evaluation (adapted from Mark *et al.*, 2000):

- Assessment of merit and worth.
- Improvement of the programme and, by extension, the organization.
- Oversight and compliance.
- Development of knowledge.

Policy evaluation can be further separated into process and outcome evaluations. Specific questions asked in process evaluations include whether (adapted from Smith and Larimer, 2009):

- the policy operates according to the relevant rules/laws/obligations;
- the target population is being served by the policy;
- the process matches the goals of the policy.

Three specific questions are often asked in outcome evaluations (adapted from Smith and Larimer, 2009):

- Is the policy having an impact?
- How much of an impact is the policy having?
- If the policy is failing to have an impact, why has it foundered?

Challenges with the policy cycle

While the policy cycle has been in use for several decades and provides a structured approach to addressing complex problems, there remain a number of concerns. The primary concern with the policy cycle is that the model assumes rational behaviour throughout each stage and that governments will follow the model. In reality – and often for very valid reasons – politics, elections, emergencies and events lead to the cycle being circumvented or even marooned. For example, often policies are developed hastily in order to seize political windows of opportunity. Deviations from the structured approach are made in the interests of expediency. In addition, policies are frequently developed by political staff, programme managers or third parties (e.g. non-government organizations, industry) who present decision-makers with policy solutions developed outside of the cycle. Lastly, the sequenced, step-wise nature of the cycle rarely coincides with reality. Decision-makers routinely consider multiple, interconnected policies simultaneously such that policy changes are made ad hoc, often with little analysis or empirical evidence.

Howard (2005) interviewed senior Australian public servants about their experiences with the policy cycle. He hoped to discover if the cycle adequately represents contemporary policy making, if policy frameworks are beneficial to public servants and if the model (as proposed by Bridgeman and Davis, 2000) correctly articulates the policy-making process. Among his discoveries, he found that the ordering of the policy cycle is sometimes altered. For example, solutions are occasionally presented before problems are identified or staff must work backward to justify the policy. He also ascertained that elected officials tend to focus on the announcement and implementation of policies and may consider the other aspects of the policy cycle as unnecessary 'process'. He concludes that the policy cycle is never as linear as the model suggests and, indeed, is often iterative in practice.

Considerations in selecting public policies

Selecting public policies may be as straightforward as establishing catch limits for a partic- ular fish stock where population trends, life history and efforts expended on catching the stock are known. However, most public policies are complex with direct effects that can only be approximately anticipated and indirect effects that may be completely unpredicted. As such, policy development requires careful consideration of the development of alterna- tive policies. This section, therefore, outlines a number of considerations when developing public policies. In the real world of politics, interest groups and seizing windows of oppor- tunity, policies are rarely 'bench tested' to the degree necessary to make informed choices. The considerations provided below, however, should guide policy developers towards some of the questions they should ask before bringing alternatives up to decision-makers.

Selecting policies primarily focuses on a number of aspects to inform different alterna- tives (solutions) that may include:

- the environmental, geographic and socioeconomic aspects requiring intervention;
- available existing policy measures that could be applied to generate the desired changes;
- the expected effects of policy change;
- who would gain and how much would be gained if the policy measure is implemented;
- who would lose and how much would be lost if the policy measure is implemented;
- the expected duration of the effects of the policy measure;
- the budgetary implications of the policy measure;
- how the policy measure would be financed;
- whether the policy measure is legal or consistent with existing law;
- whether the policy measure requires legislative change or amendment of an interna- tional convention;
- whether those affected by the policy will understand its intent and consequences;
- whether the policy will be measured to see if it is working.

While policy decisions are not simple rational discussions, high-level decision criteria can generally be categorized as relating to effectiveness, efficiency or equity of policy tools. Despite strong criteria, however, even the most rational and effective option can still be abandoned in light of political or administrative resistance. As a result, political support and administrative feasibility are significant criteria for successful policies. Depending on the issue, additional criteria may be relevant. Among these criteria are the scientific or techni- cal effects of an action, the social or cultural acceptability, the legal implications and the uncertainty involved (Loomis and Helfland, 2001).

One policy tool may not fit all situations. The achievement of deeper structural or economic changes likely requires a multidimensional approach that integrates a number of different policy tools. In considering instrument selection and design issues it is critical to establish and build on a solid foundation of information. The review process should precede policy action, and is critical to orienting and monitoring an approach based on incentives and the use of transparent metrics. Some considerations in the choice and design of policy instruments for environmental management and related examples can be found in Table 4.2.

While there are many considerations in selecting the appropriate policy instrument to manage a particular issue, assessment criteria generally fall under a few broad categories. These include effectiveness, efficiency, fairness of costs and benefits and political accept-ability. While different instruments have their strengths and weaknesses, it should be emphasized that the particular circumstances and pre-existing conditions of a management issue will determine the fit and success of a policy tool. This is particularly important in marine environments given the number of states that potentially can be involved in a policy issue. Focusing on the policy objectives, rather than favouring a particular tool, may be the best route to achieving desired outcomes.

Table 4.2 Application of selection criteria for different categories of policy instruments

Criteria	Environmental regulations	Using markets	Creating markets	Informational instruments
Effectiveness (immediacy and certainty)	High	Moderate	Moderate–low	Low
Efficiency (financial costs)	Generally high administrative and enforcement costs	Low implementation, but business compliance and revenue collection costs	Set-up and allocation costs complex and involved	High costs for research and advertising, low cost for transparency mechanisms
Fairness (of costs and benefits)	Resistance depends on regulatory target, positive public image for addressing industrial violations and 'polluter pays' principle	Low resistance as long as costs are proportionally distributed, and ability to pay is equivalent	Equity considerations for access limitations to public goods, (e.g. parks), local impacts may still be high, under regional trading scheme	Low resistance, positive associations with transparency increasing accountability, can provide short-term advantage to those who do not participate
Political considerations	Government resistance due to needs for enforcement and administration; industry push back on costs and limitations to technological innovation	Often favoured, but depend on precedent for charges, risk public reprisal for 'just another tax'; economic disincentives often met with resistance where industry based on profiting from externalities	Theoretically appealing, but complicated to implement and requires considerable intergovernmental cooperation to apply trading schemes across borders	Often viewed as a soft policy option, but low cost and positive messaging and accountability held in high regard

Source: Adapted from Sterner and Coria (2012)

Effectiveness

Considering the potential efficacy of public policies at the development stage will inform alternatives and is a key component of the solutions when presented to decision-makers. Effectiveness is not simply a question of whether the policy will achieve the desired outcome, but considers when policies will become effective, the degree of risk to those affected by the policy, certainty that the policy will achieve its intended objectives, buy-in from those affected by the policy and what constitutes 'success'. For example, contrasting effectiveness between regulatory instruments and market-based approaches will require consideration of whether the immediate results, but higher costs to industry using regulatory instruments, are more desirable than delayed results and economic flexibility using MBIs (see Chapter 3).

Analyses of effectiveness borrow measurement and evaluation techniques from other disciplines and industry sectors. Also, they may utilize evidence-based techniques, and consider direct and indirect responses on other policies and programmes. As discussed previously, effectiveness is often separated into institutional and impact effectiveness. *Institutional effectiveness* (output effectiveness) is defined as having effective policies/programmes in place and *impact effectiveness* (outcome effectiveness) is whether the policies/programmes are meeting their objectives. For example, Kenya has sufficient policies to address the management of marine oil pollution (institutional effectiveness) but whether these policies avert oil pollution (impact effectiveness) is unknown (Ohowa, 2009).

Vital to evaluating policy effectiveness is the identification of performance criteria (indicators) to support both the evaluation of alternative solutions prior to committing to a policy and to evaluate the efficacy of the policy once implemented. Performance criteria should be selected that can be routinely monitored and that correlate with one or more goals or objectives of the policy (Mazur, 2010). Table 4.3 outlines the primary types of performance criteria for policies and Table 4.4 illustrates key considerations for the selection of performance criteria.

The European Environment Agency (2005) has developed the following perspectives on policy evaluation:

- Governance can make or break the success of a policy: policies will be ineffective unless suitable governance exists.
- By tackling problems at the source, economic instruments can be fruitful ingredients in the policy mix: effective policies utilize economic incentives to discourage the production of externalities (e.g. pollution) that are costly to address.
- In assessing a policy's goal achievement, it is important to distinguish between different types of goals: policy evaluation must recognize that multiple goals and objectives exist and that some goals will be met while others will be unattained.
- Data limitations are demanding, but not insurmountable: policy evaluation is frequently limited by a lack of data. Nonetheless, sufficient data can often be obtained to understand the policy.
- Effectiveness evaluations are complex and require multidisciplinary efforts.
- Effectiveness evaluations contribute to capacity-building and shared policy learning.

Table 4.3 Main types of performance criteria for policies and marine examples

Type of criteria	Example
Benchmarks	Comparison with a documented best-case performance related to the same variable within another entity or jurisdiction
	Example: number of fisheries in a jurisdiction that are certified by the Marine Stewardship Council
Thresholds	The value of a key variable that will elicit a fundamental and irreversible change in the behaviour of a system. The policy is evaluated based on its role in making the system move toward or away from the threshold in any given period
	Example: maximum sustainable yield of a fishery
Principles	A broadly defined and often formally accepted rule. Principles should have relevant performance measures
	Example: all ports should have spill response and countermeasures programmes in place
Standards	Minimum requirements for properties, procedures or environmental quality
	Example: bilge water quality standards
Policy-specific targets	Determined in a political and/or technical process taking past performance and desirable outcomes into account
	Example: setting targets for offshore hydrocarbon production

Source: modified from Pintér et al. (2000)

Table 4.4 Key considerations for the selection of performance criteria

Transparency	Can be understood by laypersons
Policy relevance	Relates to the policy objective(s)
Analytical relevance	Accurately measures the problem
Responsiveness	Responds to changes in management
Time horizon	When results can be expected
Accountability	Identifies responsibility
Robustness	Cannot be manipulated/massaged
Measurability	Can be measured over the life of the policy

Source: modified from Jesinghaus (1999)

Efficiency

Financial costs, as well as other non-market goods and services that influence the overall level of social well-being, are another consideration in selecting a policy. Strategies that achieve their goals at a minimum cost maximize limited resources for addressing other problems. Cost dimensions can be considered as they apply to society, government agencies and targeted individuals and organizations. The financial spectrum of instruments ranges from low-cost voluntary programmes, to regulatory options with high administrative costs. The limitations of defining environmental and associated health costs in economic terms have made efficiency arguments – that is, weighing regulatory expenses against overall costs to the economy – controversial (see Box 4.1). Often, these negative costs have largely been externalized or ignored, and critiques of prescriptive regulatory policies have focused on the real financial costs of development, administration, enforcement and potential limitations to business innovation. Externalities may not only give rise to an inefficient use of the resource, but they can also result in costs being imposed on others that are widely perceived as unjust (Stavins, 2003).

Economic or MBIs are often positively associated with lower financial costs. However, there are still efficiency considerations for selecting the appropriate MBI (Chapter 3), be it quantity, price or market enhancement policy. Allocation of permits or licences has been applied to diverse types of resources, ranging from fisheries to the control of water pollution. Implementing a quantity-based trading instrument requires substantial upfront development costs to negotiate the contracts of exchange and to absorb legal, brokerage and insurance fees. Trading systems can also involve significant time and effort on the part of governments – with due consideration of the potential efficiency and equity implications involved. In particular, it has been a complex and protracted exercise for policy-makers looking to establish a domestic permit trading system for greenhouse gas emissions. For a price or market enhancement approach, trading costs are avoided, but compliance costs still apply, as do costs associated with revenue collection or subsidy allocation by government. Efficiency implications are also associated with evasion under a price-based approach. This is opposed to a trading system, where the tax or charge is commoditized and converted into a tradable 'asset' that can be exchanged (Johnstone, 2003).

Box 4.1 The Weitzman theory

In 1974 Martin Weitzman explored whether it would be better to diminish certain forms of pollution by setting emission standards (direct regulation of quantities) or by charging the appropriate pollution taxes (indirect control by prices). He demonstrated that where the value of marginal benefits and costs associated with abatement are uncertain, but where economic costs are likely to increase faster than environmental benefits, a smaller overall welfare loss is likely to result from a pricing mechanism than from a strict quantity constraint. Termed the 'efficiency rule' Weitzman recognized that if there is a risk that policy-makers are unable to determine the optimal pollution level, it is best to regulate a quantity target (even with an increase in abatement costs). However, if environmental impacts are escalating but the increased compliance costs are of greater concern, then a price approach is a more efficient choice. As a result, the most efficient policy for managing hazardous pollutant discharges or safeguarding threatened species or habitats, would likely be a form of quota or access restriction. Meanwhile, tax assessments would be an appropriate instrument for dealing with a relatively benign pollutant or slowly accumulating environmental threat.

Four decades later, most economists continue to support Weitzman's conclusions.

Source: Weitzman (1974)

Fairness of costs and benefits

Fairness can be a significant issue in selecting a policy tool. Public policies can have disparate effects on different people, varying by location, ethnicity, income or occupation. The distributional consequences of a policy typically indicate a great deal about who supports and who opposes a policy. Although 'polluter pays' is a generally accepted principle, coercive strategies are often difficult to implement. When a small group or sector bears the cost of a policy, it has significant motivation to resist implementation, while the majority that benefits are often widely distributed and without a coordinated voice. There will be trade-offs between costs and benefits with any policy; government bureaucracy, however, may not be the most responsive to changing conditions affecting the cost–benefit relationship. How policies proportionally affect vulnerable or affluent constituencies can also influence the public's perception of a policy's social equality. Voluntary instruments typically have less resistance from targeted sectors and are relatively easy to implement, but may not be viewed seriously and consequently suffer from 'free riders'. Larger environmental concerns, such as widespread pollution, are typically best mitigated by economic or regulatory approaches. MBIs (see Chapter 3) have been controversial in the allocation of the wealth associated with these resources. Although MBIs typically do not privatize resources as conventional wisdom suggests, to some degree they do privatize access to (and use of) those resources. And since access rights can be very valuable, these rights may represent a substantial amount of wealth. While the ability to reclaim and protect common resources can motivate sustainable behaviour, the ethical issues raised by their distribution among competing claimants remain a sizable challenge (McCay, 2000).

Political considerations

The political environment, as expressed through the desires of elected and non-elected politicians, drives most public policy analysis and activities. As a result, public policy failures are a frequent occurrence. Government mandates can have a major impact on instrument choice, as can the perceptions of public acceptability. As policy trends often follow political agendas, the economics of voluntary or non-regulatory approaches and streamlining or deregulating existing policies have frequently been favoured. Despite strong evidence that they could be effective, many economic disincentives are seen as difficult to implement due to constituent push back (Greiner et al., 2000). Additional regulation is often met with resistance due to increased political and administrative burdens for implementation and enforcement.

Specific political considerations in the development of policies include:

- Ensuring that key influencers do not affect outcomes that favour certain groups over others.
- Ensuring that the instrument does not create equity issues and benefits 'free riders'.
- Ensuring that policies are understood and accepted by the public. Certain instruments, including taxes, subsidies and trading schemes, can be controversial. So too are more complex combinations of policy instruments. In contrast, the public generally understands simpler instruments such as 'polluter pays'.
- Ensuring bureaucracy capacity to implement a policy.
- Maintaining relationships with sub-national levels of government. For example, regional governments often prefer MBIs over state control in instituting pollution standards.

- Maintaining relationships with other states. Cross-border policies often face additional political and diplomatic considerations. In addition, international agreements often restrict the use of certain policy instruments.
- Maintaining relationships with the media in order to accurately and effectively convey the intent of the policy.

Other evaluation criteria

Effectiveness, efficiency, equity and political considerations presented earlier in this chapter offer some general criteria for evaluating policies. Yet there are various other assessment processes that advance detailed lists of evaluation criteria for policy selection. Some additional criteria and associated examples for selecting environmental policies can be found in Table 4.5.

Methods and tools for selecting and analysing policies

Once the criteria have been analysed for each alternative, a decision still has to be made. Policy decisions are often described as being logical, ordered and merit based. While there are a number of analytical tools that are helpful for synthesizing relevant information and determining the most appropriate and strategic policy approach, there are many paths in the selection process.

In the 'rational model', ideal policy development is based on a hierarchical normative process (Fischer et al., 2006). It often begins by clearly identifying the issue that needs to be addressed and stating the objectives of the policy. Consideration of the historical background and the severity of the problem are diagnosed at this juncture. Depending on priorities, alternative options and public expectations, government may choose not to address the problem. If government action is required, existing policies should be assessed to avoid overlap. This involves directing policies at the appropriate level of government (national or regional) and engaging staff regarding the operational responsibilities for administration and enforcement. Consultation with the public and stakeholders is advisable since it provides a sounding board for them to propose alternatives, identify impacts and ensure their concerns are addressed. Policy selection may also have legal implications and needs, such as identifying the constitutional and statutory basis for the policy as well as clearly defining the legal entities to which it applies. Alternative examples and case studies of responses from other jurisdictions with similar issues can provide additional insight on the best course of action. Cost–benefit analysis or another policy selection tool may be used to support the final decision. Finally, establishing a clear scope and communication strategy to ensure the objectives are clear, consistent, understandable and accessible will ensure the success of a policy during implementation and beyond.

Although an efficient process of a rational decision is an appealing means of selecting the best policy option, policy changes rarely emerge from an orderly cost–benefit analysis. Critics argue that merit-based criteria and bureaucratic process do not adequately explain reality: government decision-making does not always follow a strictly logical progression, as some issues become 'hot' overnight, resulting in large changes rather than measured improvements. As such, a number of alternative policy decision models have been developed. These are discussed in Table 4.6.

Policy selection processes often involve decisions made with a high degree of uncertainty

Table 4.5 Criteria and considerations in the selection and design of policy instruments

Ownership/rights	Policy instruments need to be sensitive to potential compensation if rights are affected.
Diffuse/point source problem	Certain policy instruments (e.g. tradeable permits or load-based licensing systems) require knowledge of who is polluting in order to be effective. Otherwise, other instruments (e.g. ambient standards) may be more appropriate.
Single-issue multiple benefits	Instances where addressing single issues (e.g. water quality) result in multiple benefits are well served by combinations of instruments or flexible instruments.
Available information	Market-based instruments generally require reliable data on sustainable yields and an understanding of issues such as compliance costs. Regulation and financial incentives may be preferable in information-poor environments.
Proportional cost of tool	Taxes, fines and permit prices must be sufficiently large such that they change behaviour and incent change.
Intended environmental outcome	The instrument should drive behaviours towards clear, science-based, quality or quantity standards.
Efficiency gains	Instruments should attempt to realize efficiency gains, as entitlements tend to move to producers with the highest marginal returns.
Ongoing incentives	Economic instruments, such as permit systems, that provide incentives for self-management by industry, are typically preferred to those that are dependent on funding for administration and enforcement.
Timing	Instruments that can be quickly enacted are preferable to those that may take some time to implement if an environmental threat is imminent.
Flexibility	Instruments may need to be adaptable to respond to new knowledge produced by scientific research and monitoring.
Equity aspects	Instruments should be reviewed for any differential impacts on those affected by the policy. Considerations should include charges for 'public rights'; differential treatment of similar entities; flat charges or levies that act regressively, impacting those least able to pay; and prohibitively large charges or subsidies for industry restructuring.
Transaction costs	For MBIs, consideration of the effects of transaction costs on efficiency may be required.
Community acceptance	The instrument must be perceived as legitimate if it is to be effective. Some instruments, including taxes and levies may not garner public support if the public believes their purpose is to benefit certain groups or are tax grab.
Transparency	Transparency improves social acceptance of the instruments and also improves policy outcomes by minimizing the risk of fraud, corruption and mismanagement of public funds, and supports a level playing field for business.
Administrative feasibility and costs	The instrument selected ideally should be the most cost effective to administer. Considerations in determining cost effectiveness should include enforcement and monitoring costs as well as the capacity of the government to administer the programme.
Dependability or certainty	There should be certainty that the instrument will drive the desired change even when knowledge about likely responses is uncertain.
Precaution	Instruments should reflect the precautionary approach where a lack of scientific information cannot be used to justify inaction or weaken the policy.

Source: adapted from Robinson and Ryan (2002)

Table 4.6 Common models/methods/approaches/theories used to describe how public policy decisions are made

Model	Description
Rationalist	Decisions are made with the best information for the best interests of society
Bounded rationalist	Complete rationality cannot be assumed in policy-making
Incrementalism	Policy is slowly improved over time
Group theory	Groups compete to influence decisions
Elite theory	Elites influence government in their own self-interest
Public choice theory	Governments operate in their own self-interest
Normative	Decisions are made based on value judgements rather than empirical information
Valuative	Policies are examined to determine their value

Source: adapted from Fischer et al. (2006)

and expediency, and based on ill-informed political instincts. An alternative to the rational selection model likens the policy stream to a 'primeval soup' where the primary players – experts, researchers, policy analysts and politicians – create a messy pool where ideas are floated, exchanged and combined. The method for decision-making involves convoluted discussions and debate that result in particular ideas rising or sinking in popularity. While many ideas are proposed, the ones that survive meet some criteria or mandate of the contemporary policy environment. The most basic criteria for successful policies in this decision-making model are those that are technically feasible, politically acceptable and forward looking.

Ultimately, policy decisions are, however, in the hands of the decision-maker. There are a number of technical methods, however, for synthesizing policy options that aid in selecting the optimal policy. It may be necessary to conduct more than one form of analysis to assess whether the preferred policy option depends on the decision method chosen. These methods are outlined in the following section.

Dominance

Under the dominance rule, policy options, when compared to alternatives that are not superior in any factor, are eliminated. If a policy option does not meet or exceed set criteria, then logically it should be eliminated from further consideration. Nonetheless, there should be consensus on the ranking of decision criteria, for rarely is one option completely dominated by another. Typically, an objective method cannot be used to find the optimal policy choice. In fact, if one alternative clearly dominates the others on every decision criterion, it may be suspected that the criteria were inadequate or purposefully designed to make the dominant alternative the obvious choice. Dominance can be useful for eliminating substandard alternatives and reducing the number of options requiring further consideration, or it may be instructive in highlighting deficiencies in the decision criteria (Loomis and Helfland, 2001).

Criteria ranking

Following on dominance comparisons, criteria ranking prioritizes one or more criteria. Recognizing that some criteria are more important than others does not mean that other criteria are irrelevant; however, they should receive less weight in determining the appropriate policy. Additional criteria may be used in a secondary decision process or as a tie breaker between comparable alternatives. Criteria ranking requires very clear priorities and is generally a less flexible approach (Loomis and Helfland, 2001).

Cost–benefit analysis

Using a cost–benefit criterion, economic efficiency becomes the guiding benchmark. The best policy should maximize the net benefits to society. Typically, this is without regard to who gains or loses. As the gains outweigh the losses, then it is possible for the losses to be compensated. Weitzman (1974) emphasizes the importance of including natural capital (the environment) in accounting for economic growth (see Box 4.1). However, using environmental resources can cause negative environmental externalities – essentially, intangible costs such as the depletion of natural stocks or degradation of the environment. In order to fully analyse the cost benefits of a management policy that involves natural capital, intangible variables must be assigned values or 'shadow prices'. As an example, in the cost–benefit analysis of permitting an offshore wind farm operation, the lost intangible value associated with the scenic views must be priced and factored in as a cost. Cost–benefit analyses have been criticized as disguising subjective value assessments as economic equations; yet they have proven useful in establishing practical estimates for environmental damages and abatement costs. In addition, Schulz and Schulz (1991) propose that cost–benefit estimates can assist in:

- making the economic dimension of environmental degradation clearer;
- making the environmental debate more objective;
- directing scarce financial resources to the most urgent environmental issues;
- making polluters aware of the costs of their actions;
- further developing statistical measures of welfare.

In recent decades, many studies have estimated the value of a wide variety of ecosystem services. From an economic perspective, the value of ecosystem services is infinite for humanity would not exist without them. As one might suspect, the cost–benefit approach has a few structural limitations stemming from the general problem of putting a value on a public good (Costanza et al., 1987; Navrud and Pruckner, 1997). Measurement of the tangible benefits associated with any policy involves assessing the rate of change in ecosystem services from existing levels and the benefits from the service in terms of damage avoided. Although estimating the value of ecosystem services can help illuminate the extent of potential environmental damage, estimating the terms is inherently uncertain. Calculating the monetary values for damages due to degradation of coastal water quality, for example, requires multiple uncertainties in estimating the damages to shoreline property values, fishery losses, health and recreation, etc., over both the short and long terms (Freeman 1985; Pearce, 1993; Pearce and Moran, 1994; OECD 1995). Cost–benefit analysis also has challenges when considering the problems of (adapted from Hanley, 1992):

- differences between citizen and consumer values;
- detailing the complexity of ecosystems;
- irreversibility and uniqueness;
- intergenerational equity and discounting.

While acknowledging the many conceptual and empirical problems inherent in producing such estimates, the economic valuation process has proven to be an instructional exercise with multiple benefits including:

- making the range of potential values of ecosystem services more apparent;
- establishing a first approximation of the relative magnitude of global ecosystem services;
- setting up a framework for their further analysis;
- flagging those areas most in need of additional research;
- stimulating additional research and debate.

Costanza et al. (1987) presented one of the first international syntheses of ecosystem valuation. They estimated values for ecosystem services per unit area by biome, multiplied this value by the total area of each biome and then summed over all services and biomes. Most of the problems and uncertainties encountered in their work indicated that estimates typically represent a minimum value that would probably increase with additional effort in studying and valuing a broader range of ecosystem services, with the incorporation of more realistic representations of ecosystem dynamics and interdependence and as ecosystem services become more stressed and 'scarce' in the future.

Willingness to pay for environmental quality – and thus the preferred trade-off between environmental and other goods – depends upon social preference and the need to assay the preferences of future generations. In practice, these issues can only be made explicit through the political process. Assessment of benefits can nevertheless make these processes more informed and transparent.

National studies are another form of cost–benefit analysis. These weigh the consequences of additional taxes and regulation with gains in market efficiency and increases in net revenue in order to create a more comprehensive policy picture. National studies often employ independent third-party analysis to evaluate management issues. In marine environs, such an analysis might study the potential impacts and policy options for limiting coastal water pollution to target levels. Comprehensive environmental policy plans provide target values for emission, waste and resource extraction up to a future date and estimate the costs and macroeconomic effects of several different policy options. In depth analysis would consider the significance of legislation, regulation, charges and taxes and increased investment. Scenarios typically compare the measures required and impacts of achieving stringent targets, stabilizing emissions or the remaining status quo – taking no action (Loomis and Helfland, 2001).

Multi-criteria decision analysis

It has long been recognized that policy decisions are informed by socioeconomic, environmental, cultural, political and ethical considerations. Accordingly, there has been considerable effort to develop tools to integrate disparate data sets so that options can be developed and ranked, either qualitatively or quantitatively. Multi-criteria decision analysis (MCDA) methods are a systematic approach to combine different data inputs with stakeholder input and benefit/cost analyses to rank policy alternatives (Huang et al., 2011). MCDA quantifies public feedback and non-monetary factors to enable the comparison of policy options. For example, someone shopping for a new car will have certain preferences (e.g. price, reliability, style, fuel economy) that are more important. Some of these approaches can be quantified (e.g. fuel economy) while others (e.g. style) are subjective. Through the quantification of subjectives, MCDA approaches attempt to find the best alternative.

These techniques have been widely applied to many fields – reducing contaminants entering aquatic ecosystems, optimizing water and coastal resources, sustainable energy and transport policy, to name but a few. The three primary MCDA approaches include multi-attribute utility theory (MAUT), outranking and the analytic hierarchy process (AHP). These approaches build on criteria ranking (see above) by assigning weights to each variable and summing the weighted criteria for each option. Each approach produces an overall score for each option based on values for alternatives multiplied by weights; however, there are some differences between the approaches, which are discussed below.

Multi-attribute utility approach

Like other MCDA approaches, MAUT assigns weights to each variable and sums the weighted criteria for each option. However, MAUT includes a utility function that scales the importance or impact of each attribute from 0 to 1, with 0 representing the worst preference and 1 the best (Loomis and Helfland, 2001). While the absolute numbers for weights are not important, the relative values are, highlighting the subjective component of MAUT. Identifying and assigning weights can be controversial and results are sensitive to differences in opinion about the relative importance of each criterion. Sensitivity analysis, where various uncertain parameters are changed to see if the results change, can assist in this process. If the order of the policy options changed with only minor adjustments to the weighting then certainty in assigning the weights is more important. In addition, MAUT methods are challenged by both uncertain or fuzzy information and by information expressed in quantities other than a ration or interval scale.

Outranking approaches

Outranking approaches are based on 'voting theory' where the alternative a is deemed better than alternative b if the number of votes (or criteria) indicating that alternative a is better than alternative b is larger than the number of votes indicating the opposite. A single, correct alternative is not developed using outranking approaches: rather the output is the degree of dominance of one alternative over another. Outranking approaches accept incomplete value information and are suitable in situations where there are a finite number of discrete alternatives from which to choose and where the input information is descriptive in nature. Popular outranking approaches include PROMETHEE (preference ranking organizational method for enrichment evaluation) and ELECRE (elimination and choice expressing reality).

Analytic hierarchy process

AHP uses pairwise comparisons to solicit whether criterion *a* is preferential over criterion *b*. These criteria are then weighted according to importance and summed. Similar to outranking approaches, AHP does not identify a 'correct' alternative but one that reflects the preferences of the participants in the process. The approach is suited to situations where the inputs are composed of expert knowledge and subjective preferences. In addition, qualitative criteria can be applied in the evaluation.

Screening

Depending on the situation, there can be a set of criteria that present an obvious requirement for any policy. Legal obligations or threshold values present some examples of where a screening approach may assist in determining the preferred policy approach. Screening ensures that selected policies acknowledge and meet the key requirements, although in some uncertain situations achieving targets or meeting thresholds may require additional analysis. If no, or only a few, options exist that meet the screening criteria, more alternatives may be required. Also, screens do not provide methods to decide among multiple options that pass the screen.

Cost-effectiveness analysis

Cost-effectiveness analysis (also known as least cost analysis) is used to identify the most expedient financial course of action for achieving an objective. Once options for achieving the target are identified, the avenue with the lowest present value of costs is selected as the most cost-effective choice. Rather than attempting to identify and value the benefits, it is implicitly assumed that they outweigh the cost. Cost-effectiveness analysis is appropriate in situations where valid and reliable estimation of the benefits of alternative options is not feasible. For example, cost-effectiveness analysis is suited to conditions where clear and defensible environmental goals exist that can be measured in terms of biophysical units such as minimum water quality standards. It can also be used to identify the most effective option for a fixed amount of funding that has been allocated to achieve a policy objective. One drawback of cost-effectiveness analysis is that it does not identify the benefits of specific actions or gauge the willingness of society to pay for improvements in environmental quality; both these are important considerations in many decision contexts. Nevertheless, cost-effectiveness analysis is, where practical, the preferred tool for decision support.

Counterfactual analysis

Counterfactual analysis is a type of *impact analysis*. Simply put, it attempts to determine what would have happened in the absence of a policy. In doing so, it attributes cause and effect between interventions and outcomes. This is often necessary in policy analysis because a simple comparison of outcomes before and after a policy's implementation does not demonstrate the effect of the policy. Counterfactual analysis can exploit experimental design (e.g. comparing outcomes against control or comparison groups not affected by a policy) or hypothetical prediction (e.g. interviews with experts to determine outcomes without policy intervention).

Counterfactual analysis follows a logical process with the following steps (Figure 4.2): construction of a base scenario, which is a description of the current system without policy (WoP) intervention (i.e. a state that is assumed to represent the situation if the policy

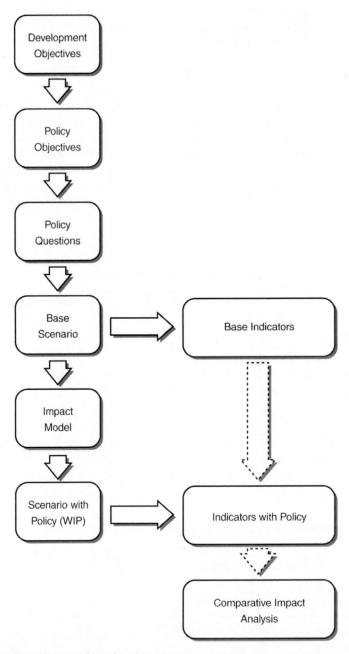

Figure 4.2 Flowchart of counterfactual analysis process
Source: adapted from Bellù, and Pansini (2009)

measure is not implemented). This will be the reference scenario, also called a benchmark or baseline, for the impact analysis of policies. The reference scenario is then described using some indicators, chosen on the basis of the type of policy that one would analyse. If, for example, the policy measure aims at protecting marine mammals and endangered marine species, one would use indicators such as the cetacean stranding rate, population growth rate, their habitat requirements and their distribution. These become reference indicators for the reference scenario. After building and describing the reference scenario, the analysis then focuses on the construction of one scenario that integrates the expected impacts of the policy option. This is the scenario with policy (WiP). If more than one policy option has to be analysed, the analyst can build different scenarios 'with' policy. The scenarios WiP are usually built as a modification of the WoP scenario.

Quantitative versus qualitative analysis

Policy analysis employs a range of qualitative and quantitative research methods to analyse and compare existing policies and possible modifications to them. It is not possible to neatly decouple qualitative and quantitative analysis, as many approaches have mixed natures, however, qualitative approaches for policy impact analysis essentially make use of non-numerical information. Qualitative research often gains a general sense of phenomena and forms theories that can be tested using quantitative research. This research includes value judgements of key informants, non-structured interviews of stakeholders, focus groups, panel discussions etc. These approaches typically compare qualitative outcomes relating to public perception and can be very detailed and context specific in order to feed the decision-making process. Many of these methods derive from ethnography, a discipline that uses fieldwork to provide descriptive studies of human societies.

In contrast, quantitative approaches are rooted in empirical methods of investigation. These focus on unbiased measurement of numerical data, usually gathered by means of surveys based on structured questionnaires with close-ended questions; or the data can be garnered from other statistical sources (e.g. national census, national accounts, custom data, international databases). Researchers treat information by means of mathematical methods in order to derive expected responses to variables affected by policy interventions. Quantitative approaches may deal with physical data (e.g. physical quantities of inputs and outputs, number of tonnes landed by a fishery, number of oil spill incidents) and provide decision-makers with information on policy impacts in physical terms.

A subset of quantitative analysis, monetary evaluations are often executed using currency as a common unit of measure. Physical quantities are converted into monetary values by means of prices expressed in monetary units. In these cases, decision-makers are provided with summary information expressed in monetary terms or in percentage variations of monetary variables (e.g. income or expenditure of various groups, GDP growth rates, value-added variations, budgetary implications) (Costanza et al., 1987).

The use of environmental indicators

Indicators can bridge qualitative and quantitative methods and link economic and environmental developments. This bridge facilitates the integration of environmental and economic policies (see Box 4.2). Environmental indicators are necessary for monitoring the discrepancy between actual developments and targets to meet environmental objectives,

---

Box 4.2 Constructing a green GDP

Including the costs of environmental degradation to better reflect measurements of national welfare, such as GDP, has involved numerous alternative 'green' calculations. While there has been significant support for the green GDP concept, critics assert that indicators, such as the GDP, require a standard judgement of environmental quality. For many reasons, however, these indicators may not capture all the dimensions of national situations. Some critics also argue that the current GDP measure is aimed at monitoring factual market activity rather than general welfare levels, and as such is not the appropriate type of indicator for including measures on the state of the environment. Consequently, the treatment of environmental spending and degradation will probably not be explicitly included in GDP calculations. Despite imperfections, knowledge of environmental policy objectives makes it possible to calculate the cost of the measures needed to achieve such objectives, and hence to construct a 'greener' GDP.

Source: Navrud and Pruckner (1997)

---

and, once calculated, to inform policy adjustments. Recording market reductions in output due to environmental degradation, spending on abatement, waste disposal and environmental charges, may give a more accurate comparison of economic costs associated with environmental policies than survey-based results (Schafer and Stahmer, 1989; Dahl, 2000).

Available information on the state of the environment may be translated into measures of flows of environmental services and estimates of damage caused by degradation of environmental resources. The use of technology, such as satellite monitoring, can be used as part of a comprehensive approach linking physical stocks of resources to national balance sheets and resource use to national accounts. Given that opportunity costs can be calculated by using targeted abatement costs, these prices can be connected to information on actual pollution or resource stocks in physical units. Still, many practical difficulties regarding both the actual measurement of physical stocks and their conversion into monetary values exist (Pearce, 1993).

Role of science in policy-making

The role of science in policy and politics is a complex topic that is explored in a growing body of scholarly literature (Ezrahi, 1980; Coppock, 1985; Barke and Jenkins-Smith, 1993; Dietz and Stern, 1998; Andresen et al., 2000; Hagendijk and Irwin, 2006; Pielke, 2008). Effective environmental policy considers a wide range of inputs, but depends on science for a number of functions. Scientific knowledge is employed at many stages in the policy-development process – to identify emerging problems, contribute to policy formation, evaluate impacts and assess the state of natural resources, and provide a rational, quantitative framework for resolving conflicts. In an ideal and healthy democracy, policy-makers and citizens need accurate, relevant and unbiased science to decide how best to proceed on issues. One of the fundamental roles that science plays in policy development is providing 'raw material' to inform policy decisions (Margules and Pressey, 2000).

Scientists and science can and often do play useful roles in policy and democratic politics, but they can also confuse, muddle or otherwise impede sound decision-making. Science is essential for effective leadership; however, there is a well-documented need to improve the flow of information in both directions between scientists and decision-makers. Scientific information for policy is required to be accurate, relevant, credible and politically legitimate; yet political interference, competing interpretations, fluctuations in funding, the influence of economic imperatives, the pace of new information and even an absence of basic research can challenge the role of science in the policy process.

The successful and effective use of science advice in decision-making has been characterized by decisions where:

- there has been full consideration of the best available science;
- the scientific knowledge used is sound (systematic, peer reviewed, independent, free from bias or bias is revealed);
- the full diversity of scientific thought from relevant disciplines is considered;
- risks and uncertainties are explicitly taken into account;
- there is an acknowledgment that the 'best available' science does not encompass all possible knowledge;
- decisions are transparent and open – all pertinent information, data, assumptions, values and interests are identified and accessible;
- a rationale for major decisions is provided/available to all interested parties;
- adaptive management principles have been incorporated and the decision is subject to ongoing review;
- formal institutional processes exist to ensure accountability.

Scientific advice is often modified by social, economic and political considerations (Margules and Pressey, 2000). Recognizing this, and that there are major inconsistencies in how many scientists treat 'facts' versus 'values', it is important to state explicitly that there are choices and options outside the linear model available for how scientists can participate in policy and politics. On one end of the spectrum is the pure scientist, who plays a role that is as independent and isolated from policy and politics as possible. For a pure scientist, research is published in academic journals and no effort is made to help resolve society's current policy challenges. At the other end of the spectrum is the issue advocate, a scientist who uses his or her science and scientific credentials to advocate for the policies he or she supports.

Chapter summary

- Policy analysis encompasses all aspects of how problems are identified, decisions are made, policies are implemented, and outputs and outcomes are measured. Policy analysis is the business of informing decision-making.
- Policy analysis may be prospective, where it attempts to forecast the likely outcomes for various policy options and uses existing data or retrospective, where it is used to assess whether the policy is meeting its goals and objectives.
- The process by which problems are identified, policies are created, outcomes evaluated and policies are revised is the termed the policy cycle.

References

Andresen, S., Skodvin, T., Underdal, A. and Wettestad, J. (2000) *Science and Politics in International Environmental Regimes: Between Integrity and Involvement*, Manchester University Press, New York.

Bardach, E. (2011) *A Practical Guide for Policy Analysis: The Eightfold Path to More Effective Problem Solving*, CQ Press, Washington, DC.

Barke, R. P. and Jenkins-Smith, H. C. (1993) 'Politics and scientific expertise: Scientists, risk perception, and nuclear waste policy', *Risk Analysis*, 13, 425–439.

Bellù, L. G. and Pansini, R. V. (2009) *Quantitative Socio-Economic Policy Impact Analysis: A Methodological Introduction*, FAO, Rome.

Bridgeman, P. and Davis, G. (2000) *Australian Policy Handbook*, Allen & Unwin, Sydney.

Coppock, R. (1985) 'Interactions between scientists and public officials: A comparison of the use of science in regulatory programs in the United States and West Germany', *Policy Sciences*, 18, 371–390.

Costanza, R., d'Arge, R., de Groot, R., Farber, S., Grasso, M., Hannon, B., Limburg, K., Naeem, S., O'Neill, R. V., Paruelo, J., Raskin, R. G., Sutton, P. and van den Belt, M. (1987) 'The value of the world's ecosystem services and natural capital', *Nature*, 387, 253–260.

Dahl, A. L. (2000) 'Using indicators to measure sustainability: Recent methodological and conceptual developments', *Marine and Freshwater Research*, 51, 427–433.

Dietz, T. and Stern, P. C. (1998) 'Science, values, and biodiversity', *BioScience*, 28, 441–444.

EEA (European Environment Agency) (2005) *Policy Effectiveness Evaluation: The Effectiveness of Urban Wastewater Treatment and Packaging Waste Management Systems*, www.eea.europa.eu/publications/brochure_2006_0305_111039, accessed 19 April 2013.

Ezrahi, Y. (1980) 'Utopian and pragmatic rationalism: The political context of scientific advice', *Minerva*, 18 (1), 111–131.

Fischer, F., Miller, G. J. and Sidney, M. S. (2006) *Handbook of Public Policy Analysis: Theory, Methods, and Politics*, Marcel Dekker, New York.

Freeman, A. M. (1985) 'Supply uncertainty, option price and option value', *Land Economics*, 6, 176–181.

Greiner, R., Young, M. D., McDonald, A. D. and Brooks, M. (2000) 'Incentive instruments for the sustainable use of marine resources', *Ocean & Coastal Management*, 43 (1), 29–50.

Hagendijk, R., and Irwin, A. (2006) 'Public deliberation and governance: Engaging with science and technology in contemporary Europe', *Minerva*, 44, 167–184.

Hanley, N. (1992) 'Are there environmental limits to cost benefit analysis?', *Environmental and Resource Economics*, 2 (1), 33–59.

Howard, C. (2005) 'The policy cycle: A model of post-Machiavellian policy making?', *Australian Journal of Public Administration*, 64 (3), 3–13.

Howlett, M. and Ramesh, M. (1995) *Studying Public Policy: Policy Cycles and Policy Subsystems*, Oxford University Press, Toronto.

Huang, I. B., Keisler, J. and Linkov, I. (2011) 'Multi-criteria decision analysis in environmental sciences: Ten years of applications and trends', *Science of the Total Environment*, 409, 3578–3594.

Jesinghaus, J. (1999) *Indicators for Decision-Making*, European Commission, Joint Research Centre, Brussels.

Johnstone, N. (2003) *The Use of Tradable Permits in Combination with other Environmental Policy Instruments*, OECD, Paris.

Lasswell, H. (1951) 'The policy orientation', in D. Lerner and H. Lasswell (eds) *The Policy Sciences*, Stanford University Press, Stanford, CA.

Loomis, J. and Helfland, G. (2001) *Environmental Policy Analysis for Decision Making*, Kluwer Academic Publishers, Dordrecht, The Netherlands.

Margules, C. R. and Pressey, R. (2000) 'Systematic conservation planning', *Nature*, 405, 243–253.

Mark, M., Henry, G. and Julnes, G. (2000) *Evaluation: An Integrated Framework for Understanding,*

Guiding, and Improving Public and Non Profit Policies and Programs, Jossey-Baso Inc., San Francisco, CA.

May, J. V. and Wildavsky, A. (1978) *The Policy Cycle*, Sage, London.

Mazur, E. (2010) 'Outcome performance measures of environmental compliance assurance', *OECD Environment Working Papers*, 18 (18).

McCay, B. J. (2000) 'Resistance to changes in property rights or, why not ITQs?', in R. Shotton (ed.) *Use of Property Rights in Fisheries Management*, FAO Fisheries Department, Rome.

Munger, M. (2000) *Policy Analysis*, W.W. Norton and Co, New York.

Navrud, S. and G. J. Pruckner (1997) 'Environmental valuation – to use or not to use? A comparative study of the United States and Europe', *Environmental and Resource Economics*, 10, 1–26.

Norris, P. (2011) 'Cultural explanations of electoral reform: A policy cycle model', *West European Politics*, 34 (3), 531–550.

OECD (organization for Economic Co-operation and Development) (1995) *The Economic Appraisal of Environmental Projects and Policies. A Practical Guide*, OECD, Paris.

Ohowa, B. O. (2009) 'Evaluation of the effectiveness of the regulatory regime in the management of oil pollution in Kenya', *Ocean and Coastal Management*, 52 (1), 17–21.

Patton, C. and Sawicki, D. (1986) *Basic Methods of Policy Analysis and Planning*, Prentice-Hall, NJ.

Pearce, D. (1993) *Economic Values and the Natural World*, Earthscan, London.

Pearce, D. and Moran, D. (1994) *The Economic Value of Biodiversity*, Earthscan, London.

Pielke, R. A. (2008) *The Honest Broker: Making Sense of Science in Policy and Politics*, Cambridge University Press, Cambridge.

Pintér, L., Zahedi K. and Cressman, D. (2000) *Capacity Building for Integrated Environmental Assessment and Reporting. Training Manual*, International Institute for Sustainable Development for the United Nations Environment Programme, Winnipeg.

Robinson, J. and Ryan, S. (2002) *A Review of Economic Instruments for Environmental Management in Queensland*, www.ozcoasts.gov.au/pdf/CRC/economic_instruments.pdf, accessed 19 April 2013.

Schafer, D. and Stahmer, C. (1989) 'Input-output model for the analysis of environmental protection activities', *Economic Systems Research*, 1 (2), 203–228.

Schulz, W. and Schulz, E. (1991) 'Germany', in J. P. Barde and D. W. Pearce (eds) *Valuing the Environment: Six Case Studies*, London: Earthscan.

Smith, K. and Larimer, C. (2009) *The Public Policy Theory Primer*, Westview Press, Boulder, Colorado.

Stavins, R. N. (2003) 'Experience with market-based environmental policy instruments', in K. Maler and J. R. Vincent (eds) *Handbook of Environmental Economics*, Elsevier, Amsterdam.

Sterner, T. and Coria, J. (2012) *Policy Instruments for Environmental and Natural Resource Management*, RFF Press, New York.

Theodoulou, S. Z. and Kofinis, C. (2004) *The Art of the Game: Understanding. American Public Policy Making*, Thomson/Wadsworth, Belmont, CA.

Weitzman, M. (1974) 'Prices vs. quantities', *Review of Economic Studies*, 41 (4), 477–491.

Chapter 5

Marine environmental protection policy

International efforts to address threats to marine biodiversity

Introduction

For centuries, the oceans were envisaged as immutable and immune to human activities. Fish were plentiful and the capacity for the oceans to absorb human wastes was believed to be unlimited. Hugo Grotius's principle of 'mare liberum' (free seas – see Chapter 2) perpetuated this worldview, which continues to underpin the actions of many states and their nationals to this day. The unregulated environment created by the free seas doctrine predictably led to the rapid degradation and exploitation of resources as well as the annexation of coastal seas (as territorial seas or EEZs) by coastal states (Hardin, 1968).

The free seas doctrine resulted in four centuries of neglect and non-management. Many stocks of once globally abundant fishes such as cod, herring and tuna have become ecologically and commercially extinct. Over 1 million whales were harvested during the past century, and to date, only the eastern Pacific grey whale has recovered to near pre-exploitation levels. Elevated levels of pollutants are found in most marine species, even those living in the polar regions. Ocean temperatures are rising – aggravated by the addition of greenhouse gases from the combustion of fossil fuels – with many deleterious effects. Just one of these is the tens of thousands of square kilometres of coral reefs that have bleached, with large associated die-offs in recent years. Important breeding, feeding, mating and resting areas for migratory species have been affected by human activities. This is merely a brief summary of the continuing degradation of marine systems. Those interested in detailed accounts of the effect of human activities on marine environments should read the comprehensive works by Norse (1993), NRC (1995) and Roberts (2013).

It was not until the 1960s that appeals from individuals such as Rachel Carson (1962) and Jacques Cousteau addressed human impacts on ocean health. Around this time, organizations such as Greenpeace also began campaigning on specific issues (e.g. whaling), creating the genesis of an 'ocean ethic' that later embodied pollution, habitat loss, introduced species and, more recently, climate change.

Unfortunately, as time goes by and new generations of people interact with the oceans, our memories and expectations of the 'natural state' of the oceans also mutate. This generational change in our perception of the state of the oceans has been captured in two memorable aphorisms – 'the shifting baseline syndrome' and 'fishing down marine food webs' (Pauly et al., 1998). The first saying captures the idea that though the oceans are progressively being degraded, each human generation comes to accept the deteriorated state as the norm. The second saying reflects the reality that ocean resources are providing smaller and smaller organisms; fisheries are trying to capitalize on smaller members of species once dominated by larger populations, and smaller species once ignored or under-valued by fishers. The changing history of our views of the oceans and especially of the history of fishing fleets have been documented by Roberts (2007 – Chapter 6). We are indisputably and ever more rapidly reducing the biodiversity of our oceans by reducing the number of species, impacting habitats and their communities and, indeed, destroying whole ecosystems.

This chapter reviews the present state of, and impacts to, the global marine environment. Following this is a discussion on the primary international mechanisms that govern the protection and preservation of the oceans. Important regional conventions and examples of bilateral and multilateral agreements are also reviewed.

Threats to marine biodiversity

There are many ways to collate threats to marine biodiversity, but threats can be broadly categorized as a result of overharvesting, pollution, habitat loss, introduced species and global climate change/ocean acidification (see e.g. NRC, 1995; Gray, 1997; Thorne-Miller and Earle, 1999). The following sections provide a brief discussion of these impacts on the marine environment. Many marine areas have a range of biota rivalling or exceeding that of tropical forests. However, the diversity of life in our oceans is now being dramatically altered by rapidly increasing and potentially irreversible human activities. Although there are differing views of present and potential threats to coastal and marine biodiversity, those shown in Table 5.1 are among the most important.

Overharvesting

The unsustainable harvest of marine populations is perhaps the most serious threat to marine environments worldwide. Overharvesting is not a new phenomenon in the oceans. Many traditional cultures either removed the available species from their local marine environment and moved onto harvesting other areas or had to develop methods of regulating the timing and amount of harvest from certain areas in order to avoid overexploitation. With the advent of the industrial revolution came the increasing mechanization of fish harvesting. Species such as large whales and offshore pelagic fish – previously difficult to catch – were now accessible in a commons environment owned by no one. Most whale species have been reduced to levels where they are considered endangered or threatened. Most populations of palatable fish stocks have been seriously depleted, and currently there is evidence that humans have, and will continue to, fish down food webs (e.g. Pauley et al., 1998 – Chapter 6).

Pollution

Unquestionably, pollution – from myriad sources – has impacted every marine system on earth. Indigenous human populations in Arctic areas have the highest contaminant levels of any people on earth, a result of ingesting marine fish and mammals that bioaccumulate toxins due to their high trophic levels. It has been assumed that pollutants reached the oceans primarily in runoff from rivers. In some coastal areas this will indeed be the case, but, overall, transport of pollutants through the atmosphere to the oceans is more significant. The types of pollutants are diverse: artificial radionuclides, petroleum hydrocarbons,

Table 5.1 Examples of threats to marine biodiversity

Risk or speed of degradation of biodiversity	Threatening process
High Low	Physical habitat destruction (e.g. reclamation, dredging) Blast fishing using explosives, meting* (either can annihilate a coral reef) Toxic pollution (e.g. chemical spills) Chemical fishing (e.g. cyanide) Introduction of exotic organisms Biological invasions Overexploitation/overfishing Indirect pollution (pesticides, herbicides in runoff) Disease/parasite infection Depletion of habitat critical to life history stages Incidental take/bycatch Destruction of adjoining watersheds Impacts of adjacent land-use practices (e.g. aquaculture) Effluent discharge (sewage, pulp/paper) Natural events (cyclones, tsunamis) Direct marine pollution, ocean dumping Sea dumping of dredge spoil Downstream impacts from dams, dykes, etc Net/debris entanglement Siltation Noise pollution Thermal pollution Climatic change – rising sea temperatures Sea-level rise Salinity changes Indigenous take

Note: *Meting is an emerging threat that involves the indiscriminate removal of all organisms from reefs using metal crow bars to rip away coral cover to harvest invertebrates such as abalone and clams. The 'high–low' scale on the left side of this table is approximate only; it seeks only to indicate that some threatening processes have a higher risk and/or speed of impact on marine diversity than others. The relative order of the various threatening processes on this scale is open to conjecture.
Source: Roff and Zacharias (2011)

chlorinated hydrocarbons, metals, carcinogens, mutagens, pesticides, excess nutrients causing nuisance and toxic algal blooms, endocrine disrupters, physical debris and so on. Furthermore, the persistence and longevity of pollutants in the marine environment, and their ecosystem-level effects on marine biota and ultimately humans, are of growing concern.

Habitat loss

Habitat loss – due to the destruction of larger vascular vegetation upon which many species depend for food and shelter – is probably the most serious threat to biodiversity in terrestrial environments. Loss of marine habitat is primarily a concern in coastal nearshore and inter-tidal marine environments. Increasing pressure on coastal systems has come from a combination of: shipping and its attendant infrastructure, modification of natural coast-lines, bottom contact fishing (e.g. trawling), recreational activities and increased land runoff – including nutrients and suspended solids. The types of habitats in these areas that can be 'lost' include marine macrophytes (kelp), mangroves, sea grasses, corals and other biotic communities (e.g. sponges, sea pens, sea fans, aphotic corals). Abiotic habitats can be lost as well, such as intertidal and estuarine mud flats and other areas that are dredged or subject to dumping. Habitat loss in deeper marine environments and the pelagic ocean is a more vague construct as these habitats are primarily composed of either oceanographic (e.g. currents, gyres, fronts) or physiographic (e.g. seabed composition) structures and processes that are more resilient to human activities – or less immediately impacted. Loss of marine habitat is significant not only from an ecological perspective, but increasingly from a socioe-conomic perspective as well. The interaction of human effects and natural marine processes is most evident in coastal waters where strategies to prevent habitat loss or restore habitats are integrated within coastal zone management initiatives (see Chapter 10).

Introduced species

Species introductions (also termed invasive, exotic and non-native species) likely have occurred since humans first used the oceans for exploration and trade. There is evidence that many species believed to be native were actually introduced through marine trans-portation prior to the industrialized era. More recently, transport in the ballast water of ships appears to be the main mode of travel (vector), and impacts are generally observed mainly in coastal waters and estuaries. The introduction of larger species such as the green crab (Carcinus maenas), the alga Calaupera taxifolia and the comb 'jellyfish' Mnemiopsis leidyi has been well publicized. Yet most species introductions are less obvious, and are found in the phytoplankton and zooplankton. Some of these invasive species can have dramatic local socioeconomic effects. Different species of jellyfish have had substantial impacts on fisheries and even coastal human recreation. Outbreaks of jellyfish have now been reported world-wide; along with species invasions, overharvesting (removing jellyfish predators), nutrient enrichment (increasing primary production resulting in increases in available food) and local water temperature increases are the probable culprits.

Global climate change

There is no doubt that the earth's climate changes over time and that cyclical changes tran-spired long before humans dominated the planet. Global climate changes have been responsible for mass extinctions in the past and the earth's climate will continue to vary, likely resulting in future mass extinctions. During the Quaternary period, sea levels deviated as much as 85m, which inhibited the evolution of established marine communities in coastal and shelf environments. Unquestionably, humans have impacted global tempera-tures, and since the 1980s there has been considerable debate on differentiating the natural and anthropogenic contributions to climate change. Recent evidence, however, concludes that climate change since the 1900s is human-induced (Hansen et al., 2012).

Human activities that influence change include the release of CO_2 through the burning of fossil fuels and large-scale deforestation – which decreases the removal of CO_2 from the atmosphere. Changes in ocean temperatures impact marine biological communities in various ways, including changes in geographical distribution, behaviour and life history (e.g. reproduction, growth and dispersal). Coral reefs, for example, are particularly vulnerable to ocean warming, for even minor temperature increases may cause many coral species to lose their algal symbiotic zooxantheallae (termed 'bleaching') that provide energy through photosynthesis. There is also evidence that the ranges of some species have shifted towards the north and south poles as temperatures increase in equatorial and temperate waters.

History of international marine environmental law and policy

The first marine environmental treaties were established to prevent the over-exploitation of fish stocks. Early treaties between Great Britain and France (1839) and the North Sea states (1882, Belgium, Denmark, France, Germany, Netherlands, UK) set the stage for negotiations between states to address marine environmental issues (Table 2.2). Treaties ensued on fur seals in the North Pacific (1911), whaling (1946) and fishing gear restrictions (1946). The first convention to specifically mention the conservation of living resources was the first the Law of the Sea Convention in 1958 (Chapter 2). Other conventions followed in order to arrest and clean up oil pollution (1971), prohibit dumping (1972) and address regional marine environmental issues (1974 – Baltic Sea).

Marine environmental law is part of the broader subject of international law, which has been formed through ICJ decisions, treaties and common practice (see Chapter 2). It was not until the 1972 UN Conference on the Human Environment (UNCHE) in Stockholm, Sweden, however, that several key principles in international environmental law were formally recognized through a global consensus. Attended by representatives from 113 states, the Conference produced the 'Stockholm Declaration', in which 26 principles were ensconced. The declaration recognized the impacts of humans on the natural environment, some of the causes of these impacts (e.g. underdevelopment, over population, lack of education) and the need for global cooperation to address these issues. While all the principles are broadly applicable to marine environments, Principle 7 specifically obliges states to prevent marine pollution. In addition, the UNCHE produced an Action Plan with 109 resolutions and one declaration. Perhaps the most significant achievement of the UNCHE was the establishment of the United Nations Environment Programme (UNEP). The UNCHE subsequently became the template for UN environment conferences in 1992, 2002 and 2012.

The 1987 World Commission on Environment and Development (WCED) was the next major advancement in international environmental cooperation. Known as the 'Brundtland Commission', its charge was to unite countries to jointly pursue sustainable development. The key output of the Commission was to define sustainable development as 'development that meets the needs of the present without compromising the ability of future generations to meet their own needs' (WCED, 1987). In addition, the report recommended that 12 per cent of each state should be protected. The Commission's report, published as *Our Common Future* (WCED, 1987) contained a chapter outlining the threats to marine areas and opportunities for improved oceans management. Specifically, the Commission proposed measures to:

- strengthen capacity for national action, especially in developing countries;
- improve fisheries management;
- reinforce cooperation in semi-enclosed and regional seas;

- strengthen control of ocean disposal of hazardous and nuclear wastes;
- advance the Law of the Sea.

While the WCED's work did not establish new international law, many of the concepts in *Our Common Future* were incorporated into the 1992 UN Conference on Environment and Development (UNCED) (see below). It is through this avenue that the WCED became an important contributor to current international environmental law.

The following section outlines the key agreements, conventions and treaties established to manage marine biodiversity.

Conventions that pertain to marine environmental protection

The LOSC

Part XII of the LOSC (see Chapter 2) addresses the protection and preservation of the marine environment. It restates earlier marine protection commitments from the 1958 Convention for the High Seas (Articles 116–120). Promoting the 'optimum utilization' of living resources, it echoes the Convention's call for catch limits based on best available science and ensuring that fisheries do not negatively affect other species (Article 61). The LOSC affirms that states have the right to capitalize on their living marine resources in their sovereign waters but must do so in a manner that protects and preserves the marine environment (Articles 192–193). Specifically, the LOSC provides direction on three specific marine protection topics:

First, states must '...reduce and control pollution of the marine environment from any source...' and ensure that their activities do not cause damage by pollution in other jurisdictions (Article 194). The LOSC (Article 194) identifies four sources of pollutants that states are obliged to address:

- Toxic, harmful or noxious substances introduced into the marine environment by land-based sources, the atmosphere or dumping.
- Pollution from vessels.
- Pollution from installations related to offshore hydrocarbon development or seabed mining.
- Pollution from other installations.

States may not reduce pollution in their jurisdictions by transporting pollution to other jurisdictions (Article 195).

Second, states must protect and preserve rare or fragile ecosystems as well as protect the habitats of depleted, endangered or threatened marine life (Article 194). Third, states must prevent the spread of alien (introduced) species (Article 196). And fourth, the LOSC delineates a global cooperation regimen to ensure that states cooperate to protect the marine environment. In particular, states must cooperate to develop science-based rules and standards to maintain environmental quality (Articles 197, 201), notify other states that are or may be affected by pollution (Article 198), jointly develop plans to respond to environmental emergencies (Article 199), undertake joint scientific research (Article 200) and cooperate on monitoring and environmental assessments (Articles 204–206).

In addition, more-developed states are obligated to assist less-developed states to improve their scientific and research capacity related to the protection of the marine environment (including ensuring less-developed states receive priority funding for these programmes). An obligation also exists to apprise less-developed states of the effects of their activities on marine environmental quality (Articles 202–203).

The majority of Part XII, however, addresses the requirements for states to develop national (domestic) legislation to contribute to the aims of Part XII. Specific LOSC direction to states includes the requirements to adopt and enforce laws and cooperate across jurisdictions to prevent or minimize pollution. Land-based sources (Articles 207 and 213), seabed activities in sovereign areas (Articles 208 and 214), activities in the Area (Articles 209 and 215), dumping (Articles 210 and 216), vessels (Article 211) and the atmosphere (Article 212) are specific pollution-related issues mentioned in Park XII.

Additionally, the LOSC articulates the requirements for enforcement by flag states. States must ensure that vessels flying their flag comply with all necessary international and national rules and regulations established to protect the marine environment (see Chapter 7). Flag states are also required to ensure that vessels flying their flag carry the necessary documentation to demonstrate compliance and to undertake inspections and investigations, with other states if necessary, to guarantee compliance (Article 217).

Port states are entitled to undertake investigations of ships operating voluntarily within their territorial waters or EEZs where evidence suggests that a vessel operating on the high seas was in contravention of international rules with respect to Part XII of the LOSC (Article 218). Port states may also prohibit a vessel from operating in their national waters if the vessel is believed to be unseaworthy and a risk to the marine environment (Article 219). Likewise, coastal states may inspect, institute proceedings or detain a vessel transiting its territorial waters or EEZs where evidence suggests that the vessel was in contravention of international rules with respect to Part XII of the LOSC (Article 120).

Port and coastal states may exercise their rights under Part XII under the following conditions:

- Flag states and their vessels have an opportunity to present evidence during investigations or hearings (Article 223).
- Enforcement against foreign vessels may only be undertaken by warships, military aircraft or other clearly marked government vessels (Article 224), and enforcement actions must not subject ships and their crews to risk (Article 225).
- Detained foreign vessels must be released without delay and, if repairs are needed, be permitted to transit to the nearest repair facility (Article 226).
- Foreign vessels are not to be discriminated against (Article 227).
- Proceedings against foreign vessels with alleged violations outside of the port or coastal state's territorial sea are suspended if the flag state initiates proceedings. The exception is where major damage was caused in the port or coastal state's EEZ (Article 228).
- Penalties may not be imposed on any vessel beyond three years after the incident was alleged to have occurred (Article 228).
- Only monetary penalties can be levied against foreign vessels determined to have violated national or international laws within a coastal or flag state's territorial sea (Article 230).
- Flag states will be immediately notified of all measures taken against vessels flying their flag (Article 231).
- Warships or other state vessels are immune from enforcement obligations but are expected to meet the spirit and intent of Part XII (Article 236).

While the LOSC addresses pollution from land-, sea- and air-based sources, the Convention is noticeably silent on other aspects of protection of the marine environment. Habitat loss, climate change and overharvesting are ignored, but this is unsurprising given that the original intent of the Convention was to regulate seabed mining and these threats were only emerging in 1982.

The Convention on Biological Diversity

The CBD was signed in June 1992 at the United Nations Conference on Environment and Development (the Rio 'Earth Summit' or UNCED) and entered into force in December 1993 (UN, 1992). The CBD was the first global agreement specifically intended to conserve and sustainably use biodiversity, with its attendant genetic components. Biodiversity in the CBD is defined as '...variability among living organisms from all sources including, inter alia, terrestrial, marine and other aquatic ecosystems and the ecological complexes of which they are part; this includes diversity within species, between species and of ecosystems'.

The CDB also recognizes the role(s) of indigenous cultures, women and poverty reduction in biodiversity conservation. The CDB commits more-developed states to provide funding to less-developed states in order to further the aims of the Convention. Moreover, the CBD mandated a subsidiary body for the provision of scientific, technical and technological advice.

While the text of the Convention does not specifically address marine environments, it does consider the marine environment as a component of biodiversity (Article 2) and that the CBD is to be applied to the marine environment consistent with the laws of the sea (Article 22.2). However, the Parties to the Convention established 'Marine and Coastal Biodiversity' as one of its thematic programmes and marine and coastal biodiversity was an early priority for the COP. At the second COP in 1995, Decision II/10 (CBD, 1995) committed signatories to address marine biodiversity issues and provided some overall guidance on the architecture of this programme. In addition, COP II produced a global consensus on the importance of marine and coastal biodiversity. Known as the 'Jakarta Mandate on Marine and Coastal Biological Diversity', the mandate reaffirms that '...there is a critical need for the COP to address the conservation and sustainable use of marine and coastal biological diversity, and urge Parties to initiate immediate action to implement the decisions adopted on this issue' (CBD, 2000a).

The Jakarta Mandate further articulates the following six implementation principles:

- Apply the ecosystem approach.
- Apply the precautionary principle.
- Recognize the importance of science.
- Utilize a roster of experts.
- Involve local and indigenous communities and consider traditional knowledge in decision making.
- Implement programmes at the national, regional and global levels.

At the COP 4 meeting in Bratislava, Slovakia in 1998, the Parties adopted Decision IV/5 on the conservation and sustainable use of marine and coastal biological diversity, including a Multiyear Programme of Work on Marine and Coastal Biological Diversity. Coral reefs and coral reef bleaching induced by climate change were particular focuses of this COP (CBD, 1998).

At the COP 8 meeting in Curitiba, Brazil in 2006, the Parties adopted Decisions VIII/21, 22 and 23. These stressed the value of researching and protecting seabed habitats, the import of integrated marine and coastal area management, and the recognition that protected areas are an essential tool for the conservation of marine biodiversity (CBD, 2006). The meeting's delegates set a target of 10 per cent for the protection of the world's ecological regions, including coastal and marine areas.

At the COP 9 meeting in Bonn, Germany in 2008, the Parties adopted Decision IX/20 to compile and synthesize scientific information on the effects of ocean fertilization (to sequester CO_2 and address climate change) on marine biodiversity (CBD, 2008). The Parties also settled on criteria for the delineation of ecologically and biologically significant marine areas and criteria for establishing networks of marine protected areas (see Chapter 10).

A number of other agreements have been framed under the Convention, including:

- The Cartagena Protocol on Biosafety (adopted 2000, entered into force 2003) that aims to ensure the safe handling, transport and use of organisms that have been modified using modern biotechnology. It takes measures against the adverse effects on biological diversity and risks to human health (CBD, 2000b).
- The Nagoya Protocol on Access and Benefit Sharing (adopted 2010, not yet entered into force) that governs the access to and benefits derived from genetic resources. Specifically, it attempts to address the needs of those searching for genetic resources (bio-prospecting) with those who live on or govern the lands and coastal areas that contain these genetic resources (CBD, 2010).

The World Summit on Sustainable Development and the protection of the marine environment

The World Summit on Sustainable Development (WSSD) was held between 26 August and 4 September 2002. Termed the 'Earth Summit 2002' or 'Rio+10', the purpose of the Summit was to focus on global problems (e.g. poverty, conflict, disease, discrimination) inhibiting sustainable development that could be addressed through multilateral cooperation. The meeting produced the Johannesburg Declaration on Sustainable Development, which proclaimed an aspirational, non-binding suite of resolutions (framed as sustainable development outcomes) furthering the intent of the UNCHE and UNCED. The Declaration was executed through a Plan of Implementation of the WSSD.

The Plan of Implementation formulated a number of marine-related objectives. While reiterating a number of binding actions from previous international agreements, it also introduced a number of new commitments and implementation dates. Specifically, states agreed to restore depleted fish stocks by 2015 (s.31(a)), establish a global network of representative marine protected areas by 2012 (s. 32(c)) (see Chapter 10) and establish a process to report on the state of the marine environment by 2004 (s. 36(b)).

Rio+10 was followed by Rio+20 in 2012. Conference attendees adopted a document titled *The Future We Want*, which reiterates commitments from previous conferences but updated to reflect technological changes (e.g. the importance of connections to the internet) and the state of the global economy (e.g. the need to address youth unemployment). Sections 158–177 address marine issues and, consistent with the previous sections, reiterate past promises. New issues raised in the document include the threats of plastic marine debris and the possible unintended consequences of ocean fertilization.

Threat-specific conventions and the protection of the marine environment

Binding conventions

A number of legal international agreements exist to protect oceans from specific human impacts. A list of all the marine treaties and agreements are provided in Table 2.2, but the most significant conventions are listed below.

The Convention on the Prevention of Marine Pollution by Dumping of Wastes and Other Matter (1972)

After MARPOL (Chapter 7), the Convention on the Prevention of Marine Pollution by Dumping of Wastes and Other Matter (the 'London Convention' or 'London Dumping Convention' abbreviated as LDC) is perhaps the next strongest international law related to marine environmental protection. Adopted in 1972 and entering into force in 1975, the LDC is administered by the IMO. The Convention prohibits the deliberate dumping of different types of hazardous materials from ships, aircraft and platforms, including ships, aircraft and platforms themselves (IMO, 1972). The Convention does not apply to discharges from land-based sources (e.g. outfalls), waste generated as a result of the normal operation of ships, intentional placement of materials for purposes other than dumping (e.g. seawalls), and materials produced as a by-product of fish processing or seabed mining. Articles in the Convention may be contravened if the safety of humans or structures is compromised by unforeseen events (e.g. storms). The Convention contains 22 articles and 3 annexes and has been ratified by 81 states. It applies to all territorial waters and high seas but not internal waters.

In 1996, the Protocol to the Convention on the Prevention of Marine Pollution by Dumping of Wastes and Other Matter 1972 ('London Protocol') was approved and entered into force in 2006. It is intended to eventually replace the 1972 LDC (IMO, 1996). The London Protocol introduced a new approach to marine waste management and introduced the concept of the 'precautionary principle', the principle of 'polluter pays' and stressed the need to avoid transferring wastes from one type of receiving environment (e.g. the ocean) to another (e.g. the land). While the London Convention prescribed the substances and compounds (e.g. organohalogen compounds) that were prohibited from dumping, the Protocol – recognizing the pace in development of new compounds that may negatively impact marine and human health – prohibited the dumping of all materials with the exception of those listed in Annex I. Thus, the Protocol reverses the onus of proof and was a key development in the evolution of marine governance.

The London Protocol includes three annexes. It is important to note that the annexes of the London Protocol in no way relate to the annexes of the London Convention. The annexes of the London Protocol are as follows:

- Annex I specifies the wastes and other matter that may be dumped. The Protocol bans the dumping of all waste except for dredged material, sewage sludge, fish waste from processing, manmade structures, inert or inorganic geological material, bulky items composed of inert materials that will not interfere with navigation or fishing, and CO_2 for sub-seabed sequestration in geological features (i.e. not on the deep sea floor). A permit is required for any dumping activity under the Protocol.

- Annex II outlines the assessments that must be undertaken prior to consideration for dumping. Specifically, the purpose of Annex II is to prevent waste from being produced (e.g. product lifecycle management), consider alternatives to ocean dumping once waste has been produced (e.g. recycling), and ensure that the material considered for ocean dumping is free from or contains limited hazardous materials (e.g. treatment). Annex II describes the processes and options to be considered prior to issuing a dumping permit.
- Annex III outlines the arbitral procedure to be used between contracting parties.

Permits are granted by the states that have ratified the Convention. States are required to establish an appropriate authority to issue permits as well as record the location, nature and quantities of the dumping of permitted materials that must be provided to the IMO. Permits are to be issued by the port state where the material is loaded onto ships or the flag state if material is loaded onto a ship from a port state not party to the Convention.

Incineration of industrial wastes and sewage sludge is prohibited. Other wastes may be incinerated at sea under special permit. Dumping of low-level radioactive wastes is prohibited.

Signatories to the Convention agree to cooperate with the IMO and other international organizations (e.g. UNEP) to assist member states with training, to provide the necessary equipment that complies with the Convention, and to facilitate the management of wastes to avoid their disposal at sea.

International Convention for the Control and Management of Ships' Ballast Water and Sediment

Ballast is a broad term for any liquid or solid carried by ships to increase draft, change trim, regulate stability or maintain structural integrity of the ship. Modern steel ships, including cruise ships, tankers and bulk cargo carriers, use water for ballast that is pumped in or out of separate compartments depending on ship's cargo and anticipated weather and sea-state conditions. Most modern ships are built with ballast water tanks that hold between 25 and 30 per cent of their dead weight tonnage. Upwards of 100,000 tonnes of ballast water can be contained in the largest bulk carriers (Branch, 2007).

The environmental, economic and public health implications of the use of water as ballast are significant. While the world's oceans are part of a single, large water body, marine biological communities are constrained to certain geographic areas. This separation is caused by land masses, currents, unfavourable temperature regimes or biological control (competition and predation). This geographic isolation has resulted in the evolution of countless unique marine species over the past several million years. Many of these species likely would flourish in other parts of the ocean if they were able to either move there through their own locomotion or if they were advected by currents, storms or as hosts on a species that is able to move long distances (Carlton, 1996; Davidson and Simkanin, 2012).

Modern marine shipping, however, provides another means (termed a 'vector') for the transport of these species into marine areas they would otherwise have slight chance of reaching. Ships may inadvertently move species between areas when species attach to a ship's hull, are captured when ballast water is pumped into ballast tanks or reside in a ship's sea chests (piping and chambers used to bring water on board a ship for ballast and other purposes such as cooling). The speed of modern ships combined with the cleanliness of modern ballast water tanks increases the likelihood of an organism surviving its voyage in ballast tanks. Furthermore, shipping now moves over 80 per cent of the world's commodities and approximately 3 to 5 billion tonnes of ballast water internationally each year. This provides ample opportunities for the movement of species across the oceans (Endresen *et al.*, 2004). It is estimated that shipping is responsible for at least a third of documented marine invasions and that at any given moment at least 7000 different species are being carried in ships' ballast tanks around the globe (Davidson and Simkanin, 2012).

While modern ships have grates and mesh on water intakes to prevent larger species from entering ballast tanks, these preventative measures are incapable of excluding viruses, bacteria, microbes and smaller zooplankton. In addition, most marine organisms (even sessile species attached to the ocean floor) have planktonic lifecycles where eggs, larvae and juveniles are available to enter ballast tanks and, once released at the ship's destination, may settle into their sessile adult phase (Chapter 1). The environmental and economic impact of introduced species from shipping is large: perhaps the most well-known invasions are the incursion of the North American comb jelly (*Mnemiopsis leidyi*) into the Black and Azov Seas and the introduction of the Eurasian zebra mussel (*Dreissena polymorpha*) into the Laurentian (North America) Great Lakes. *Mnemiopsis* was introduced into the Black Sea in the 1980s and has reached densities of 1kg of biomass per cubic metre (400 animals per cubic metre) and contributed to the collapse of commercial fisheries in the Black Sea. The zebra mussel was accidentally introduced into the Great Lakes in 1998 and now infects approximately 40 per cent of internal waterways in the US, requiring up to $100 million annually on control measures (globallast.imo.org).

Currently, the IMO is most concerned with the following invasives:

- Cholera (*Vibrio cholera*) (various strains)
- Cladoceran water flea (*Cercopagis pengoi*)
- Mitten crab (*Eriocheir sinensis*)
- Toxic algae (red/brown/green tides) (various species)
- Round goby (*Neogobius melanostomus*)
- North American comb jelly (*Mnemiopsis leidyi*)
- North Pacific seastar (*Asterias amurensis*)
- Asian kelp (*Undaria pinnatifida*)
- European green crab (*Carcinus maenas*)

While initial IMO guidelines on the control and management of ballast water were published in 1991 and subsequently revised in 1997, the International Convention for the Control and Management of Ships' Ballast Water and Sediment was adopted 13 February 2004 (IMO, 2004). The Conference was attended by 75 states, 1 IMO associate member, 2 governmental organizations and 18 non-government organizations. The Convention will enter into force 12 months after the date on which the Convention is signed by at least 30 countries representing at least 35 per cent of the world's fleet tonnage. To date, the Convention has been signed by 37 countries representing 30.32 per cent of the world's fleet tonnage (Wright, 2012).

The purpose of the Convention is to prevent, minimize and ultimately eliminate the transfer of harmful aquatic organisms and pathogens through the control and management of ships' ballast water and sediments (Article 2.1). Key articles of the Convention include:

- Article 3: Outlines the application of the Convention and instances where the Convention does not apply.
- Article 4: Requires parties to the Convention to adhere to the standards and requirements outlined in the Annex. It also requires parties to cultivate national policies, strategies and programmes for ballast water management.
- Article 5: Requires parties to establish sediment reception facilities where the cleaning and repair of ballast water tanks occurs.
- Article 6: Requires parties to cooperate on scientific and technical research and monitoring.
- Article 7: Requires parties to survey ships flying their flag or operating under their authority to ensure compliance with regulations in the Annex.
- Article 8: Violations of the Convention are the responsibility of the flag state to investigate and pursue.
- Article 9: Port states may inspect visiting ships but inspections are limited to the verification of a valid certificate, the inspection of the ballast water record book and the examination of the ship's ballast water.
- Article 10: Port states may warn, detain or exclude a ship that is detected to have violated the Convention.

The annex contains the regulations for the control of ships' ballast water and sediments. Key regulations include:

- Regulation B-1: Requirement for each ship to have a ballast water management plan. The plan includes safety procedures for ballast water management, how the ship will meet the regulations as outlined in the Annex, and detailed procedures for the disposal of sediments at sea and on shore.
- Regulation B-2: Requirement for each ship to have on board a ballast water record book that retains records for a period of two years.
- Regulation B-3: Establishes a tiered implementation timetable – based on the construction date of the ship – for the implementation of regulations.
- Regulation B-4: Outlines standards for ballast water exchange, stating that whenever possible ballast water be exchanged at least 200nmi from shore in depths at least 200m.
- Regulation C-1: Parties, individually or jointly, may develop measures that exceed the regulations in the Annex.
- Regulation C-2: Permits port states to identify areas where uptake of ballast water should not occur.
- Regulation D-1: Requires ships exchanging ballast water to meet a 95 per cent volumetric efficiency in the exchange.
- Regulation D-2: Requires ships to discharge less than ten viable organisms per cubic metre greater than or equal to 50 micrometres in minimum dimension and less than ten viable organisms per millilitre less than 50 micrometres in minimum dimension and greater than or equal to 10 micrometres in minimum dimension. Also sets out the indicator microbes to be used (cholera, *Escherichia coli*, intestinal enterococci) as a human health standard.
- Regulation D-3: Active substances used to treat ballast water must be approved by the IMO.
- Regulation E-1: Sets out a survey schedule for ships 400 gross tonnes or larger to which the Convention applies.

- Regulation E-5: Ballast water certificates are valid up to five years.

Basel Convention on the Control of Transboundary Movements of Hazardous Wastes and their Disposal

The Basel Convention on the Control of Transboundary Movements of Hazardous Wastes and their Disposal (the 'Basel Convention') was adopted 22 March 1989 and came into force in 5 May 1992. There are currently 175 parties to the Convention (UNEP, 1989).

The Convention resulted from the growing public awareness in the 1970s and 1980s of the dumping of toxic wastes from more developed to less developed nations who often lacked the capacity to effectively store and process these wastes. Africa, in particular, was a destination for much of these wastes. Hazardous waste exports received public attention in the 1980s with 'toxic trader' ships that moved wastes to Africa, Eastern Europe and other regions. Perhaps the most famous of these was the *Khian Sea*, which in 1986 was loaded with approximately 12 million kg of municipal incinerator and waste ash containing numerous toxic compounds. Over a period of 18 months the ship was renamed twice (*Felicia, Pelicano*), refused entry into at least 11 countries and dumped approximately 1800 tonnes of ash on a Haitian beach. The ship was eventually found empty in Singapore (Jaffe, 1995).

The purpose of the Convention is to 'protect human health and the environment against the adverse effects of hazardous wastes'. This is achieved primarily through controls on the transboundary movement of hazardous wastes. Objectives of the Convention are to: minimize the production of hazardous wastes and hazardous recyclable materials; and ensure hazardous materials are disposed in an environmentally responsible manner, preferably close to the source of production.

The Convention was amended at the third meeting of the Conference of the Parties on 22 September 1995 (UNEP, 1995a). Termed the 'Ban Amendment' it prohibits the export of all hazardous wastes covered by the Convention from more developed countries (OECD, EC, Liechtenstein) to less developed countries. Of the 90 states that are signatories to the Ban Amendment, 68 must ratify it before it enters, which has not yet occurred. However, many nations subscribe to the principles of the Ban Amendment and the EU fully complies with the Amendment under its Waste Shipment Regulation.

The Basel Convention is viewed as highly successful. Specific achievements are the development of technical guidelines for specific waste issues world-wide. Even though they are non-binding from an international legal perspective they are now accepted as national policy in most states.

Non-binding agreements

Global Programme of Action for the protection of the Marine Environment from Land-based Activities

The Global Programme of Action (GPA) for the protection of the Marine Environment from Land-based Activities is a UNEP-led programme adopted 3 November 1995 by 108 states. Termed the 'Washington Declaration', the GPA is not legally binding but rather a framework to protect the health, productivity and biodiversity of coastal and marine environments from land-based sources of pollution (UNEP, 1995b; Osborn and Datta, 2006). The GPA is concerned with nine specific impacts on coastal and marine environments:

sewage, persistent organic pollutants, radioactive substances, heavy metals, oils (hydrocarbons), nutrients, sediment mobilization, litter and the physical alteration/destruction of habitats. The Programme does not identify specific policy or legislative tools that states should adopt nor does it develop global 'one size fits all' solutions to various land-based threats. Instead, the GPA is designed to assist states in their efforts to protect and restore coastal and marine environments. Five primary goals underpin the Programme:

- Identify the sources and impacts of land-based sources of marine pollution.
- Identify priority problems for which action is needed.
- Set management objectives for these problem areas.
- Develop strategies to achieve these objectives.
- Evaluate the impacts of these strategies.

Specifically, the GPA encourages states and regions to develop national programmes of action (NPAs) to identify priorities, devise action plans and monitor performance. NPAs are specifically tailored to individual states or regions and may utilize regulatory, market-based, voluntary or other means to reduce impacts on marine environments. To date, 72 countries are developing or have completed their NPAs.

The GPA was reviewed in 2001, the outcome of which is the Montreal Declaration on the Protection of the Marine Environment from Land-based Activities. The Declaration focuses attention on municipal wastewater, ecosystem approaches to coastal and ocean governance, and financing of the GPA. The second review of the Programme was completed in 2006. The review recommended that the NPAs include valuations of ecosystem goods and services, integrate freshwater and coastal management, address wastewater issues and link GPA activities with the amelioration of poverty. Nutrients, wastewater and marine litter were the principal subjects tabled at the third review in 2012.

International law related to climate change

Since oceans cover 73 per cent of the earth's surface and comprise a volume of 1.3 billion square km, it is no surprise that the oceans are important in structuring and regulating the global climate. As well as storing and exchanging gases such as CO_2, oceans store and release heat, transfer heat horizontally and vertically, and provide water for the atmosphere (Rahmstorf and Richardson, 2009). Of the 26.4 gigatonnes of CO_2 released every year into the atmosphere by human activities, approximately 40 per cent is absorbed by the oceans, making them an integral buffer against anthropogenic climate change (IPCC, 2007).

While the effects of climate change on marine environments are not fully understood, climate change has significant potential implications to coastal and marine environments, including:

- Rising sea levels (0.09–0.88m by 2100) and their impact on coastal areas and island nations.
- Melting of polar sea ice resulting in changes to marine trophic structure. In addition, the loss of sea ice reduces the albedo effect (where incoming solar radiation is reflected back into space).
- Changes to global ocean circulation patterns and the effects on terrestrial climates, marine trophic structure and commercial fisheries.

- Increased ocean acidification resulting in diminished shell formation for calcareous organisms (organisms with shells).
- Increased frequency and severity of extreme weather events due to ocean warming.
- Migration of biological communities towards polar areas as a result of increasing ocean temperatures.
- Release of seabed methane deposits as a result of increasing ocean temperatures.

The global response to both climate change mitigation (reducing the releases of anthropogenic greenhouse gases) and adaptation (preparing for the impacts of climate change) has significant marine policy implications for coastal, fisheries, public safety and biodiversity management. Accordingly, an appreciation of the international approaches to mitigating and adapting to climate change and how these approaches affect the marine environment is essential.

United Nations Framework Convention on Climate Change

While climate change was recognized by the UN General Assembly as a potential threat to the global environment and human population as early as 1989 (UN, 1989), the first standalone international agreement on climate change was the UNFCCC. It opened for signature 9 May 1992 at the UNCED in Rio de Janeiro and entered force on 21 March 1994 (UNFCC, 1992). Presently, the Convention has 195 parties.

The Convention's objective (Article 2) is 'stabilization of greenhouse gas concentrations in the atmosphere at a level that would prevent dangerous anthropogenic interference with the climate system'. Furthermore, Article 2 states that 'such a level should be achieved within a time-frame sufficient to allow ecosystems to adapt naturally to climate change, to ensure that food production is not threatened, and to enable economic development to proceed in a sustainable manner'.

Article 3 of the Convention outlines the principles intended to guide parties, which include:

- More developed states will lead in the mitigation of and adaptation to climate change (3.1).
- Less developed states, and especially those vulnerable to climate change, will be given special consideration (3.2).
- Scientific uncertainty should not be used to postpone decisions (3.3).
- Parties have a right to sustainable development and should use economic cooperation and trade to enable sustainable development and to address climate change (3.4 and 3.5).

Article 4.1 of the Convention commits all parties to the following key actions:

- Establish inventories of greenhouse gas emissions by source (4.1a).
- Address and report on anthropogenic emissions (4.1b).
- Develop and share mitigation technologies and practices (4.1c).
- Promote the conservation and enhancement of sinks and reservoirs (4.1d).
- Prepare climate change adaptation plans (4.1d).
- Incorporate climate change considerations into national policies (4.1e).

- Promote and share knowledge on climate science (4.1f, 4.1g).
- Promote, and cooperate on, education and awareness related to climate change (4.1i).

Article 4.2 of the Convention commits developed state parties listed in Annex I of the Convention to the specific strategic actions:

- Develop and implement national policies on limiting greenhouse gas emissions and protecting and enhancing greenhouse gas sinks and reservoirs (4.2a).
- Reduce the emission of greenhouse gases to 1990 levels (4.2b).
- Use best available scientific information (4.2c).

The directives in Articles 4.3–4.10 outline the specific responsibilities of developed parties in assisting less developed parties in their efforts to mitigate and adapt to climate change.
    The remainder of the Convention is summarized below:

- Article 5: Commits parties to undertake research and monitoring and to assist developing states to this end.
- Article 6: Commits parties to education, training and public awareness.
- Article 7: Establishes a conference of the parties (COP) and outlines the duties of the COP.
- Article 8: Establishes a secretariat and its functions.
- Article 9: Establishes a subsidiary body to provide scientific and technological advice.
- Article 10: Establishes a subsidiary body to implement the Convention.
- Article 11: Establishes a financial mechanism to fund grants or the transfer of technology.
- Article 12: Identifies reporting requirements for parties to the Convention.
- Articles 13–16: Provide housekeeping details common to most international agreements.

Kyoto Protocol to the United Nations Framework Convention on Climate Change

The Kyoto Protocol was adopted on 11 December 1997 and entered force on 16 February 2005. While the UNFCCC only encourages parties to stabilize and reduce greenhouse gas emissions, the Kyoto Protocol commits 37 developed parties to 5 per cent reductions to their 1990 levels of greenhouse gas emissions over the 2008–2012 period (UNFCC, 1997). Certain states have greater than 5 per cent greenhouse gas reduction targets including Iceland (10 per cent) and Australia (8 per cent).
    Less developed states are not bound by the Protocol because of their disproportionately small contributions of greenhouse gasses relative to more developed states. This principle is stated in the Protocol as 'common but differentiated responsibilities'. The US has not ratified the Protocol and Canada withdrew from the Protocol in 2011.
    The Protocol functions in two ways: first, by setting limits on the overall amounts of greenhouse gases that a state may emit, it monetizes emissions (creates an economic value). Emissions that were previously unregulated now have economic value. This creates the opportunity to buy and sell emissions domestically for the industrialized state parties to the Convention. Second, the Protocol creates market-based mechanisms where states may meet

their emissions targets by reducing emissions in other states. The intent of these mechanisms is to reduce greenhouse gas emissions in less developed states where the costs of reduction are likely to be less than in developed nations and where the benefits to less developed states are likely to be greater.

The Protocol contains three market-based mechanisms:

- Emissions trading (Article 17): Allows states that have exceeded their greenhouse gas reduction targets to sell the difference between their targets and actual emissions to other states. Known as the 'carbon market', emissions trading also allows states to consider changes to land use/land cover or changes in practices that reduce CO_2 as equivalent to direct reductions on emissions.
- Clean Development Mechanism (CDM) (Article 12): Allows states to undertake emissions reduction projects in other states in order to assist with meeting their own greenhouse gas targets (e.g. funding a switch from fossil fuels to renewable energy for electricity generation).
- Joint implementation (Article 6): Allows states with emission reduction targets to trade emissions between each other.

Greenhouse gases covered in the Protocol include:

- Carbon dioxide (CO_2).
- Methane (CH_4).
- Nitrous oxide (N_2O).
- Hydrofluorocarbons (HFCs).
- Perfluorocarbons (PFCs).
- Sulphur hexafluoride (SF_6).

A number of states have explored ocean fertilization and direct injection of CO_2 into the deep oceans as potential methods to sequester CO_2 and meet their greenhouse gas reduction targets. Ocean fertilization generally proposes the addition of biologically available iron or urea (nitrogen) to increase primary production, while deep injection proposes to inject liquefied CO_2 into the deep oceans or in geological structures (Freestone and Rayfuse, 2008). To some researchers/lawmakers, the UNFCCC's direction to consider stabilization of climate through the use of sinks suggests that ocean fertilization may be an avenue for states to meet their reduction targets. Ocean fertilization, however, may be considered a 'pollutant' under the LDC, and potentially may contravene the LOSC if fertilization potentially harms marine living resources.

Important marine regional agreements

The proliferation of international law in the twentieth century demonstrates that states could, without difficulty, enter into agreements with one another on various issues. While most states have many bilateral agreements with neighbouring jurisdictions (Table 2.2), the concept of states within a defined geographic area subscribing to an action plan to address one or more issues was rarely seen in areas outside of the EU until the 1970s.

UNEP Regional Seas Programme

The UNEP Regional Seas Programme was initiated in 1974 to address degradation of coastal and marine areas. The Programme is administered by UNEP's Oceans and Coastal Areas Programme Activity Centre (OCA/PAC) in Nairobi, Kenya. The objectives of the Programme (www.unep.org/regionalseas/default.asp) are to:

- control marine pollution and to protect and manage aquatic resources through the advancement of regional conventions, guidelines and action plans;
- understand and monitor pollution and its impacts on human health and marine resources;
- cooperate on the sustainable development of coastal and marine resources;
- support developing countries in the protection, development and management of marine and coastal resources.

The Programme considers both the socioeconomic and environmental aspects within a geographic area and is therefore one of the earliest attempts at region-based sustainable development through international cooperation.

Early Regional Seas agreements focused almost exclusively on marine pollution and dumping as well as scientific research and large-scale monitoring. While entrained seas (e.g. Mediterranean) were found to contain significant levels of various types of pollution, the open oceans were generally found to be much cleaner and, by the 1990s, pollution levels in most marine areas were stable. This suggested that the Programme might benefit from a revised focus. As a result, UNEP revised its Regional Seas policies in 1983 to focus on integrated coastal area management (see Chapter 10) with a specific emphasis on land-based sources of pollution and the human activities responsible for degradation of coastal environments. The revised Programme also focused on pollution control measures, environmental impact assessment, training, public awareness and collaboration between jurisdictions (Akiwumi and Melvasalo, 1998). Further fomenting change in the Regional Seas Programme is the Global Programme of Action for the protection of the Marine Environment from Land-based Activities (discussed previously). Its purpose is to reduce, control and/or eliminate marine degradation from land-based sources.

Regional Seas programmes are initiated by states through a formal request to UNEP. National and technical experts are tasked to develop an action plan that is enabled (in most cases) through a regional convention with associated protocols specific to particular regional issues. Regional Seas conventions and their protocols, therefore, are legal agreements. Many of the conventions have multiple protocols addressing different threats and, in many cases, the development of a Regional Seas Programme begins with the simultaneous adoption of a regional convention and subsequent protocols. The action plans are financed by participating states, although states can request that UNEP administer the funding under the Financial Rules and Regulations of the United Nations. The UNEP Environment Fund also provides seed funding to support start-up costs (Verlaan and Khan, 1996).

Currently, 143 countries participate in 13 Regional Seas programmes (or closely associated partner programmes) that represent all the populated marine coasts of the world (Figure 5.1). The Programme is unique in the UN in that it is the only UN body with regional, autonomous structures. In addition, five partner programmes consisting of the Antarctic, Arctic, Baltic Sea, Caspian Sea and North-East Atlantic Regions, while not

Figure 5.1 UNEP Regional Seas programmes

Note: Programme names are as follows:

1 Arctic	10 North-East Pacific
2 Antarctic	11 North-West Pacific
3 Baltic	12 Red Sea and Gulf of Aden
4 Black Sea	13 ROPME Sea Area
5 Caspian	14 South Asian Seas
6 South Asian Seas	15 South-East Pacific
7 Eastern Africa	16 Pacific
8 Mediterranean	17 Western Africa
9 North-East Atlantic	18 Wider Caribbean

under the auspices of UNEP, are considered part of the Regional Seas Programme. Several Regional Seas programmes merit further discussion.

UNEP Regional Seas Programme: Mediterranean

The Mediterranean Sea covers 2.5 million square km, averaging a depth of 1500m and bounded by 46,000km of coastline. With only two connections to the global oceans at the Strait of Gibraltar (14km wide) and the Suez Canal (less than 1km wide) turnover of the sea is likely greater than a century. Although the Mediterranean comprises only 0.82 per cent of the surface area and 0.32 per cent of the volume of the world's oceans, the sea is estimated to contain at least 7 per cent of all known marine species. Moreover, approximately 30 per cent of the area's fauna species are endemic to the Mediterranean (UNEP, 2012). Meanwhile, 460 million people reside in the Mediterranean's watersheds and this population is expected to grow to 520 million by 2025.

In 1974, the UNEP Governing Council endorsed the Mediterranean Sea as a high-priority area for the protection of living resources and the prevention of pollution. The Mediterranean Action Plan (MAP) was in place a year later. It called for integrated planning and development, the coordination of pollution research and monitoring, the elaboration of robust financing mechanisms and the establishment of a convention and related protocols (Vallega, 2002).

The Convention for the Protection of the Mediterranean Sea Against Pollution (the 'Barcelona Convention') was adopted 16 February 1976 and entered into force on 12 February 1978. Fourteen Mediterranean states and the EU signed the Convention. The Convention was revised on 9–10 June 1995 as the Convention for the Protection of the Marine Environment and the Coastal Region of the Mediterranean but has not yet entered into force. In addition, the MAP was revised and termed MAP Phase II with 21 participating states focusing on coastal management and biodiversity protection (Vallega, 2002; UNEP, 2005).

The Convention's primary intent is to protect the Mediterranean from pollution through a system of protocols (Article 4). While all types and sources of pollution are covered by the Convention, specific Articles address pollution occasioned by: dumping from ships and aircraft (Article 5), discharges from ships (Article 6), exploration and exploitation of the continental shelf and the seabed and its subsoil (Article 7), and land-based sources (Article 8). Cooperation on pollution emergencies (Article 9), monitoring (Article 10) and scientific cooperation (Article 11) are also treated in distinct articles.

The Convention is amended by the following seven protocols:

- Dumping (adopted 1976, entered into force 1978 revised 1995).
- Prevention and emergencies (adopted 1976, entered into force 1978; revised version adopted 2002, revised version entered into force 2004).
- Land-based pollution (adopted 1980, entered into force 1983; revised version adopted 1996, revised version entered into force 2008).
- Specially protected areas and biodiversity (adopted 1982, entered into force 1986; revised version adopted 1995, revised version entered into force 1999).
- Seabed development and offshore exploration (adopted 1994, entered into force 2011).
- Hazardous wastes (adopted 1996, entered into force 2008).
- Integrated coastal zone management (adopted 2008, entered into force 2008).

By uniting states to confront mutual problems, the Convention has achieved success and there are now 18 signatories. Key achievements include the creation of the Mediterranean Strategy for Sustainable Development, which integrates the environment and development (UNEP, 2005) and the Strategic Action Plan for the Conservation of Marine and Coastal Biodiversity in the Mediterranean (SAP BIO) (sapbio.rac-spa.org/).

UNEP Regional Seas Programme: Kuwait/ROPME

The Kuwait Regional Convention for Cooperation on the Protection of the Marine Environment from Pollution ('Kuwait Convention') was adopted on 23 April 1978 by a regional conference convened in Kuwait and attended by Bahrain, Iran, Iraq, Kuwait, Oman, Qatar, Saudi Arabia and the United Arab Emirates. The Convention covers the Persian Gulf and Gulf of Oman. The eight nations bordering these seas produce 25 per cent of the world's oil, have experienced three major wars in the past 25 years and the world's largest oil spill until the Deep Water Horizon accident (de Mora et al., 2010). Furthermore, the Persian Gulf is unique among regional seas because it is very shallow, averaging 36m depth, and much of its marine biota lives near the upper range of tolerances to salinity and temperature. The region also has significant exposure to human induced threats – including externalities of oil and gas production and transportation – alien species and overharvesting of marine resources. Another unique consideration of the Persian Gulf that increases attention on marine water quality is that a significant portion of its coastal residents obtain their water through desalinization (Hamza and Munawar, 2009).

Adopted with the Convention is the Action Plan for the Protection of the Marine Environment and the Coastal Areas of Bahrain, Iran, Iraq, Kuwait, Oman, Qatar, Saudi Arabia and the United Arab Emirates. Further bolstering the protection of the area is the Protocol concerning Regional Cooperation in Combating Pollution by Oil and other Harmful Substances in Cases of Emergency. Other protocols deal with seabed development and offshore activity (adopted 1989, entered into force 1990), land-based pollution (adopted 1990, entered into force 1993) and hazardous wastes (adopted 1998, entered into force 2003).

Article XVI of the Convention established ROPME in 1979 to implement the Action Plan and, in 1982, the establishment of the ROPME Secretariat in Kuwait. ROPME, while part of the Regional Seas Programme, operates independently of UNEP. Broadly, the Convention's aims are to prevent damage to the marine environment and protect human health within a context of sustainability.

UNEP Regional Seas Programme: West and Central Africa

The Abidjan Convention for Cooperation in the protection and Development of the Marine and Coastal Environment of the West and Central African Region ('Abidjan Convention') was adopted in 1981 and entered into force in 1984 (UNEP, 1981a). Sometimes termed 'WACAF', the Convention has 23 contracting parties and 10 additional ratifying parties. In addition, the Action Plan for the Protection and Development of the Marine Environment and Coastal Areas of the West and Central African Region and the Protocol Concerning Cooperation in Combating Pollution in Cases of Emergency were also adopted and entered into force at the same time as the Convention.

The primary objectives of the Action Plan are to:

- undertake environmental assessment and scientific research into coastal and marine areas (Article 4.1);
- promote methods for sustainable development (Article 4.2);
- foster regional agreements for the protection and development of marine and coastal areas (Article 4.3);
- cultivate financial and institutional arrangements for implementing the Action Plan (Article 4.4).

Between 1982 and 1998 the WACAF region received $560,000 to administer the Action Plan. As UNEP contributions were scaled down, states were unable to fund their participation in the Action Plan and by 1993 only two projects were completed: a workshop on land-based pollutants and the development of integrated coastal zone management strategies for three states (Peart et al., 1999).

UNEP Regional Seas Programme: South-East Pacific

The Convention for the Protection of the Marine Environment and Coastal Area of the South-East Pacific (Lima Convention) and Action Plan for the Protection of the Marine Environment and Coastal Areas of the South-East Pacific were adopted 12 November 1981 and entered force in 1986 (UNEP, 1981b). Subsequent protocols and agreements include:

- Agreement on Regional Cooperation in Combating Pollution in the South-East Pacific by Hydrocarbons and other Harmful Substances in cases of Emergency (signed 1981).
- Complementary Protocol on the Agreement for Regional Cooperation in Combating Pollution in the South-East Pacific by Hydrocarbons and other Harmful Substances in Cases of Emergency (adopted 1983, entered into force 1987).
- Protocol for the Protection of the South-East Pacific Against Pollution from Land-based Sources (adopted 1983, entered into force 1986).
- Protocol for the Conservation and Management of Protected Marine and Coastal Areas of the South-East Pacific (adopted 1989, entered into force 1995).
- Protocol for the Protection of the South-East Pacific from Radioactive Pollution (adopted 1989, entered into force 1995).
- Protocol on the Regional Program for the Study of the El Niño phenomenon in the South-East Pacific (ERFEN) (adopted 1992).
- Proposed Protocol on the Assessment of the Environmental Impact on the Marine Environment and Coastal Areas the South-East Pacific (signed).
- Proposed Protocol to Prohibit Transboundary Movements of Hazardous Wastes and their Disposal to the South-East Pacific (signed).

The Convention encompasses the entire Pacific coast of South America and includes Panama, Columbia, Ecuador, Peru and Chile. Its goals are numerous: prevent pollution from all sources (Article 4), prevent, reduce and control erosion from coastal areas (Article 5), cooperate on responding to marine pollution emergencies (Article 6), monitor pollution (Article 7), undertake environmental impact assessment (Article 8), exchange information (Article 9) and cooperate on scientific investigations (Article 10). The CPPS is responsible for implementing the Convention and Action Plan, and the secretariat is headquartered in Quito, Ecuador.

UNEP Regional Seas Programme: Red Sea and Gulf of Aden

The Programme for the Environment of the Red Sea and Gulf of Aden (PERSGA) was initiated in 1974 to address coastal development, overfishing, marine transportation and marine emergencies. The programme secretariat is headquartered in Jeddah, Saudi Arabia.

The Regional Convention for the Conservation of the Red Sea and Gulf of Aden Environment ('Jeddah Convention') was adopted 15 February 1982 and entered into force 20 August 1985 (UNEP, 1982). Parties to the Convention include Djibouti, Egypt, Jordan, Saudi Arabia, Somalia, Sudan and Yemen. In addition to the Jeddah Convention, the regional conference that produced the Convention adopted the Action Plan for the Conservation of the Marine Environment and Coastal Areas in the Red Sea and Gulf of Aden and Protocol Concerning Regional Cooperation in Combating Pollution by Oil and Other Harmful Substances in Cases of Emergency. In September, 1985 PERSGA approved the following additional protocols:

- Protocol Concerning the Conservation of Biological Diversity and the Establishment of a Network of Protected Areas in the Red Sea and Gulf of Aden.
- Protocol Concerning the Protection of the Marine Environment from Land-based Activities in the Red Sea and Gulf of Aden.
- Protocol Concerning the Exchange of Personnel and Equipment in Case of Marine Emergency.

The two most recent protocols were adopted in 2005: Protection of Marine Environment from Land Based Activities and Biodiversity and Establishment of Marine Protected Areas (www.persga.org).

UNEP Regional Seas Programme: Wider Caribbean

The Convention for the Protection and Development of the Marine Environment of the Wider Caribbean Region ('Cartagena Convention') was adopted 24 March 1983 and entered into force 11 October 1986 (UNEP, 1983). Twenty-eight island and continental states are party to the Convention, which builds on UNEP's Caribbean Environment Program (CEP) and Caribbean Action Plan adopted in 1981. The Convention included the Protocol concerning Cooperation in Combating Oil Spills in the Wider Caribbean Region, and is administered by the CEP in Kingston, Jamaica. It is supplemented by the Specially Protected Areas and Wildlife Protocol (1990), the Land-based Sources of Marine Pollution Protocol (1999) and the Specially Protected Areas and Wildlife (2000).

UNEP Regional Seas Programme: Eastern Africa

The Convention for the Protection, Management and Development of the Marine and Coastal Environment of the Eastern African Region ('Nairobi Convention') was adopted in Nairobi on 21 June 1985 and entered into force on 30 May 1996 (UNEP, 1985). Other plans and protocols adopted concurrently and entering into force at the same time were the Action Plan for the Protection, Management and Development of the Marine and Coastal Environment of the Eastern African Region ('EAP'), Protocol Concerning Protected Areas and Wild Fauna and Flora in the Eastern African Region and the Protocol Concerning Cooperation in Combating Marine Pollution in Cases of Emergency in the Eastern African Region. Six states were required to ratify or accede to the Convention before it came into force. UNEP led the initial development of the programme.

The Convention is a comprehensive compact covering the protection, management and development of the marine and coastal environment. Nine items comprised a priority list, including: pollution-related commitments included controlling pollution from ships (Article 5), dumping (Article 6), land-based sources (Article 7), seabed activities (Article 8), airborne sources (Article 9) and in cases of emergencies (Article 11). Additional commitments are related to specially protected areas (Article 10), environmental damage from engineering activities (Article 12), environmental assessment (Article 13) and scientific cooperation (Article 14).

Despite its sensible intentions, the Convention has been criticized for focusing attention on pollution prevention and single-species management at the expense of integrated coastal zone management. A Regional Trust Fund to support implementation of the Action Plan was established in 1985; however, over the following ten years the $700,000 in contributions from the parties was only 25 per cent of the total amount of the initial commitment (Peart *et al.*, 1999). The Swedish and Belgian governments also financially supported implementation.

The 2010 Conference of Plenipotentiaries and the Sixth Conference of Parties (COP6) to the Nairobi Convention adopted a replacement for the Convention titled the Amended Nairobi Convention for the Protection, Management and Development of the Marine and Coastal Environment of the Western Indian Ocean. Additionally, the Protocol for the Protection of the Marine and Coastal Environment of the Western Indian Ocean from Land-based Sources and Activities was framed, the contracting parties being Comoros, France, Kenya, Madagascar, Mauritius, Mozambique, Seychelles, Somalia, Tanzania and the Republic of South Africa.

UNEP Regional Seas Programme: South Pacific

The Convention for the Protection of Natural Resources and Environment of the South Pacific Region ('Noumea Convention') was adopted 24 November 1986 and entered into force 22 August 1990 (SPREP, 1986). The Convention encompasses the territorial waters and EEZs of 24 state parties, which includes 7500 islands and approximately a third of the world's languages (Table 5.2). The Convention also contains protocols for dumping and pollution emergencies. It is administered by the Secretariat of the Pacific Community, head-quartered in Anse Vata, Noumea.

Key Articles of the Convention include:

- Article 3: Any party may add areas under its jurisdiction within the Pacific Ocean between the Tropic of Cancer and 60 degrees south latitude and between 130 degrees east longitude and 120 degrees west longitude.
- Article 4: Parties will form bilateral or multilateral agreements – including regional or sub-regional agreements – for the protection, development and management of the marine and coastal environment. Protected areas (Article 14), pollution emergencies (Article 15) and environmental impact assessment (Article 16) are incorporated in these agreements.

Table 5.2 State parties to the Convention for the Protection of Natural Resources and Environment of the South Pacific Region (Noumea Convention), including overseas parties

State parties		
American Samoa	Australia	Cook Islands
Federated States of Micronesia	French Polynesia	Fiji
Guam	Kiribati	Marshall Islands
Nauru	New Caledonia	New Zealand
Niue	Northern Mariana Islands	Palau
Papua New Guinea	Pitcairn Islands	Samoa
Solomon Islands	Tokelau	Tonga
Tuvalu	Vanuatu	Wallis and Futuna
Overseas parties		
France	United States	

- Articles 5, 8–11: Parties will work to prevent, reduce and control pollution and ensure sound environmental management and development.
- Article 13: Parties will work to control coastal erosion as a result of development.
- Articles 17–18: Parties will share information, cooperate on research programmes and provide assistance to island developing states and territories.

The Convention also set the foundation for establishing the South Pacific Regional Environmental Programme (SPREP) in 1993. The Programme was renamed the Pacific Regional Environment Programme in 2004, however, it is still referred to as SPREP when abbreviated. SPREP has four thematic areas: biodiversity and ecosystems management; climate change; environmental monitoring and governance; and waste management and pollution control.

UNEP Regional Seas Programme: Black Sea

The 20 states located in the Black Sea watershed input a combined 350 cubic km of river water annually into the Black Sea. The sea is in turn connected to the world's oceans through the shallow, 35km-wide Istanbul Strait. As a result, the Black Sea is highly entrained and therefore sensitive to a number of human impacts. Eutrophication/nutrient enrichment, chemical and oil pollution, overfishing, habitat loss, changes in biodiversity and introduced species are all significant threats to the area. The unique nature of the Black Sea called for a convention focusing on its issues; yet those states – Bulgaria, Georgia, Romania, Russian Federation, Turkey and Ukraine – surrounding its waters needed to remain abiding to the 1976 Barcelona Convention.

The Convention on the Protection of the Black Sea Against Pollution ('Bucharest Convention') was adopted 21 April 1992 and entered into force 15 January 1994 (BSC, 1992). Parties to the Convention include Bulgaria, Georgia, Romania, Russian Federation, Turkey and Ukraine, and the Convention is supported by a permanent secretariat based in Istanbul, Turkey. It contains three protocols on the control of land-based pollution, dumping and pollution emergencies. The primary purpose of the Convention is to reduce pollution in the Black Sea and the Programme developed the 1996 Strategic Action Plan for the Environmental Protection and Rehabilitation of the Black Sea that was updated in 2009 (BSC, 2009).

Chapter summary

- Threats to biodiversity can be broadly categorized as a result of overharvesting, pollution, habitat loss, introduced species, ocean acidification and global climate change.
- The first international marine treaties were established to address the exploitation of fish stocks and marine mammals.
- The LOSC affirms that states have the rights to exploit their living marine resources in their sovereign areas. It also requires states to protect and preserve the marine environment.
- The LOSC requires that states: control pollution of the marine environment from any source; preserve rare or fragile ecosystems as well as protect habitats of depleted, endangered or threatened marine species; and prevent the spread of alien (introduced) species.
- 143 states participate in 13 Regional Seas programmes that address the degradation of coastal and marine areas.

References

Akiwumi, P. and Melvasalo, T. (1998) 'UNEP's regional seas programme: Approach, experience and future plans', *Marine Policy*, 22 (3), 229–234.

BSC (Black Sea Commission) (1992) *Convention on the Protection of the Black Sea Against Pollution*, Istanbul, Turkey, www.blacksea-commission.org/_convention-fulltext.asp, accessed 10 December 2012.

BSC (2009) *Strategic Action Plan for the Environmental Protection and Rehabilitation of the Black Sea*, Istanbul, Turkey, www.blacksea-commission.org/_bssap2009.asp, accessed 10 December 2012.

Branch, A. E. (2007) *Elements of Shipping*, Routledge, Milton Park.

Carlton, J. T. (1996) 'Marine bioinvasions: The alteration of marine ecosystems by nonindigenous species', *Oceanography*, 9 (1), 36–43.

Carson, R. (1962) *Silent Spring*, Houghton Mifflin, NY.

CBD (Convention on Biological Diversity) (1995) *Conservation and Sustainable Use of Marine and Coastal Biological Diversity*, Montreal, Canada, www.cbd.int/decision/cop/?id=7083, accessed 7 December 2012.

CBD (1998) *COP 4 Decision IV/5: Conservation and Sustainable Use of Marine and Coastal Biological Diversity, including a Programme of Work*, www.cbd.int/decision/cop/?id=7128, accessed 7 December 2012.

CBD (2000a) *The Jakarta Mandate – From Global Consensus to Global Work*, Montreal, Canada, www.cbd.int/doc/publications/jm-brochure-en.pdf, accessed 7 December 2012.

CBD (2000b) *Cartagena Protocol on Biosafety*, Montreal, Canada, bch.cbd.int/protocol/text/, accessed 7 December 2012.

CBD (2006) *COP 8 Decision VIII/22: Marine and Coastal Biological Diversity: Enhancing the Implementation of Integrated Marine and Coastal Area Management*, Montreal, Canada, www.cbd.int/decision/cop/?id=11036, accessed 7 December 2012.

CBD (2008) *COP 9 Decision IX/20: Marine and Coastal Biodiversity*, Montreal, Canada, www.cbd.int/decision/cop/?id=11663, accessed 7 December 2012.

CBD (2010) *Nagoya Protocol on Access and Benefit-sharing*, www.cbd.int/abs/, accessed 7 December 2012.

Davidson, I. C. and Simkanin, C. (2012) 'The biology of ballast water invasions 25 years later', *Biological Invasions*, 14, 9–13.

de Mora, S., Tolosa, I., Fowler, S. W., Villeneuve, J.-P., Cassi, R. and Cattini, C. (2010) 'Distribution of petroleum hydrocarbons and organochlorinated contaminants in marine biota and coastal sediments from the ROPME Sea Area during 2005', *Marine Pollution Bulletin*, 60 (12), 2323–2349.

Endresen, O., Behrens, H. L., Brynestad, S., Andersen, A. B. and Skjong, R (2004) 'Challenges in global ballast water management', *Marine Pollution Bulletin*, 48 (7–8), 615–623.

Freestone, D. and Rayfuse, R. (2008) 'Ocean iron fertilization and international law', *Marine Ecology Progress Series*, 364 (213–218), 227–233.

Gray, J. S. (1997) 'Marine biodiversity: Patterns, threats and conservation needs', *Biodiversity and Conservation*, 6, 153–175.

Hamza, W. and Munawar, M. (2009) 'Protecting and managing the Arabian Gulf: Past, present and future', *Aquatic Ecosystem Health & Management*, 12 (4), 429–439.

Hansen, J., Sato, M. and Ruedy, R. (2012) 'Perception of climate change', *Proceedings of the National Academy of Sciences*, 10.1073/pnas.1205276109.

Hardin, G. (1968) 'Tragedy of the commons', *Science*, 162 (3859), 1243–1248.

IMO (International Maritime Organization) (1972) *Convention on the Prevention of Marine Pollution by Dumping of Wastes and Other Matter*, IMO, London, treaties.un.org/doc/Publication/UNTS/Volume%201046/volume-1046-I-15749-English.pdf, accessed 7 December 2012.

IMO (1996) 1996 Protocol to the Convention on the Prevention of Marine Pollution by Dumping of Wastes and Other Matter, 1972, IMO, London, www.imo.org/OurWork/Environment/

SpecialProgrammesAndInitiatives/Pages/London-Convention-and-Protocol.aspx, accessed 25 October 2013.

IMO (2004) *International Convention for the Control and Management of Ships' Ballast Water and Sediments*, IMO, London, www.ecolex.org/server2.php/libcat/docs/TRE/Multilateral/En/TRE001412.pdf, accessed 7 December 2012.

IPCC (Intergovernmental Panel on Climate Change) (2007) *Climate Change 2007: The Physical Science Basis*, Contribution of Working Group I to the Fourth Assessment Report of the Intergovernmental Panel on Climate Change, Cambridge University Press, Cambridge.

Jaffe, D. (1995) 'The international effort to control the transboundary movement of hazardous waste: The Basel and Bamako Conventions', *ILSA Journal of International & Comparative Law*, 123, 123–137.

Norse, E. A. (ed.) (1993) *Global Marine Biodiversity: A Strategy for Building Conservation into Decision Making*, Island Press, Washington, DC.

NRC (National Research Council) (1995) *Understanding Marine Biodiversity*, National Academy Press, Washington, DC.

Osborn, D. and Datta, A. (2006) 'Institutional and policy cocktails for protecting coastal and marine environments from land-based sources of pollution', *Ocean & Coastal Management*, 49 (9–10), 576–596.

Pauly, D., Christensen, V., Dalsgaard, J., Froese, R. and Torres F. (1998) 'Fishing down marine food webs', *Science*, 279 (5352), 860–863.

Peart, R., Rich, J. and Masia, M. (1999) 'Sub-Saharan African multilateral programmes for coastal management: How well have they performed?', *Ocean & Coastal Management*, 42 (9), 791–813.

Rahmstorf, S. and Richardson, K. (2009) *Our Threatened Oceans*, Haus Publishing, London.

Roberts, C. (2007) *The Unnatural History of the Sea*, Island Press, Washington, DC.

Roberts, C. (2013) *The Ocean of Life: The Fate of Man and the Sea*, Penguin Books, New York.

Roff, J. C. and Zacharias, M. A. (2011) *Marine Conservation Ecology*, Earthscan, London.

SPREP (Secretariat of the Pacific Regional Environment Programme) (1986) *Convention for the Protection of the Natural Resources and Environment of the South Pacific Region*, Apia, Samoa www.sprep.org/attachments/NoumeConventintextATS.pdf, accessed 8 December 2012.

Thorne-Miller, B. and Earle, S. A. (1999) *The Living Ocean: Understanding and Protecting Marine Biodiversity*, Island Press, Washington, DC.

UN (United Nations) (1989) 'Protection of global climate for present and future generations of mankind', United Nations General Assembly, A/RES/44/207, www.un.org/documents/ga/res/44/a44r207.htm, accessed 25 October 2013.

UN (1992) *Convention on Biological Diversity*, treaties.un.org/doc/Treaties/1992/06/19920605%2008-44%20PM/Ch_XXVII_08p.pdf, accessed 7 December 2012.

UNEP (United nations Environment Programme) (1981a) *Convention for Cooperation in the Protection and Development of the Marine and Coastal environment of the West and Central African Region and protocol*, Nairobi, Kenya, abidjanconvention.org/index.php?option=com_content&view=arti-cle&id=51&Itemid=169, accessed 7 December 2012.

UNEP (1981b) *Convention for the Protection of the Marine Environment and Coastal Area of the South-East Pacific*, Nairobi, Kenya, sedac.ciesin.org/entri/texts/marine.environment.coastal.south.east.pacific.1981.html, accessed 8 December 2012.

UNEP (1982) *Regional Convention for the Conservation of the Red Sea and Gulf of Aden*, Nairobi, Kenya, www.unep.ch/regionalseas/main/persga/convtext.html, accessed 7 December 2012.

UNEP (1983) *Convention for the Protection and Development of the Marine Environment in the Wider Caribbean Region*. Nairobi, Kenya, www.cep.unep.org/cartagena-convention/cartagena-conven-tion.pdf, accessed 8 December 2012.

UNEP (1985) *Convention for the Protection, Management and Development of the Marine and Coastal Environment of the Eastern African Region*, Nairobi, Kenya, www.unep.org/NairobiConvention/docs/English_Nairobi_Convention_Text.pdf, accessed 8 December 2012.

UNEP (1989) *Basel Convention on the Control of Transboundary Movements of Hazardous Wastes and their Disposal*, Nairobi, Kenya, treaties.un.org/doc/Treaties/1992/05/19920505%2012-51%20PM/Ch_XXVII_03p.pdf, accessed 7 December 2012.

UNEP (1995a) *Amendment to the Basel Convention on the Control of Transboundary Movements of Hazardous Wastes and their Disposal*, Nairobi, Kenya, treaties.un.org/doc/Treaties/1995/09/19950922%2012-54%20PM/Ch_XXVII_03_ap.pdf, accessed 7 December 2012.

UNEP (1995b) *Global Programme of Action for the protection of the Marine Environment from Land-based Activities*, Nairobi, Kenya, www.gpa.unep.org/downloads/doc_download/55-gpa-full-text.html, accessed 7 December 2012.

UNEP (2005) *Mediterranean Strategy for Sustainable Development: A Framework for Environmental Sustainability and Shared Prosperity*, Nairobi, Kenya, 195.97.36.231/acrobatfiles/05WG270_Inf13_Eng.pdf, accessed 7 December 2012.

UNEP (2012) *State of the Mediterranean Marine and Coastal Environment 2012 Highlights for Policy Makers*, Nairobi, Kenya 195.97.36.231/publications/UNEPMAP_SoEhighlights_2012_ENG.pdf, accessed 7 December 2012.

UNFCC (United Nations Framework Convention on Climate Change) (1992) *United Nations Framework Convention on Climate Change*, Bonn, Germany, treaties.un.org/doc/Treaties/1994/03/19940321%2004-56%20AM/Ch_XXVII_07p.pdf, accessed 8 December 2012.

UNFCC (1997) *Kyoto Protocol to the UN Framework Convention on Climate Change*, Bonn, Germany, unfccc.int/resource/docs/convkp/kpeng.pdf, accessed 8 December 2012.

Vallega, A. (2002) 'The regional approach to the ocean, the ocean regions, and ocean regionalisation? A post-modern dilemma', *Ocean & Coastal Management*, 45 (11–12), 721.

Verlaan, P. A. and Khan, A. S. (1996) 'Paying to protect the commons: Lessons from the Regional Seas Programme', *Ocean and Coastal Management*, 31 (2–3), 83–104.

WCED (World Commission on Environment and Development) (1987) *Our Common Future*, Oxford University Press, Oxford.

Wright, D. A. (2012) 'Logistics of compliance assessment and enforcement of the 2004 ballast water convention', *Journal of Marine Engineering and Technology*, 11 (1), 17–23.

Chapter 6

International fisheries policy
Sustaining global fisheries for the long term

Introduction

The importance of marine fisheries is often overlooked. From a Western perspective, seafood is simply another source of protein for consumers to choose. However, currently over 1 billion people depend on seafood as their primary source of protein, and 3 billion people rely on seafood for at least 15 per cent of their protein (Davies and Rangeley, 2010; FAO, 2011). Fishing recently has been estimated to employ 34 million people in full- or part-time jobs, producing 80 million tonnes of seafood in 2009 (FAO, 2011). To say fishing is 'big business' is a colossal underestimation: the first-sale value of the world's fisheries was estimated at $84.9 billion in 2004 (Sinclair et al., 2002; Davies and Rangeley, 2010).

However, the sustainability of marine capture fisheries has recently been shown to be at risk. The FAO currently reports that 28 per cent of the world's fisheries are now overexploited, depleted or recovering. Indeed, 57.4 per cent are fully exploited and only 20 per cent underexploited or moderately exploited (FAO, 2009b). Global catches peaked in 1996 and have since declined by approximately 10 per cent (FAO, 2011). For context, in 1974, the corresponding percentages were 10, 50 and 40 per cent respectively. The Commission of the European Communities has determined that, for the European fish stocks where stock status is known, only a third are currently managed for sustained future production (CEC, 2008). In the US, 47 per cent of the 188 stocks or stock complexes with sufficient information to determine whether overfishing is occurring, are either overfished or currently being overfished (NMFS, 2009). In New Zealand, 29 per cent of the 101 stocks or stock complexes in which status can be determined are below their management targets (New Zealand Ministry of Fisheries, 2009).

The status of high-seas stocks is even less well known. Of the 17 Atlantic fish stocks managed under the North Atlantic Fisheries Organization (NAFO), six are collapsed and only three are considered to be sustainable (NAFO, 2008). Atlantic cod biomass is estimated at 6 per cent of historical levels and North Sea cod stocks are depressed to such a degree that for years fishers have been unable to harvest up to their allowable catches.

Through an examination of regionally based on-average catches from 2005–2009, the following regions contribute the largest proportions of the global catch (FAO, 2011):

- Northwest Pacific (25 per cent);
- Southeast Pacific (16 per cent);
- Western Central Pacific (14 per cent);
- Northeast Atlantic (11 per cent);
- Eastern Indian Ocean (7 per cent).

A closer look at certain aspects of marine capture fisheries shows that 50–70 per cent of pelagic predators (i.e. tunas, swordfish) have been removed by fishing (Myers and Worm, 2003). Fishing pressure continues to shift towards lower trophic levels as apex predators decline; this has been termed 'fishing down food webs' (Pauly et al., 1998). Furthermore, the global fishing fleet is far larger than what necessity dictates, especially since technological innovations increase the ability to catch fish. This surplus fishing capacity is underwritten by $30–40 billion in annual subsidies by most fishing nations, thus providing no incentive to reduce fishing effort (Sumaila et al., 2010). In addition, evidence suggests that fishing effort from port-based fisheries has increased over the past three decades at upwards of 3 per cent per decade (Watson et al., 2012). As such, more fishers are chasing fewer fish, and if one was to imagine the global fishing fleet were to be a country, it would be the 18th largest oil-consuming nation on earth (Tyedmers et al., 2005).

Another serious problem is the unintended harvest of species. An estimated 20 per cent of the global fisheries catch constitutes unwanted bycatch that is discarded. Shrimp fisheries produce the largest bycatch and small pelagic fisheries the least. Bycatch also consists of the incidental take of endangered marine mammals, turtles and seabirds although recent advances in gear technology are reducing these impacts. Lastly, approximately 35 per cent of the global annual catch is captured on non-tropical continental shelves, which occupy only 5 per cent of the total ocean area. It is estimated that current harvest levels on non-tropical continental shelves require 36 per cent of the total primary productivity in these areas to sustain these fisheries (Frid et al., 2005).

This chapter first provides a brief overview of how marine fisheries are managed. This is followed by a discussion on the laws and policies of international fisheries and the organizations that oversee marine fisheries.

An overview of fisheries management

There was little perceived need for fisheries management prior to the twentieth century. Marine fisheries (including marine mammals) were thought to be inexhaustible so there was little impetus to allocate resources towards their management. Most fisheries (including nearshore ones) were 'commons' resources that were open to anyone with the equipment to exploit them. With the advent of the internal combustion engine fisheries easily accessible from land were quickly depleted and states realized the need for management. A suite of national and international legislation and conventions was developed to enforce fisheries regulations. These included the International Convention for the Regulation of Whaling (1946) and the Law of the Sea (1958) (Chapter 2). A number of other conventions (discussed in Chapter 5), including MARPOL (1973/78) and the London Dumping Convention (1972) were designed to protect marine environments – and therefore fisheries – from land- and marine-based pollutants.

The purpose of this section, however, is not to chronicle the plight of specific fisheries or dwell on the failure of fisheries management. Rather, this section presents some of the principles of fisheries management and how it has evolved since its inception in the early twentieth century. While humans have utilized almost all marine taxa for food, fuel and medicine, fishing efforts have traditionally targeted finfish (e.g. class/order Agnatha, Chondrichthyes, Osteichthyes), crustaceans (e.g. class/order Amphipoda, Decapoda, Euphausiacea, Mysidacea), mollusks (e.g. class/order Bivalvia, Cephalopoda, Gastropoda) and marine mammals (e.g. class/order Odontoceti, Mysticeti, Pinnipedia, Sirenia). Fisheries for marine algae are important in many regions, and many species from other marine taxa (e.g. Echinodermata) are also intensely harvested (see Chapter 1).

Fisheries employ a variety of techniques and can be categorized in a number of ways (Box 6.1). However, four types of fisheries are generally recognized: 'food', 'industrial', 'shellfish' and 'recreational'. Food fisheries are generally operated by and for local peoples and – while declining over the past 30 years – they continue to be vital to both developed and developing nations. Food fisheries generally harvest nearshore finfish and shellfish, with some remaining subsistence whaling by indigenous peoples in Arctic areas. Industrial fisheries have been the fastest growing in recent decades, due to improvements in vessels, fishing gear and technology (e.g. satellite images, global positioning systems). Industrial fisheries are the dominant harvesters on the high seas and are increasingly harvesting pelagic invertebrates

---

Box 6.1 Types of fisheries based on gear types

- **Trolling**: Use of one or more baited fishing lines often pulled slowly through the water column behind a boat. Trolling targets pelagic species such as albacore tuna.
- **Drift gillnets**: A gillnet is a panel of netting suspended vertically in the water by floats, with weights along the bottom. Fish are entangled in the net. Drift gillnet gear is anchored to a vessel, and drifts along with the current. It is usually used to target swordfish and some shark species.
- **Harpoon**: A larger spear, which may contain explosives, that is thrown or launched at individual animals. The harpoon fishery mainly targets swordfish and marine mammals.
- **Pelagic longline**: Pelagic longline gear consists of a main horizontal line that has shorter lines with baited hooks attached to it. The gear is used at various depths and at different times of day, depending on the species being targeted.
- **Cable longline**: This gear consists of heavy wire for the mainline instead of monofilament. It has been used to target mako and blue sharks.
- **Coastal purse seine**: A purse seine is an encircling net that is closed by means of a purse line threaded through rings on the bottom of the net. This gear is effective in catching schooled tunas. 'Coastal' purse seiners are smaller vessels that fish close to the shore. They mainly harvest coastal pelagic species (sardines, anchovies, mackerel), but they also fish for bluefin and other tunas when they are present.
- **Large purse seine**: Large purse seine gear is used in major high-seas fisheries.
- **Recreational fisheries**: The recreational fisheries for resident and migratory species consist of private vessels and charter vessels using hook-and-line gear.
- **Bottom trawl**: A fishing net is dragged across the seafloor (benthic trawling) or immediately above the seafloor (demersal trawling). Bottom trawling targets shrimp, squid and various types of groundfish. Also termed 'bottom-contact fishing'.
- **Midwater trawl**: A fishing net is suspended in the pelagic zone targeting tunas and small, schooling fish.
- **Dynamite**: Explosives are used to concuss vertebrate fish to facilitate harvesting. Most frequently applied in tropical coral reef environments.
- **Poison**: Poison, most often sodium cyanide or bleach, is used to kill or stun fish. Most frequently applied in tropical coral reef environments.

(e.g. squid, krill). Shellfish fisheries, by contrast, target invertebrates that are primarily on or within the benthos. Finally, recreational fisheries operate throughout the world and there are limited statistics on the total amount of such landings.

To date, fisheries management has essentially been limited to assessment of populations or stocks of individual species. The basic goal of fishery management is to estimate the amount of fish that can be harvested (total allowable catch – TAC) so as to impose catch limits (Box 6.2), while maintaining viable fish populations. These estimates may be modified by political, economic and social considerations. Overly conservative management can result in under-utilized fisheries production (underharvesting), while overly liberal management may result in overharvesting and severely reduced populations.

Fisheries management is generally composed of four components: the fish(es) that are harvested; the species that are not harvested; the environment or habitat where the fish live; and the human use and interactions with the fish and their habitat (Lackey and Nielsen, 1980). Each of these aspects of fisheries management has its own set of theories, concepts and methods to address these issues. While the ideal ecological unit of fisheries

---

Box 6.2 Catch limits

Total allowable catch is a management measure that limits the total output from a fishery by setting the maximum weight or number of fish that can be harvested over a period of time. TAC-based management requires that landings be monitored and that fishing operations stop when the TAC for the fishery is met. A TAC is based on stock assessments (discussed below) and other indicators of biological productivity, usually derived from both fishery-dependent (catch) and fishery-independent (biological survey) data. Data collected from fishermen, processors or dockside sampling can be combined with at-sea observations and independent fishery survey cruises to provide information about the total biomass, age distribution and number of fish harvested. Typically, the TAC is determined on an annual basis, but then partitioned across seasons. To the extent that a TAC is well estimated and enforced, it can control total fishing mortality on a stock (e.g. Pacific halibut).

Trip limits and bag limits are measures that pace landings by limiting the amount of harvest of a species in any given trip. Trip limits are applied in commercial fisheries when there is interest in spacing out the landings over time or a desire to specify maximum landings sizes, and they are usually accompanied by a limit on the frequency of landings.

Individual fishing quotas (IFQs) are a fishery management tool that allocates a certain portion of the TAC to individual vessels, fishermen or other eligible recipients, based on initial qualifying criteria (see Chapter 10). Although used throughout the world, the tool is used in the US to manage Alaska halibut and sablefish, wreckfish and surf clam/ocean quahog fisheries.

Individual vessel quotas (IVQs) are used in a number of fisheries worldwide, including some Canadian and Norwegian fisheries. IVQs are similar to IFQs, except that they divide the TAC among vessels registered in a fishery, rather than among individuals (see Chapter 10).

management is a fish population, for practical reasons fisheries are managed on the basis of stocks (akin to subpopulations). Sometimes more than one species is included in a stock because they are harvested together as though they were one species. In other cases, different species may be managed together for convenience.

Since the early twentieth century, fisheries have generally been managed using the concept of maximum sustainable yield. Early efforts by Baranov (1918) used growth rates of fish combined with their natural mortality rate in order to identify the peak biomass of a cohort. Russell (1931) incorporated recruitment and mortality from fishing and natural causes. Graham (1939) developed models that suggested that low rates of harvesting yielded smaller catches of larger fish while high rates of harvesting yielded similar catches of smaller fish. Therefore, Graham concluded that harvesting less than maximum was an inefficient use of fish and harvesting more than the maximum was an inefficient use of effort. For these reasons MSY was adopted as the primary fisheries management tool from the 1940s until the 1970s. In addition to MSY, many fisheries managers believed the stocks could not become extinct as the cost of harvesting the few remaining fish would be such that fisheries would shift to other, more readily available species. Economic theory, however, did not reflect reality as fishers continued to pursue stocks to ecological and economic extinction (Taylor, 1951).

With the failure of MSY as a fisheries management tool, the concept of optimum sustained yield (OSY) was developed. OSY is essentially MSY modified to reflect relevant economic, ecological or social considerations and is more of a management construct than an empirically derived number. The vagueness of OSY, however, has led to many difficulties and debates of how the concept should be applied.

Fisheries managers have a number of tools and techniques to manage a stock. These techniques can be broadly separated into input controls, an indirect form of control since they do not limit the catch, and output controls, which directly limit the catch.

Input controls are designed to limit either the number of people fishing or the efficiency of fishing. Input controls, often adopted when a fishery is first managed, include restrictions on gear types, number and size of vessels, the area fished, the time fished or the numbers of fishers. They can be applied to both commercial and sport fisheries, and may be applied to an entire fishery or to segments of it. Two popular input controls are licenses and license endorsements. They may be used to certify fishermen or vessels, or used as a management measure to limit the number and types of fishers or vessels participating in the fishery. The licensing system is designed to limit fishing capacity and effort, but their effect on either is indirect. For instance, if licenses do not stipulate a maximum vessel size or other limits on fishing power or capacity, the capacity of the fleet can drift upwards as small vessels are replaced with larger ones. The problem arises because size is only one dimension of fishing power. Also, attempts to control size can lead to adaptations that are inefficient or unseaworthy.

Output controls directly limit catch and therefore a significant component of fishing mortality (which also includes mortality from bycatch, ghost fishing and habitat degradation due to fishing). Output controls can be used to set catch limits for an entire fleet or fishery, such as a TAC (see Box 6.2 for this and other types of limits). They can also be used to set catch limits for specific vessels (trip limits, IVQs), owners or operators (IFQs). In this case the sum of the catch limits for individuals or vessels equals the TAC for the entire fishery. Output controls rely on the ability to monitor total catch. This can be achieved by either (1) measuring total landed catch with reliable landings records, port-sampling data and some estimates of discarded or unreported catch, or (2) measuring the actual total catch with at-sea observer coverage or verifiable logbook data.

One of the most important aspects of fisheries management is the ability to describe the condition or status of a stock, also known as *stock assessment*. Not only do stock assessments provide information on the size, age structure and health of a stock, but they form recommendations on the management of the stock. There are two components of stock assessment. The first is to study the biology and life history of the species and the second is to understand the effects and impacts of fishing on the stock. For many stocks throughout the globe there is little information available on the life history of the species so fisheries management decisions are often made based on information obtained from the landing (catch) of fish from the stock. This is a classic dilemma in fisheries management: since fisheries are often established without baseline information on the natural state and variability of the stock prior to fishing, management decisions must either be based on patterns in the landings of fish or must attempt to build a 'picture' of the stock prior to exploitation from which to base management decisions.

In a perfect world with perfect information, however, full and accurate stock assessments would be made using the following information:

- The number of fishers and types of gear used in the fishery (e.g. longline, trawl, seine);
- Annual catches by gear type;
- Annual effort expended by gear type;
- Age structure of the fish caught by gear type;
- Ratio of males to females in the catch;
- How the fish are marketed (preferred size, etc.);
- Value of fish to the different groups of fishermen;
- Timing and location of best catches.

The biological information would include:

- The age structure of the stock;
- The age at first spawning;
- Fecundity (average number of eggs each age fish can produce);
- Ratio of males to females in the stock;
- Natural mortality (the rate at which fish die from natural causes);
- Fishing mortality (the rate at which fish die from being harvested);
- Growth rate of the fish;
- Spawning behaviour (time and place);
- Habitats of recently hatched fish (larvae), of juveniles and of adults;
- Migratory habits;
- Food habits for all ages of fish in the stock.

When the above information is collected by examining the landings of fishermen, it is called *fishery-dependent data*. When the information is collected by biologists through their own sampling programme, it is called *fishery-independent data*. Both methods contribute valuable information to the stock assessment.

In the real world, however, very little of the above information is available, particularly for newer fisheries that target poorly understood species (often high-seas species).

Depending on how much information is available on the stock, there are a number of ways to manage a fishery, which include estimating populations of stocks based on the ease of catching fish, protecting fish until they can spawn, ensuring a proper age structure in the stock and developing models of population survivabilty. The following paragraphs briefly discuss some of these methods.

The simplest stock assessments are made by calculating *catch per unit effort* (CPUE) using a combination of the history of landings (catches) for the stock and the level of effort expended in harvesting the stock. CPUE, therefore, is essentially an indicator of stock abundance. When new fisheries are established, CPUE tends to be high as catches are high while the effort required to catch the fish is low. As more fishers participate in the fishery, or technological improvements result in the ability to harvest more fish, CPUE generally will either stabilize in a well-managed fishery, or continue to decline in a poorly managed fishery. CPUE is a relatively simple and intuitive way to manage a fishery. Yet CPUE has some serious limitations and therefore is no longer solely used as a management tool. The primary shortcoming is that the fishery is often in decline by the time that reliable estimates of CPUE have been generated. Moreover, other problems hamper the efficiency of CPUE: insufficient information on landings, inadequate information on fishing effort and technological advances that skew comparisons with past practices.

A somewhat more reliable way to assess a stock is to determine at what age(s) fish spawn and then to structure the fishery in such a way that fish are not harvested before they can recruit (reproduce). The objective is to protect fish until they are old enough to spawn. The harvest of fish before they can recruit and replace themselves is termed *recruitment overfishing* and has had serious consequences on many fisheries. Once spawning age(s) is determined, the fishery can be managed through specifications on gear type (e.g. mesh size in nets) or size limits. Nevertheless, protection against recruitment fishing does not protect a stock from being overharvested for more fish can still be removed than can be recruited. Recent evidence has shown that larger fish are disproportionately more fecund than smaller fish in relation to body sizes.

Another method to estimate stocks is to develop a mortality and *spawning potential ratio* (SPR). A SPR is the number of eggs that could be produced by an average recruit over its lifetime when the stock is fished, divided by the number of eggs that could be produced by an average recruit over its lifetime when the stock is unfished. In other words, SPR compares the spawning ability of a stock in the fished condition to the stock's spawning ability in the unfished condition. SPR can also be calculated using either the biomass (weight) of the entire adult stock, the biomass of mature females in the stock or the biomass of the eggs they produce. These measures are called *spawning stock biomass* (SSB), and when they are put on a per-recruit basis they are called *spawning stock biomass per recruit* (SSBR). SPRs are based on knowledge of the age structure of a stock collected from either estimating ages from the lengths and weights of a fish or from the examination of the ear bones from fish, termed otoliths, which – like trees – develop annual growth rings that can be counted.

Current approaches and difficulties with fisheries management

Many definitions of 'fishery management' exist but the FAO defines fishery management as:

> The integrated process of information gathering, analysis, planning, consultation, decision-making, allocation of resources and formulation and implementation, with

enforcement as necessary, of regulations or rules which govern fisheries activities in order to ensure the continued productivity of the resources and the accomplishment of other fisheries objectives.

(FAO, 1997)

The human history of fishing is relatively straightforward and tends to repeat itself. As humans begin to overexploit a species, they (a) exploit the same species somewhere else, (b) exploit less preferred species locally and (c) increase local resource production through aquaculture (Lotze, 2004). Thus, if fishery resources are limited and effort is uncontrolled, mortality will increase until the fishery becomes economically non-viable or the population collapses.

The previous section provided an overview of how fisheries function and the methods used to set catch limits. This section explores approaches to how fisheries are currently managed and the consequences and limitations of these approaches.

Single-species approach to fisheries

Throughout history, most fisheries have operated under the principle of single-species management, where each target stock or population is managed independently of other targeted stocks, which may overlap in space and time (Larkin, 1977). Thus, fishery management advice under this model is provided on a stock-by-stock basis – a model under which most species continue to be managed today. Under moderate fishing pressure, the single-species approach is an acceptable management strategy; it has been demonstrated to work well for terrestrial wildlife management and, if properly applied, for the management of certain marine species (e.g. whales – see below). However, as discussed in the previous section, global fishing pressures are currently at a level where species-by-species management has the potential to seriously undermine the structure and function of entire marine ecosystems.

The purpose of this section is to review the single-species approach, its strengths and weaknesses, and show how this approach has been improved over the past two decades to encompass communities and ecosystems.

The purpose of single-species fishery management is to control fishing effort in order to avoid overexploitation of a stock and to ensure that fishers are provided with a suitable return on their investment. One alternative to single-species management is a situation where fishers make minimal or no profits. Meanwhile the fishery becomes biologically overexploited through overfishing, resulting in fish being caught at a size before they have realized their full growth and spawning potential. Therefore the goal of single-species management is to invoke technical conservation measures to prevent overexploitation. This is actualized by protecting young or spawning fish or making the fishery sufficiently inefficient that the zero profit level is reached before the stock is overexploited (Larkin, 1977; Cochrane, 2002).

The fundamental objective of traditional single-species management is to manage a fishery to MSY, which assumes that every fish stock generates 'surplus production' and that this production can be reduced (by fishing) to a biomass necessary in order to maintain the stock at some sustainable level. As such, traditional applications of MSY usually result in biomass reductions of the target stock of 30–50 per cent of unfished levels (Mace, 2001, 2004). Given that managing multiple stocks to MSY in the same geographic (food web) area will inevitably affect food-web structure, fisheries managers have begun to realize that MSY is an upper threshold that should not be exceeded. Hence they have stipulated a fishing mortality less than MSY, often re-expressed as 'F_{MSY}'.

A further improvement on the single-species approach is *multi-species fishery management*, where decisions are made in consideration of other target species under harvest in the same area, and may include ecotrophic considerations in management decisions. Regardless of the improvements to the MSY concept since its introduction in the 1930s, the following assumptions underpin single-species management (summarized by Babcock and Pikitch, 2004):

- The objective of management is to maximize the long-term average yields of the fishery.
- A population biomass level exists that will maximize the long-term average yield.
- Fish growth, natural mortality and fecundity are constant and do not change over time, irrespective of the abundance of other species, environmental changes or the effects of fishing.
- The total fishing mortality can be controlled by regulation of the fishery.

Current fishery management practices in most countries and international agencies are based on the single-species approach and generally follow the model below (adapted from Hilborn, 2004):

- Single-species stock assessments are undertaken for each stock or stock complex to set maximum sustained yield or some other maximum threshold such as F_{MSY}.
- Regulations that determine allowable time, area, gear and catch limits are recommended by fishery managers. These are ultimately established through a political process.
- A centralized management agency responsible for science, decision-making and enforcement is responsible for operating the fishery.
- Stakeholders are involved in decision-making, either through policy, legal or political means.

In reality the traditional assumptions that underpin the single-species management model have resulted in the following characteristics (adapted from Pavlikakis and Tsihrintzis, 2000; Marasco et al., 2007):

- A disproportionate focus on short-term economic objectives and the maintenance of a single species.
- Removal of humans as ecosystem components from fishery models and decision-making.
- Political, economic and social values being either discounted or ignored.
- Needs of commercial fishing stakeholders taking precedence over the public interest.
- Science and management occurring at local scales, where scaling up to regional, national or international interests is difficult.
- A focus on population ecology rather than community ecology in conjunction with oceanography, fisheries economics, ecology and fisheries biology.
- Modern tools, such as geographic information systems, oceanographic models and economic techniques not being applied.

Single-species fishery management is primarily based on developing stock assessment models to establish reference points that define overfishing limits based on (spawning) stock biomass and fishing mortality (Link, 2005). When these limits are exceeded, protocols are triggered to reduce fishing, either through input controls (e.g. number and size of vessels, time allotted to fish, gear restrictions) or output controls (e.g. amount of fish that can be caught).

Properly applied, the single-species approach is a viable model for the management of certain fisheries and can provide significant economic and ecological value to the sustainable management of a fishery (Marasco et al., 2007). Single-species approaches can work where stock status can be determined, life histories are well known, the effects of environment changes on stocks are understood and compliance with regulations can be enforced. Notable examples of species that can be managed under this approach include certain marine mammals, such as those covered by the International Whaling Commission's Revised Management Procedure (see below), and nearshore shellfish stocks where stock status and environmental conditions can be easily determined.

Most fisheries, however, do not meet the conditions for being managed under the single-species approach. In certain circumstances, single-species approaches will lead to overexploitation of a stock or economic failure of the fishery (Mace, 2004). Failures of the approach have generally not been the fault of the science and management, but rather data limitations and a failure of political will (Marasco et al., 2007). For example, a review of the scientific advice provided to European decision-makers on 18 fish stocks in 2002 found that scientists provided the correct advice 53 per cent of the time, provided the wrong advice resulting in a detriment to the fishery 23 per cent of the time and 24 per cent of the advice consisted of 'false alarms' where catch reductions were recommended but later found to be unnecessary (Frid et al., 2005). Thus, scientific advice on single-stocks was either correct or neutral to the health of the stock 77 per cent of the time.

Perhaps the most contentious aspect of the single-species approach is the reliance on MSY or MSY variants in order to set fishery harvests. Many jurisdictions continue to use stock assessment models to attempt to determine MSY or the rate of fishing morality F_{MSY} that a stock can sustain (Symes, 2007). As discussed earlier in this chapter, nearly 30 per cent of the world's fisheries are overfished, suggesting that attempting to manage to MSY is failing. This failure is a result of either how MSY is calculated, uncertainty in the knowledge of how marine ecosystems function, or political interference that sets allocations above MSY. Many stock assessment models used to calculate MSY result in typical biomass reductions of 50–70 per cent below unfished levels, which has been demonstrated to be biologically unfeasible for slow-growing, late-maturing species such as sharks and rays (Hirshfield, 2005). As such, setting catches below MSY results in larger stock sizes that should, in turn, cause fewer stock collapses and more favourable outcomes (Hillborn, 2004). In addition, traditional application of the MSY concept has led to presumptions that all mature female fish are of equal importance. In actuality, in fact older females have a disproportionate spawning potential and thus may be critical to maintain population viability. The MSY concept, moreover, ignores ecosystem effects of high harvest rates on fast-growing, high-biomass species, where populations appear to be resilient to high levels of fishing (Hirshfield, 2005).

Much of the difficulty in managing to MSY is related to uncertainty in single-species stock assessment models. Perceived uncertainty has led governments to ignore these models. In a few critical cases, incorrect assessment leading to rapid fishery declines has jolted confidence in the models. The models have also failed to consider ecosystem and food web effects that potentially negatively affect stocks. Another difficulty with the models is that they have been ineffectual in providing assistance to governments attempting to design regulatory systems to achieve fishery targets (Pauley et al., 2002). Uncertainty has been further exacerbated by funding limitations: models have lacked input data, scientific analyses and applications across spatial and temporal scales.

There are a number of recent and proposed improvements to the single-species model, some of which are now enshrined in national and international law. Most suggested improvements to the model fall under the guise of establishing marine protected areas or applying ecosystem approaches (see Chapter 10) and improving fishery management regulations. While useful, they are generally used in specific circumstances and have limited scope to address the broader challenges of ocean conservation and management (Day et al., 2008). A broader consideration of fundamental improvements to the approach falls under the following types (adapted from Hilborn, 2004; Symes, 2007):

- Eliminating the approximately $30 billion in annual subsidies for fishing fleets.
- Reducing target fishing moralities through adopting a more precautionary approach to setting harvest levels.
- Establishing no fishing areas over a significant (20 per cent or more) area of the world's oceans either through marine protected areas (MPAs) or other means.
- Eliminating destructive fishing practices, including prohibiting bottom trawling of previously untrawled areas and reducing bycatch.
- Weakening the current command and control fisheries management model to a system of co-management where responsibilities are shared between stakeholders.
- Establishing new forms of marine tenure where individuals or groups of fishers are guaranteed a specific share of a future catch thus removing incentives to overcapitalize and permit alignment of economic interests with long-term conservation goals.
- Establishing incentive-based versus regulatory-based management approaches.

Additional improvements to the single-species approach may include consideration of the species of interest within the broader ecosystem context. Examples of recent efforts toward these ends include: modifying stock assessment models to account for density dependence for the target species that arises from predation from another species (community effects); considering time-dependent disease and predation effects when modelling natural mortality; incorporating perceived regime shifts in defining biological reference points; and incorporating considerations for low productivity or endangered species that may be affected by harvest of the target species in stock models (Marasco et al., 2007).

In conclusion, the single-species approach, however well intentioned and applied, is subject to the political whims of governments, which are rarely willing to fundamentally reform fisheries policies (Symes, 2007). Improvements on the single-species approach – including more precautionary re-expressions of MSY that also account for uncertainty (e.g. recruitment failure) in the dynamics of stocks and populations – are a step in the right direction; but these have not yet been proven to result in more sustainable fisheries management outcomes. This is most likely due to political interference and illegal fishing. Similarly, the multispecies approach has also shown little promise that it is a sustainable tool for fisheries management. Since 1986, the US northeast multispecies groundfish management plan has consisted of 24 target species managed collectively. A 20-year review of the plan concluded that indicators of population and community health (e.g. spawning stock biomass) were not improving as projected, and that illegal harvest was responsible for 12–24 per cent of total harvest (King and Sutinen, 2010). There is no shortage of national and international fishery conservation laws, conventions, agreements and policies that should be sufficient to control overfishing and ensure that community and ecosystem considerations are incorporated into decision-making (see below). However, for reasons articulated above, fisheries continue to be overexploited, thus requiring a new approach to managing both fisheries and oceans.

Guerry (2005) eloquently summarizes the reasons for the failure of traditional species-based approaches as follows: fragmented ocean governance where the fishery as a commons resource has led to jurisdictions competing for the resources; inability to maintain ecosystem elements, such as water quality or spawning habitat necessary to sustain successful fisheries; inability to manage diverse, non-fisheries related impacts, including pollution, habitat loss, overharvesting, climate change and introduced species; and lack of recognition of connections between ecosystem structure, function and services.

The ecosystem approach to fisheries

The ecosystem approach to fisheries (EAF) stems from the realization that single species (and multispecies) approaches to fisheries management have, regardless of the reasons, resulted in a situation where most of the world's fish stocks are fully exploited or overfished (FAO, 2009b). The term EAF is used in this book because, after much debate, the FAO concluded that EAF is a more appropriate term than ecosystem-based fisheries management since the EAF explicitly considers ecosystem processes in the formulation of management actions. This section explores the ecosystem approach to management as it applies to fisheries.

While the term EAF is relatively new, EAF is fundamentally a re-expression of the principles of sustainable development, which have been enshrined in international conventions since the 1972 UNCHE (Scandol et al., 2005 – Chapter 5). General concepts and definitions of the EAF converge around the need either to alter existing practices or to develop a new fishery management paradigm that explicitly recognizes the interrelationships and dependences within food webs and human activities (Pitcher et al., 2009). EAF is predicated on the assumption that an improved understanding and management of stock interactions, stock–prey relationships and stock-habitat requirements will result in more accurate fishery assessment models (Christie et al., 2007). Activities and concepts that roll up into the EAF include: establishing a sustainable fishery resource for future generations; inclusion of humans in ecosystems; emphasizing ecosystem sustainability over ecosystem products; understanding the dynamic nature of marine systems; establishing clear management goals and objectives; and precautionary and adaptive management.

The overall goal of the EAF, then, is to achieve ecologically sound resource conservation that is responsive to the reality of ecosystem processes (Marasco et al., 2007). The EAF objectives therefore include (modified from Pikitch et al., 2004):

- avoiding degradation of, and potentially restoring, ecosystems as measured by indicators of environmental quality and system status;
- maintaining ecosystem structure, process and function at the community and ecosystem level;

- obtaining and maintaining long-term socioeconomic benefits without compromising the sustainability of the ecosystem;
- generating knowledge of ecosystem processes sufficient to understand the likely consequences of human actions.

Elements of the EAF are summarized as follows (adapted from Marasco *et al.*, 2007 and Sissenwine and Murawski, 2004):

- ensuring that broader societal goals are taken into account and balance social objectives;
- employing geographic (spatial) representation;
- recognizing the importance of climatic-oceanic conditions;
- emphasizing food web interactions and pursuing ecosystem modelling and research;
- incorporating improved habitat information (for target and non-target species);
- expanding monitoring and ecosystem assessments;
- acknowledging and responding to higher levels of uncertainty;
- employing adaptive management;
- accounting for ecosystem knowledge and uncertainties;
- considering multiple external influences.

Perspectives on the EAF range from it being simply an incremental extension of current fisheries management approaches to a complete redesign of marine management (Pitcher *et al.*, 2009). Regardless of which perspective is applied, the EAF necessitates that short-term socioeconomic benefits will be reduced due to decreased harvests. This is due to the application of the precautionary principle and consideration of the ecosystem effects of fishing (Marasco *et al.*, 2007). Furthermore, uncertainty cannot be used as an excuse to maintain the status quo.

There have been many criticisms of the EAF. Perhaps the most significant issue with the EAF is that, given humanity's dismal record with single-species management, the likelihood of successfully implementing the EAF is low, especially given that the same fishery scientists, politicians and stakeholders responsible for single-species declines are now in charge of implementing the EAF (Mace, 2004; Murawski, 2007). Compounding this problem is the lack of a general theory of the functioning of marine ecosystems, which limits the ability to explain and predict even simplified single-species impacts on marine systems. As such, the expectation that the EAF will be able to cope with the increased uncertainty due to the additional need to model community and ecosystem considerations will severely limit the usefulness of the tool (Valdermarsen and Suuronen, 2003; Curry *et al.*, 2005). Others argue that improvements in single-species approaches, such as the application of F_{MSY} negate the need to implement the EAF. While these criticisms are valid, many of the primary issues to be addressed under the EAF, including bycatch, indirect effects of harvesting, and interactions between biological and physical components of ecosystems are already being addressed (Sissenwine and Murawski, 2004).

With a few exceptions that are discussed below, the EAF to date has consisted of a sporadic application of marine protected areas, limits on destructive fishing practices and efforts to limit bycatch (Shelton, 2009). The EAF, being relatively new, has not yet realized its potential primarily because the sustainability objectives have not yet been clearly articulated for most fisheries, the MSY (or F_{MSY}) concept has not been quantitatively put into the context of ecosystem objectives, and therefore setting the EAF-informed MSYs has not been undertaken for most fisheries. In addition, the realization that the EAF, if applied properly and with a precautionary focus, will almost always lead to reduced fishery targets, has led to bureaucratic inertia by governments and stakeholder groups unwilling to accept reduced catch allocations (Shelton, 2009).

A number of perspectives exist on how the EAF should be implemented. Some view the EAF as simply a recalculation of MSYs for target fisheries that considers direct and indirect impacts of the fishery on the local and regional ecosystem. Adjusting MSYs to limit known fishery impacts (e.g. bycatch, benthic disturbance) is straightforward and would likely result in catch limits at or below F_{MSY} (Link, 2005; Shelton, 2009). This approach is similar to 'weak stock management' used in the singe-species approach. Here all fisheries are regulated at levels to prevent any one stock from overexploitation and to produce single-species MSYs for target fisheries (Hilborn, 2004). Compared to more ambitious applications of the EAF, this approach requires minimal new data collection and avoids the need for additional scientific studies in order to fully understand the relationships between reducing a stock and impacts on the broader ecosystem. One disadvantage of simply recalculating MSYs to reduce known impacts is that significant community and ecosystem impacts may still occur under this management scenario. This entails that the EAF was advanced primarily because these types of limited approaches have not resulted in sustainable fisheries or stable ecosystems.

A more robust application of the EAF exists that expressly captures ecosystem interactions. It involves developing ecological objectives for target and non-target stocks and populations. The application considers diversity, productivity and resilience at the genetic, population and community levels (Rogers et al., 2007). However, developing objectives requires sufficient understanding of fish population dynamics and the structural and functional aspects of life history, which are required to define ecosystem overfishing in a way that relates to setting MSYs in the single-species context. Hence, setting objectives will likely be one of the more difficult aspects of implementation of the EAF (Link, 2005; Rogers et al., 2007).

The most comprehensive application of the EAF involves the integration of the target stock with community dynamics and environmental variables before determining catch limits and management measures. While this approach requires a level of inventory and scientific understanding that may not currently exist, it does attempt to identify the surplus production required to satisfy ecosystem needs (Goodman et al., 2002). Additional challenges of this application of the EAF are addressing the high levels of uncertainty as a result of integrating biological and oceanographic processes into stock assessment models. These are likely to have limited predictive power, given most currently available data and understanding of marine ecosystem function (Jennings and Revill, 2007). For example, Georges Bank, off the east coast of North America, is one of the world's most studied ecosystems, yet the proper application of the EAF has been hampered by a lack of sufficient data (Froese et al., 2008).

Regardless of what 'flavour' of the EAF is selected for a particular application, it faces a number of implementation challenges. Among the things that the EAF will need to develop are long-term ecosystem-related objectives, meaningful indicators that clearly demonstrate when a threshold has been exceeded, and a stronger scientific foundation to inform EAF decisions (Cury et al., 2005).

The science gaps to fully operationalize the EAF can be summarized as follows (modified from Frid et al., 2006):

- Understanding the relationships between hydrographic regimes and fish stock dynamics.
- Understanding and inventorying habitat distributions.
- Establishing 'design rules' for marine protected areas.
- Understanding ecological dependencies within marine communities.
- Developing predictive capabilities in complex systems.
- Incorporating uncertainty into management advice and direction.
- Understanding the genetics of target and non-target stocks, populations and species.
- Understanding the response of fishers to management measures.

Regardless of how the EAF is used, a number of implementation tools have been identified, many of which are currently in practice and discussed elsewhere in this book (modified from Pikitch *et al.*, 2004):

- Marine habitat delineation at multiple spatial and temporal scales.
- Identification and protection of essential fish habitat.
- Marine protected areas.
- Marine spatial planning and zoning.
- Impacts of fisheries on endangered species.
- The reduction of bycatch.
- Managing target species.
- Precautionary approach.
- Adaptive management.
- New analytical models and management tools.
- Integrated management plans.

As discussed previously, while dozens of publications and conferences have explored the principles and implementation of the EAF, there have been few examples of the EAF actually being applied to real-world management decisions. Pitcher *et al.* (2009) evaluate the performance of the EAF using a number of criteria. Only two states were rated as 'good' (Norway, US), while four countries (Iceland, South Africa, Canada, Australia) were rated as 'acceptable', leaving over half of the countries with failing grades.

However, there are a number of the EAF successes, including Alaskan groundfish management and CCAMLR (Chapter 8). CCAMLR has demonstrated that the EAF and the precautionary approach can be applied in high-seas environments. Under the Treaty, the scientific committee has developed a number of innovative methods to manage prey species in order to protect dependant predators, limit bycatch and develop precautionary-based protocols prior to the exploitation of new fisheries. Advice from the CCAMLR scientific committee is almost always followed and the quality of science is high (Agnew, 1997; Constable, 2004). Continuing challenges to implement the EAF under the Treaty exist, ranging from data limitations, controlling fishing effort and incorporating climate change into ecosystem and stock assessments.

Another example of a successful implementation of the EAF is the management of groundfish in the Northeast Pacific. Although four species of crabs are considered to be overexploited, there are currently no overfished groundfish stocks (Dew and Austring, 2007). As part of the EAF approach in this region a full ecosystem assessment was developed (PICES, 2004) and significant research has been undertaken to address particular challenges. This research, for instance, has explained how threatened Steller sea lions have been impacted by the combined effects of fishing, predation, competition and ocean productivity in the Aleutian Islands and the eastern Gulf of Alaska (Christensen *et al.*, 2007). Much of the success in applying the EAF techniques in this region has been ascribed to the fact that the North-East Pacific is primarily a bottom-up driven ecosystem. Food chains are shorter and inferences on ecosystem health and setting catch limits can be at least partially predicted using primary production estimates (Ware and Thomson, 2005).

In conclusion, the EAF concept has been part of the suite of fishery management tools for nearly two decades. In particular, it has been successfully applied to ecosystems with certain characteristics. Those that are generally bottom-up structured and those where their EEZs fall within developed nations (CCAMLR excepted) have fared well. Success has also come where there is genuine stakeholder and political will to move away from traditional, single-species management and towards a more holistic approach to ocean conservation. The degree and speed of uptake of the EAF will depend on the willingness of fishing industries and governments to bear the high short-term costs associated with reducing fishing efforts and moving towards sustainability.

History of international fisheries law and policy

International efforts to regulate fishing are not new: France and Britain's first collective attempt to manage fisheries in the North Sea was in 1839. This was followed by the multi-lateral 1881 North Sea Fisheries Conference, focused on the orderly exploitation of cod, haddock and plaice stocks. While overfishing was understood to be occurring at the time of the Conference, it was not formally part of the Conference. Instead, the Conference addressed dispute resolution and avoiding gear conflicts.

The International Council for the Exploration of the Sea (ICES) was established in 1902 and by 1934, size limits, closed areas to assist fisheries and mesh sizes had been recommended.

The first conservation-based international agreement was the North Pacific Fur Seal Convention (signed 7 July 1911 and entered into force on 14 December 1911). In this Convention Great Britain (for Canada), Japan, Russia and the US agreed to halt the pelagic (sea-based) harvesting of northern fur seals and sea otters (Table 2.2).

In 1937, in recognition of the rapid depletion of European inshore fisheries since World War I, the British government convened the first of three conferences on overfishing. The conference to implement the 1934 ICES recommendations was held, but due to the World War II, the convention was never brought into force. The second convention in 1943 also failed to reach consensus on any issues. However, in 1946, the London International Overfishing Conference was convened and attended by 12 European nations to address overfishing in the North Sea and waters adjacent to the British Isles. The resulting Convention for the Regulation of Meshes of Fishing Nets and the Size Limits of Fish (the 'Overfishing Convention') applied only to the North East Atlantic and did not enter into force until 1953. By this time, however, the Overfishing Convention was inadequate to address emerging fisheries issues (e.g. herring) and the parties (now numbering 14) to the Convention signed the North East Atlantic Fisheries Convention in 1959 that, in turn, established the North East Atlantic Fisheries Commission (NEAFC), a body extant today (see below).

A companion convention to the North West Atlantic, the International Convention for the Northwest Atlantic Fisheries was enacted in 1950. This established the International

Commission for the Northwest Atlantic Fisheries (ICNAF) with the authority under Article VIII to prescribe 'an overall catch limit for any species of fish'.

The widespread introduction in the mid-1970s of EEZs and the adoption of the LOSC in 1982 provided a new framework for the better management of marine resources. More recently, advances in fishing technology (e.g. cold storage, GPS, sonar) have opened up the previously inaccessible 60 per cent of the oceans that are outside national EEZs. The high seas' contribution to overall marine catches increased from 9 per cent in the 1950s to 15 per cent in 2003. This increase necessitated the establishment of regional fisheries management organizations that are discussed below (Cullis-Suzuki and Pauly, 2010). Furthermore, the ability to access the high seas has resulted in the growth of *illegal, unreported and unregulated* (IUU) fishing (Sumaila *et al.*, 2006). It is estimated that IUU fishing may comprise up to 30 per cent of the global fisheries catch (Borit and Olsen, 2012).

Organizations that oversee international fisheries management

The Food and Agriculture Organization of the United Nations

In 1945 the UN established the FAO as a specialized agency to improve agricultural productivity. Better productivity would ensure the world's population would have access to sufficient and nutritious food, improve rural quality of life and contribute to the world's economy. Currently, the FAO is headquartered in Rome and has 191 member states, 2 associate members and 1 member organization (the EU). The FAO has staff located in over 130 states and is governed by the FAO Council, which oversees the activities of the FAO and sets programmes and budgets.

The FAO's Fisheries and Aquaculture Department is one of seven FAO departments. It was established to improve the conservation and utilization of the world's fisheries in inland and marine waters. The Department is broadly responsible for the following activities:

- Reporting on the status of the world's wild fisheries and aquaculture operations.
- Collecting, analysing and summarizing environmental and socioeconomic information on the world's wild fisheries and aquaculture operations.
- Supporting national and regional (multinational) fisheries management organizations through the provision of scientific, policy, technical and legal advice.

The Fisheries and Aquaculture Department is overseen by the Committee on Fisheries (COFI), which is a subsidiary body of the FAO Council. COFI reviews and approves the fisheries and aquaculture related work of the FAO and makes recommendations on pertinent fisheries and aquaculture issues. COFI also provides a venue for the negotiation of agreements.

Key long-term fisheries and aquaculture related FAO activities include:

- Implementing the Code of Conduct for Responsible Fisheries (see below).
- Implementing the various international plans of action on the management of sharks, incidental seabird catches, fishing overcapacity and IUU fishing (see below).
- Administering the global partnerships for responsible fisheries (FishCode) to improve the socioeconomic and nutritional status of developing nation populations.
- Administering the Species Identification and Data Programme (FAO FishFinder) to improve fish identification.

- Administering GLOBEFISH, which provides analysis and information on the international fish trade and promoted market access, including managing the FAO fishery country profiles and FAO fishing areas (Figures 6.1 and 6.2).
- Developing the Global Record of Fishing Vessels, Refrigerated Transport Vessels and Supply Vessels (Global Record) in order to provide a single database of global fishing vessels to deter and eliminate IUU fishing.

Today, the FAO's fisheries and aquaculture programmes are globally respected and represent the world's best available information on global catch statistics and trends. Rather than denunciations of methodologies, criticisms of FAO fishery programmes have been directed at points of interpretation and underlying assumptions concerning the reporting of catch statistics (Pauly and Froese, 2012).

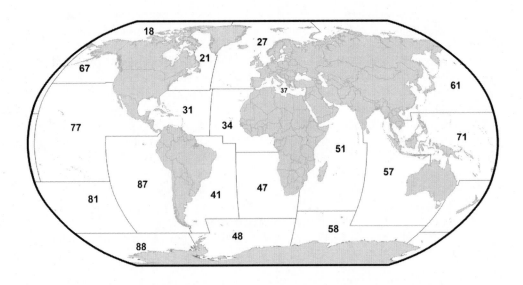

Figure 6.1 FAO major fishing areas

Note: Fishing area names are as follows:

Atlantic Ocean and adjacent seas
18 Arctic Sea
21 Atlantic, Northwest
27 Atlantic, Northeast
31 Atlantic, Western Central
24 Atlantic, Eastern Central
37 Mediterranean and Black Sea
41 Atlantic, Southwest
47 Atlantic, Southeast
Indian Ocean
51 Indian Ocean
57 Indian Ocean

Pacific Ocean
61 Pacific, Northwest
67 Pacific, Northeast
71 Pacific, Western Central
77 Pacific, Eastern Central
81 Pacific, Southwest
87 Pacific, Southeast
Southern Ocean
48 Atlantic, Antarctic
58 Indian Ocean, Antarctic
88 Pacific, Antarctic

Source: www.fao.org/fishery/area/search/en

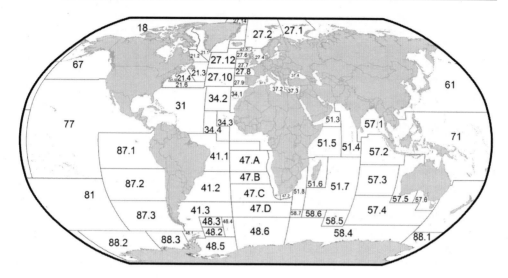

Figure 6.2 FAO major fishing subareas
Note: See Figure 6.1 for key to area names.
Source: www.fao.org/fishery/area/search/en

Regional fisheries bodies and regional fisheries management organizations

Globally applicable agreements are important for setting the overall direction for fisheries policy with respect to access, conservation, equity and the beneficial use of fisheries. However, given the diversity of marine ecosystems and the states that depend on them, global agreements have limited use in establishing management regimes for individual regions.

This limitation was recognized during the LOSC negotiations and, while the freedom of fishing is guaranteed to all nations (including landlocked as per Articles 87 and 116), these rights are constrained by both the need to conserve resources on the high seas (Articles 117–120) and the requirement to establish regional or sub-regional organizations to manage stocks found in the EEZs of two or more coastal states (Articles 63–67). Furthermore, the IJC concluded in the 1974 *Fisheries Jurisdiction Case* that the freedom to fish doctrine was '...replaced by a recognition of a duty to have due regard of other states and the needs of conservation for the benefit of all' (Henriksen, 2009). Subsequent agreements, most notably Articles 8–13 of the 1995 UN Fish Stocks Agreement (discussed below), direct signatories to establish *regional fisheries management organizations* (RFMOs) as the means to achieve fisheries objectives outlined in the LOSC, UNCED and UNCBD (Chapter 5).

While treaties on the management of 'fisheries' (including marine mammals) date back to 1911, the International Pacific Halibut Commission (IPHC) (see Table 2.2), established in 1923, was the first true regional fisheries body. This was followed in the 1950s with the establishment of a number of tuna-based regional fisheries bodies (Table 6.1) to provide scientific advice and management. While regional fisheries bodies established before the 1995 UN Fish Stocks Agreement may have had both an advisory and management

Table 6.1 List of current RFMOs grouped by purpose and listed by year of establishment

Name by type	Year of establishment	Coverage of fish stocks	Membership	Decisions binding on parties
Regional management				
International Pacific Halibut Commission (IPHC)	1923	Halibut	Closed	Yes
Asia-Pacific Fisheries Commission (APFIC)	1948	All	Open	No
General Fisheries Council for the Mediterranean (GFCM)	1949	All	Open	Yes
Inter-American Tropical Tuna Commission (I-ATTC)	1949	Tuna	Closed	No
Permanent Commission on the South Pacific (CPPS)	1952	All	Closed	Yes
International Commission for the Conservation of Atlantic Tunas (ICCAT)	1966	Tuna	Open	Yes
International Baltic Sea Fisheries Commission (IBSFC)	1973	All	Closed	Yes
Northwest Atlantic Fisheries Organization (NAFO)	1978	All	Closed	Yes
North East Atlantic Fisheries Commission (NEAFC)	1980	All	Closed	Yes
Commission for the Conservation of Atlantic Marine Living Resources (CCAMLR)	1980	All	Open	Yes
North Atlantic Salmon Conservation Organization (NASCO)	1983	Anadromous	States of origin	No
Pacific Salmon Commission (PSC)	1985	Anadromous	States of origin	Yes
North Atlantic Marine Mammal Commission (NAMMCO)	1992	Marine mammals	Closed	No
North Pacific Anadromous Fish Commission (NPAFC)	1993	Anadromous	States of origin	No
Commission for the Conservation of Southern Bluefin Tuna (CCSBT)	1994	Tuna	Closed	Yes
Convention on the Conservation and Management of the Pollock Resources in the Central Bering Sea	1995	Pollock	Open	Yes
Indian Ocean Tuna Commission (IOTC)	1996	Tuna	Open	Yes
South East Atlantic Fisheries Organization (SEAFO)	2003	All (except tuna)	Open	Yes
Western and Central Pacific Fisheries Commission (WCPFC)	2004	Tuna and other highly migratory	Open	Yes
South Indian Ocean Fisheries Agreement (SIOFA)	2006	All except tuna		Yes

Table 6.1 continued

Name by type	Year of establishment	Coverage of fish stocks	Membership	Decisions binding on parties
Fishery Committee of the West Central Gulf of Guinea (FCWC)	2006	All	Open	Yes
South Pacific Regional Fisheries Management Organization (SPRFMO)	2012	All	Open	Yes
Coordination and development				
Committee for the Eastern Central Atlantic Fisheries (CECAF)	1967	All	Closed	No
Western Central Atlantic Fisheries Commission (WECAFC)	1973	All	Open	No
South Pacific Forum Fisheries Agency (FFA)	1979	Tuna	Closed	No
Latin American Organization for the Development of Fisheries (OLDEPESCA)	1984	All	Closed	No
Sub-Regional Commission on Fisheries (SRCF)	1985	All	Closed	No
Regional Convention on Fisheries Cooperation among African States Bordering the Atlantic Ocean (ASBAO)	1995	All	Closed	No
Scientific research				
International Council for the Exploration of the Sea (ICES)	1902	All	Closed	No
North Pacific Marine Science Organization (PICES)	1992	All	Closed	No

Note: RFMOs are defined as organizations established through international treaties with a primary focus on fisheries issues. The International Whaling Commission is sometimes considered a RFMO and is discussed below.
Source: Adapted from Sydnes (2001)

mandate, more recently 'regional fisheries bodies' are assumed to have an advisory mandate. RFMOs, meanwhile, set fisheries quotas binding on state parties to the agreements (Drankier, 2012).

In the simplest sense, RFMOs are legally mandated fisheries management bodies operating in EEZs, ABNJ or both, depending on the nature of the agreement (Figure 6.3). RFMOs may include 'regional economic integration organizations' such as the EU and semi-sovereign states such as Taiwan. Besides fisheries management, RFMOs gather and exchange information, address broader resource management issues of significant import and nurture scientific cooperation. RFMOs also provide a forum for international cooperation on the management of fisheries found in the EEZs of two or more nations or in the high seas.

Most RFMOs follow the current organizational model of international agreements and have a decision-making body and secretariat. The decision-making body may be a commission, committee, council, meeting of ministers or any other forum selected by signatories.

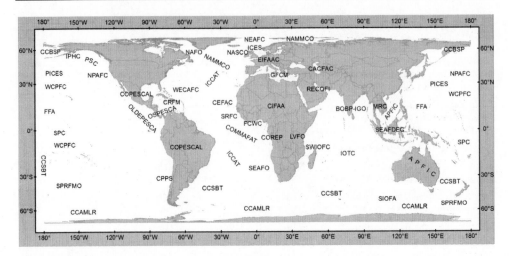

Figure 6.3 Regional fisheries management organizations
Note: Acronyms are listed in Table 6.1
Source: www.fao.org/fishery/topic/14908/en

RFMOs lack supranational powers, meaning that states have to agree to be bound by RFMO decisions. This is one reason why some question if RFMOs are meeting their conservation mandate (Cullis-Suzuki and Pauly, 2010). Certain RFMOs, particularly those established to manage tuna fisheries, have been criticized for ignoring scientific advice and setting unsustainable harvest levels (McCarthy, 2013).

International agreements affecting fisheries

Fisheries treaties between states have existed since the 1880s. Considering the number of commercially important fish stocks that straddle one or more EEZs and the need to manage valuable high-seas stocks it is no surprise that they are prevalent throughout the world. The LOSC formalized the rights of developing and landlocked states to access both high-seas fisheries as well as surplus production in coastal states' EEZs. This created urgency to develop new international agreements that operated within the jurisdictional framework established by the LOSC.

International fisheries agreements negotiated since the LOSC entered into force in 1982 include the following 'soft law' agreements: the FAO Code of Conduct for Responsible Fisheries and its associated FAO international plans of action (IPOAs); and the Jakarta Mandate developed by parties to the CBD.

'Hard law' agreements include (Juda, 2002): the FAO Agreement to Promote Compliance with International Conservation and Management Measures by Fishing Vessels on the High Seas; and the Agreement for the Implementation of the Provisions of the United Nations Convention on the Law of the Sea of 10 December 1982 Relating to the Conservation and Management of Straddling Fish Stocks and Highly Migratory Fish Stocks.

The following sections discuss the key global fisheries agreements negotiated since World War II. There is a particular focus on the management of fisheries under the LOSC. Though many of the following agreements include discussions on aquaculture, since these activities largely occur within states' internal waters they are not discussed here.

International Convention for the Regulation of Whaling

Mechanization and technical improvements to the world's whaling fleets throughout the nineteenth and twentieth centuries resulted in the near extinction of many cetacean species. For example, the blue whale (*Balaenoptera musculus*) was reduced to less than 1 per cent of its historical population levels and perhaps as few as 5000 individuals existed prior to the 1986 moratorium on their harvest. Similarly, the western Pacific gray whale (*Eschrichtius robustus*) population continues to be approximately 130 animals and remains listed as 'critically endangered' by the IUCN (www.iucn.org/wgwap/).

Recognizing that many whale (cetacean) populations had become precariously low as a result of commercial whaling, the International Convention for the Regulation of Whaling (ICRW) was convened in Washington DC in 1946 to conserve whale stocks to ensure the development of an orderly whaling industry (Box 6.3).

Box 6.3 Whale species covered by the ICRW

Of the 120+ species of cetaceans, currently only 21 species of 'great whales' are covered by the ICRW. Cetaceans covered by the Convention either reside in ABNJ or migrate between states and, as such, are 'shared between nations'.

Baleen whales (order Mysticeti – use baleen plates to filter food):

- Blue whale (*Balaenoptera musculus*)
- Bowhead whale (*Balaena mysticetus*)
- Bryde's whale (*Balaenoptera edeni, B. brydei*)
- Fin whale (*Balaenoptera physalus*)
- Gray whale (*Eschrichtius robustus*)
- Humpback whale (*Megaptera novaeangliae*)
- Minke whale (*Balaenoptera acutorostrata, B. bonaerensis*)
- Pygmy right whale (*Caperea marginata*)
- Right whale (*Eubalaena glacialis, E. australis*)
- Sei whale (*Balaenoptera borealis*)

Toothed whales (order Odontoceti)

- Cuvier's beaked whale (*Ziphius cavirostris*)
- Shepherd's beaked whale (*Tasmacetus shepherdi*)
- Baird's beaked whale (*Berardius bairdii*)
- Arnoux's whale (*Berardius arnuxii*)
- Southern bottlenose whale (*Hyperoodon planifrons*)
- Northern bottlenose whale (*Hyperoodon ampullatus*).
- Killer whale (*Orcinus orca*)
- Pilot whale (*Globicephala melaena*) (*G. macrorhynchus*).
- Sperm whale (*Physeter macrocephalus*)

The Convention was formally amended only once, in 1956, through a protocol that designated helicopters or aircraft as 'whale catchers'. The annual catch limits by species and geographic area are published annually in the 'Schedule' to the Convention. The Schedule also prescribes capture methods, the locations of capture and rules for IWC sanctuaries and reporting requirements for whaling states. The Schedule is reviewed and approved annually at the meetings of the Commission.

There are currently 89 member states party to the Convention as well as a number of non-governmental organizations who may attend IWC meetings. The Convention established the International Whaling Commission whose secretariat is headquartered in Cambridge, England. The Commission is comprised of the Scientific, Technical, Finance and Administration, and Conservation Committees. The Scientific Committee recommends research necessary for the regulation of whaling, reports on the status and trends of whale stocks and recommends actions to increase whale populations. The work of the Scientific Committee is used by the Commission to develop regulations for the member states (Zacharias et al., 2006).

Changes to the Schedule require the consent of three quarters of the member states. These modifications become effective 90 days after they are approved at the annual meeting of the Commission. A change to the Schedule is non-binding on signatory states if formally objected to within 90 days. If one state objects within the 90 period, additional time is provided for other states to object.

The IWC's original mandate to regulate whaling in order to foster the development of the whaling industry has changed significantly over the past decades. The IWC, while still retaining its original mandate, has evolved into a conservation organization much to the displeasure of many whaling states. This shift began in 1972 where the UNCHE called on the IWC to establish a ten-year moratorium on commercial whaling along with a recommendation to undertake additional research efforts. In the early 1970s this call coincided with the establishment of global anti-whaling campaigns by Greenpeace and other environmental organizations.

In 1982, the IWC, at its annual meeting, agreed to a 'Comprehensive Assessment' or a pause in commercial whaling. The goal of the Assessment was to undertake population assessments and other research such that improved catch limits could be set to avoid the further decline of harvested populations. The Comprehensive Assessment, better known as the 'moratorium' was later defined by the IWC Scientific Committee as '...an in-depth evaluation of the status of all whale stocks in the light of management objectives and procedures...that...would include the examination of current stock size, recent population trends, carrying capacity and productivity' (www.iwcoffice.org/conservation/estimate.htm). Stated differently, the moratorium (Paragraph 10e of the Schedule) sets commercial whaling catch limits to zero for all species in all areas, irrespective of their conservation status while a comprehensive assessment of stock status is undertaken. Thus, the moratorium, adopted in 1982 and implemented in 1985, is a de facto global whale sanctuary. The moratorium applies only to commercial harvesting activities and a number of nations, including Japan, Norway and Iceland have continued whaling under scientific research permits where the products from this 'research whaling' have been sold commercially. Norway resumed commercial whaling in 1994 and Iceland in 2006.

The moratorium is intended to be replaced by the Revised Management Procedure (RMP), a science-based harvest framework within the broader Revised Management Scheme (RMS). The RMP is a conservative management approach that takes scientific uncertainty explicitly into account. It would allow limited sustainable whaling on populations meeting certain conditions. No catches would be allowed for any populations below 54 per cent of their unexploited level, with the ultimate general aim of stabilizing these populations at approximately 72 per cent of their unexploited level (Young, 2002).

Currently, the IWC is nearly evenly divided between pro- and anti-whaling states. This, of course, makes any IWC business nearly impossible. The Commission is currently examining opportunities to move forward with its business and have proposed resolutions that would severely curtail whaling under scientific permit but would allow limited commercial whaling.

Convention on Fishing and Conservation of the Living Resources of the High Seas (1958)

As we have seen, the primary purpose of the four 1958 Geneva Conventions, culminating in the 1982 LOSC was to codify the geographical limits of, and obligations in, the territorial seas. Yet, the Conventions also laid the groundwork for future fisheries agreements. In particular, the 1958 Conventions bestowed full control of fisheries within a state's territorial seas and continental shelf with no duties of conservation. Beyond territorial waters, the 'freedom of the high seas' doctrine also applied to fishing and the only stipulation (albeit large) was that nations must consider the interests of other nations when fishing in the high seas.

The 1958 Convention on Fishing and Conservation of the Living Resources of the High Seas (the 'Convention on Fishing') was signed on 19 April 1958 and entered into force on 20 March 1966 (UN, 1958). The Convention, for the first time in international law, recognized that fisheries could and were being overexploited. The Convention also recognized that international cooperation was the appropriate vehicle for the sustainable management of high-seas fish stocks. Key articles of the Convention include:

- Article I: All states have the right to fish in the high seas and must cooperate with other states to conserve the living resources of the high seas.
- Article II: Defines the term 'conservation of the living resources of the high seas' to mean obtaining the optimum sustainable yield from these resources.
- Article III: Each state will adopt its own conservation measures when fishing high-seas stocks that are not fished by other nations.
- Article IV: When nationals of two or more states are fishing the same stock on the high seas, states shall enter into negotiations to develop conservation measures for the stock under consideration.
- Article V: Conservation measures for fish stocks developed under Articles III and IV will apply to nationals of other states who intend to fish the same stocks.
- Article VI: Coastal states are entitled to participate in negotiations regarding high-seas fish stocks adjacent to their territorial seas even if nationals of the coastal state are not fishing the stock under consideration.
- Article VII: Coastal states may unilaterally develop conservation measures for high-seas fish stocks adjacent to their territorial seas if a negotiated agreement with other states cannot be completed within six months.
- Article VIII: Non-coastal states whose nationals are not fishing a particular fish stock may request states whose nationals are fishing a particular stock to develop conservation measures as per Articles III and IV.

- Articles IX–XII: Dispute resolution procedures including the establishment of five-member special commissions to address disputes.

While the Convention was the first attempt at managing fish stocks in the high seas it was widely perceived to have a number of shortcomings. It seemed to favour the rights of fishing nations over coastal states and it was unable to accommodate the growing number of nations claiming sovereignty out to 200nmi. Also, in lieu of the ability of coastal states to enforce compliance, the dispute resolution mechanisms appeared hopelessly convoluted. As a result, only 38 states have ratified the Convention. In addition, the concept of managing to MSY has failed, resulting in continued overexploitation of fisheries.

Lastly, the ICJ ruled in 1974 that Iceland could claim a 12nmi zone seaward of the coastline where the state fully controlled fishing rights. The court also recognized that Iceland had 'preferential fishing rights' beyond the 12nmi seaward boundary to an unspecified distance so that its fisheries could be preserved (ICJ, 1975). This decision necessitated the establishment of marine jurisdictional arrangements and was pivotal in the development of the LOSC.

Fisheries management and the Convention on International Trade in Endangered Species of Wild Fauna and Flora (1973)

The Convention on International Trade in Endangered Species of Wild Fauna and Flora (CITES) was signed on 3 March 1973 and entered into force on 1 July 1975. It was amended on 22 June 1979 (the 'Bonn Amendment'). The Convention currently has 175 parties (CITES, 1973). The purpose of CITES is to regulate the trade in wild animals and plants to prevent their overexploitation. The Convention functions by assigning species of concern to one of the following three lists:

- Appendix I: All species threatened with extinction that are, or may be, affected by trade.
- Appendix II: All species that may become threatened with extinction that are, or may be, affected by trade.
- Appendix III: All species that any state identifies as being subject to regulation within its jurisdiction for the purpose of preventing or restricting exploitation, and needing the cooperation of other Parties in the control of trade.

More than 30,000 species are currently listed in the CITES Appendices. While CITES has no authority over the operation of marine fisheries, the Convention has the ability to contribute to fisheries management objectives by listing species of commercial importance. Listed marine species on Appendix I of current or past commercial importance include all marine turtles, all beaked whales, all dugongs, all coelacanths and a number of other seals, dolphins, porpoises, sturgeon and fish. Appendix II lists thousands of marine species, including the Caribbean queen conch (*Strombus gigas*), whale shark (*Rhincodon typus*), great white shark (*Carcharodon carcharias*) and basking shark (*Cetorhinus maximus*) (Cochrane and Doulman, 2005). Proposals for CITES to add additional commercially important species, including threatened or endangered shark, coral and tuna species have not yet been successful. Pressure for CITES to list these species continues to increase.

Fisheries management and the United Nations Convention on the Law of the Sea

It is important to note that while the LOSC is not a prescriptive convention on fisheries management it has fundamentally changed the way global fisheries are managed. For example, the establishment of the 200nmi EEZs resulted in approximately 95 per cent of the world's fisheries harvest to fall within national jurisdiction, either within internal waters, the territorial sea or archipelagic waters. The LOSC sets the legal framework for the conservation, management and research with respect to marine living resources and is the foundation for subsequent international fisheries management agreements (Cochrane and Doulman, 2005). Broadly, the LOSC attempts to balance several considerations, including: coastal states' sovereign rights to manage their fisheries against providing opportunities for other states to fish in their territorial seas and EEZs; optimum utilization against long-term sustainability; and freedom to fish the high seas against the requirement to cooperate with other nations to avoid overexploitation.

The specific LOSC Articles relevant to fisheries management are summarized below:

- Part II (Territorial Sea and Contiguous Zone)
  - Article 19: Fishing activities are not considered innocent passage and may only be conducted with the agreement of the coastal state.
- Part IX (Enclosed or Semi-enclosed Seas)
  - Article 123: Coordinate the management, conservation, exploration and exploitation of the living resources of the sea.
- Part V (EEZs), Articles 61–68
  - Article 61: The coastal state determines the allowable catch of the living resources in its EEZ and is required to ensure that these resources are not overexploited.
  - Article 62: Coastal states will promote the objective of optimum utilization of living resources within their EEZs and provide other states the opportunity to harvest surplus catch. Nationals of other states fishing within another state's EEZ must abide by the laws and regulations of the coastal state.
  - Article 63: States will cooperate on the conservation and development of fish stocks that overlap two or more states' EEZs.
  - Article 64: States that fish for highly migratory species (listed in Annex I), whether in or out of a state's EEZ, must cooperate through an international organization to ensure the conservation and optimum utilization of the stock.
  - Article 65: States will cooperate to conserve marine mammals.
  - Article 66: States in whose rivers anadromous stocks originate have the primary interest in, and responsibility for, such stocks.
  - Article 67: States in whose waters catadromous species spend the greater part of their lifecycle shall have responsibility for the management of these species and shall ensure the ingress and egress of migrating fish.
- Part VII (High Seas), Section 2, Articles 116–120
  - Article 116: All states have the right for their nationals to engage in fishing on the high seas.
  - Article 117: All states have a duty to the conservation of the living resources of the high seas.
  - Article 118: States shall cooperate with each other in the conservation and

management of living resources in the areas of the high seas including the establishment of sub-regional or regional fisheries organizations.

- – Article 119: States will manage for MSY on the high seas based on best scientific evidence. States will share information and not discriminate against fishers from any state.
- – Article 120: States will cooperate to conserve marine mammals on the high seas.
- Annex I Highly Migratory Species
  - – Albacore tuna: *Thunnus alalunga*
  - – Bluefin tuna: *Thunnus thynnus*
  - – Bigeye tuna: *Thunnus obesus*
  - – Skipjack tuna: *Katsuwonus pelamis*
  - – Yellowfin tuna: *Thunnus albacares*
  - – Blackfin tuna: *Thunnus atlanticus*
  - – Little tuna: *Euthynnus alletteratus*; *Euthynnus affinis*
  - – Southern bluefin tuna: *Thunnus maccoyii*
  - – Frigate mackerel: *Auxis thazard*; *Auxis rochei*
  - – Pomfrets: Family *Bramidae*
  - – Marlins: *Tetrapturus angustirostris*; *Tetrapturus belone*; *Tetrapturus pfluegeri*; *Tetrapturus albidus*; *Tetrapturus audax*; *Tetrapturus georgei*; *Makaira mazara*; *Makaira indica*; *Makaira nigricans*
  - – Sail-fishes: *Istiophorus platypterus*; *Istiophorus albicans*
  - – Swordfish: *Xiphias gladius*
  - – Sauries: *Scomberesox saurus*; *Cololabis saira*; *Cololabis adocetus*; *Scomberesox saurus scombroides*
  - – Dolphin: *Coryphaena hippurus*; *Coryphaena equiselis*
  - – Oceanic sharks: *Hexanchus griseus*; *Cetorhinus maximus*; Family *Alopiidae*; *Rhincodon typus*; Family *Carcharhinidae*; Family *Sphyrnidae*; Family *Isurida*
  - – Cetaceans: Family *Physeteridae*; Family *Balaenopteridae*; Family *Balaenidae*; Family *Eschrichtiidae*; Family *Monodontidae*; Family *Ziphiidae*; Family *Delphinidae*.

The LOSC has been complemented by the Agreement to Promote Compliance with International Conservation and Management Measure by Fishing Vessels on the High Seas ('Compliance Agreement' – discussed below) intended to improve the regulation of fishing vessels on the high seas by strengthening 'flag-state responsibility'. Parties to the agreement must ensure that they maintain an authorization and recording system for high-seas fishing vessels and that these vessels do not undermine international conservation and management measures. The agreement aims to deter the practice of 're-flagging' vessels with the flags of states that are unable or unwilling to enforce such measures.

Fisheries management and the UN Conference on Environment and Development

As discussed in Chapter 5, the purpose of the 1992 UNCED was to reconcile environment and development concerns in the context of disparities between nations' abilities to implement sustainable development. The conference launched Agenda 21, aimed at preparing the world for the challenges of the following century and contained a programme of action for sustainable development to be undertaken by the world's nations. Agenda 21 contained

Chapter 17, titled 'Protection of the Oceans', which addresses: integrated management and sustainable development of coastal areas; marine environmental protection; sustainable use and conservation of marine living resources of the high seas; sustainable use and conservation of marine living resources under national jurisdiction; critical uncertainties for the management of the marine environment and climate change; strengthening international, including regional, cooperation and coordination; and sustainable development of small islands (UNEP, 1992).

Specific fishery-related issues that were also raised in Chapter 17 (17.45) included:

- the need for adoption, monitoring and enforcement of effective conservation measures;
- problems of unregulated fishing;
- overcapitalization;
- vessel reflagging to circumvent controls;
- insufficiently selective gear;
- unreliable information;
- lack of sufficient cooperation among states.

Furthermore, Chapter 17.49 encourages states to manage fisheries in accordance with the LOSC and specifically to:

- give effect to the Convention's provisions with regard to straddling stocks that move between EEZs and high seas and with regard to highly migratory species;
- negotiate international instruments for the effective management and conservation of fish stocks;
- define and identify appropriate management units.

Additionally, Chapter 17 calls on states to:

- ensure that fishing activities by ships under their flag minimize incidental catch (17.50);
- take measures to monitor and control high seas-fishing efforts by ships flying their flag in order to ensure compliance with applicable conservation and management rules (17.51);
- deter reflagging of vessels by their nationals who do so to avoid compliance with high-seas conservation and management rules (17.52);
- prohibit dynamiting, poisoning and other destructive fishing practices (17.53);
- fully implement UN General Assembly Resolution 46/25 on large-scale pelagic drift net fishing (17.54);
- reduce wastage, post-harvest loss, discards and improve techniques of processing, distribution and transportation (17.58);
- cooperate within regional and global fisheries bodies and, where such bodies do not exist, establish regional organizations (17.59);
- join regional high-seas fisheries organizations if they have an interest in a high-seas fishery regulated by that organization (17.60).

In addition, UNCED resulted in the UNFCCC and CBD, which are discussed below.

Fisheries management and the United Nations Convention on Biological Diversity

As discussed in Chapter 5, the 1992 CBD came into force in 1993 and is the global umbrella convention for the protection of terrestrial and marine living resources (UN, 1992). The Convention addresses biodiversity at the genetic, species and ecosystem or seascape levels, and directs signatories to establish systems of protected areas and to protect endangered species. A 1995 meeting in Jakarta, Indonesia led to the Jakarta Ministerial Statement (the 'Jakarta Mandate') on the implementation of the CBD, which includes protection of marine biodiversity within sustainable fisheries practices (Sinclair et al., 2002). A programme of action for the Jakarta Mandate was developed in 1998 with a focus on integrated coastal and marine management, sustainable use of marine resources, establishing marine reserves (protected areas) and addressing alien species (Cochrane and Doulman, 1995).

FAO Agreement to Promote Compliance with International Conservation and Management Measures by Fishing Vessels on the High Seas

The Agreement to Promote Compliance with International Conservation and Management Measures by Fishing Vessels on the High Seas ('1993 FAO Compliance Agreement') was approved on 24 November 1993 by Resolution 15/93 of the Twenty-Seventh Session of the FAO Conference. The Agreement entered into force on 24 April 2003. The Agreement originated from discussions at the 1992 UNCED for the need to effectively implement the LOSC. The Compliance Agreement was drafted to advance the objectives for marine living resources in the LOSC and articulate clear responsibilities, particularly for flag states, on ensuring that vessels flying their flag are in compliance with regional and sub-regional high-seas fisheries agreements and that information on high-seas fishing vessels is provided to the FAO (FAO, 1995a). The Agreement closes a legal loop-hole where vessels could change their state of registration ('flag') to avoid being bound by international agreements that had been signed by their previous flag state. The intent of the Agreement is that state parties will transpose the Agreement into national legislation (Freestone et al., 2001).

Key articles of the Compliance Agreement include:

- Article II: Agreement applies to all vessels that are used or intended for fishing on the high seas.
- Article III: Parties to the Agreement commit to ensuring that vessels flying their flag do not engage in activity that undermines the effectiveness of international conservation and management measures. Parties are required to ensure that vessels flying their flag and fishing on the high seas are authorized to do so and are able to operate according to the Agreement. Parties are also required to take enforcement measures against their vessels that contravene this Agreement.
- Article IV: Parties will maintain a record of fishing vessels entitled to fly that flag and authorized to be used for fishing on the high seas.
- Article V: Parties will exchange information, including evidentiary material, relating to activities of fishing vessels. This assists the flag State in identifying those fishing vessels flying its flag reported to have engaged in activities undermining international conservation and management measures.

- Article VI: Parties will provide the FAO with information on each vessel fishing on the high seas.
- Article VII: Parties will support developing nations in implementing this Agreement.
- Article VIII: Parties will encourage non-parties to comply with the Agreement.

1995 United Nations Agreement for the Implementation of the Provisions of the United Nations Convention on the Law of the Sea of 10 December 1982 relating to the Conservation and Management of Straddling Fish Stocks and Highly Migratory Fish Stocks

The Agreement for the Implementation of the Provisions of the United Nations Convention on the Law of the Sea 10 December 1982 relating to the Conservation and Management of Straddling Fish Stocks and Highly Migratory Fish Stocks ('1995 UN Fish Stocks Agreement' or UNFSA) entered into force 11 December 2001 (UN, 1995). Besides ensuring the long-term conservation and sustainable use of straddling fish stocks and highly migratory fish stocks in ABNJ (Article 2), the Agreement provides specific language related to the protection of the marine environment (Articles 5 and 6). As with the FAO Code of Conduct, the intent of the Agreement is that state parties will transpose its objectives into national legislation (Freestone et al., 2001). The Agreement contains the same general measures as the FAO Code of Conduct discussed below, and is intended to (modified from Freestone et al., 2001):

- ensure consistent conservation and management measures for stocks (straddling and highly migratory) found in both EEZs and the high seas;
- apply the precautionary approach to straddling and high migratory stocks;
- articulate duties of the flag state with respect to the operation of their vessels on the high seas;
- articulate the duties and responsibilities of RFMOs;
- confer enforcement powers on non-flag and port states to regulate fishing vessels and provide a means for dispute resolution.

Key articles of the agreement include:

- Article 3: Agreement applies to straddling fish stocks and highly migratory stocks outside of national jurisdiction (Article 3.1), requires states to apply conservation principles outlined in Article 5 (Article 3.2), and commits signatories to assist developing states in implementing the Agreement (Article 3.3).
- Article 5: Requires signatories to: sustainably manage straddling and highly migratory fish stocks and promote optimum utilization (Article 5.a); base management decisions on best available science and pursue maximum sustained yield (Article 5.b); apply the precautionary approach (Article 5.c); consider the ecosystem effects of fishing and manage other fish stocks in an appropriate manner (Article 5.d, e); minimize pollution, waste, discards, catch by lost or abandoned gear, and catch of non-target species (Article 5.f); protect marine biodiversity (Article 5.g); prevent overfishing and excess fishing capacity (Article 5.h); consider artisanal and subsistence fishers (Article 5.i); collect and share data (Article 5.j); conduct scientific research (Article 5.k); and implement monitoring and compliance (Article 5.l).

- Article 6: Requires signatories to adopt the precautionary approach for straddling fish stocks and highly migratory fish stocks in order to protect the living marine resources and preserve the marine environment.
- Article 7: Commits signatories to cooperate on management measures for straddling fish stocks and highly migratory fish stocks. It also provides specific direction on how states should cooperate with each other.
- Article 8: Commits signatories to cooperate and resolve disputes through sub-regional or regional fisheries management organizations or other arrangements.
- Articles 9–13: Provides further specifics on the operation of sub-regional or regional fisheries management organizations or arrangements.
- Article 14: Commits states to collect and provide information and cooperation in scientific research.

FAO Code of Conduct for Responsible Fisheries

The FAO Code of Conduct for Responsible Fisheries was adopted on 31 October 1995 and originated from a recommendation from the 1992 International Conference on Responsible Fishing, held in Cancun (Mexico). Unlike other fisheries agreements discussed in this Chapter, the Code is voluntary and provides the principles and standards applicable to the conservation, management and development of all fisheries. The Code also addresses the capture, processing and trade of fish and fishery products. Fishing operations, aquaculture, fisheries research and the integration of fisheries into coastal area management fall under the Code's umbrella as well (FAO, 1995b).

While voluntary, the Code is an ambitious effort to integrate a number of binding principles from the LOSC, the FAO Compliance Agreement and the UN Fish Stocks Agreement. To address all aspects of fishing, a much broader set of operating principles and best practices to address all aspects of fishing is exploited. The Code applies to inland fisheries and also addresses aquaculture development (Hoscha et al., 2011).

Article 7 (Fisheries Management) is the most pertinent Article to fisheries management and commits signatories to adopt measures for the long-term conservation and sustainable use of fisheries resources. The management scheme calls for effective compliance and enforcement, transparent science and decision-making, reductions in excess fishing capacity, protection of aquatic habitats and endangered species, minimizing bycatch and implementation of the precautionary approach. Specifically, the agreement states that there is a '. . . conscious need to avoid adverse impacts on the marine environment, preserve biodiversity, and maintain the integrity of marine ecosystems' (Valdimarrson and Metzner, 2005).

Article 8 (Fishing Operations) describes the duties of flag and port states with respect to: monitoring; compliance and enforcement; health and safety standards; search and rescue; education and training; recording keeping and reporting; vessel and gear markings; insurance; and accident reporting. This Article also directs states to: prohibit the use of dynamite, poison and other destructive fishing practices; minimize bycatch and discards; utilize methods to limit the loss of fishing gear; develop and utilize selective fishing gear; reduce energy consumption and minimize pollution; provide port-based services to minimize pollution and introduction of alien species; consider the use of artificial reefs and aggregation devices to increase stock populations.

Additional Articles address the following subject areas:

- Article 9: Aquaculture development.
- Article 10: Integration of fisheries into coastal area management.
- Article 11: Post-harvest practices and trade.
- Article 12: Fisheries research.

Importantly, the Code of Conduct introduced and reinforced key principles of sustainable fisheries management. Among these principles was use of traditional ecological knowledge (6.4), the precautionary approach (6.5 and 7.5), an ecosystem-based approach to fisheries management (6.2, 6.4, 6.6, 6.8, 7.2.2 and 7.3.3.), environmentally responsible gear and practices (6.6 and 8.5) and international cooperation (7.1.4 and 7.1.5.). Also, the Code identifies the FAO Secretariat to monitor the application of the Code and to report on developments to the Committee of Fisheries (4.2) (Juda, 2002).

The Code engendered a global discussion on fisheries management and informed the development of 'hard law' agreements including the FAO Compliance Agreement, the UN Fish Stocks Agreement and subsequent amendments to the Code of Conduct (discussed below) (Juda, 2002).

The Code was amended by the Rome Declaration on the Implementation of the Code of Conduct for Responsible Fisheries, agreed to on 11 March 1999 at a FAO Ministerial Meeting on Fisheries (FAO, 1999). The Declaration specifically recognized the following:

- Article 2: The growing amount of IUU fishing activities being carried out, including fishing vessels flying 'flags of convenience'.
- Article 3: The importance of ensuring the contribution of sustainable aquaculture to food security, income and rural development.
- Article 5: The need for further technical and financial support for certain states to apply the Code and Plans of Action.
- Article 6: The need to give greater consideration to the development of more appropriate ecosystem approaches to fisheries development and management.

To date, aspects of the Code have proved particularly challenging for both more developed and less developed states, particularly with respect to the amount and quality of information necessary to implement the Code. Hoscha et al. (2011) found, in a review of the implementation of the Code in 9 states, that its goals have been hampered by socioeconomic considerations, administrative inertia, faltering political will and short-sighted leadership. Furthermore, they found that little process had been made on combating IUU fishing and addressing fishing overcapacity.

The Code has also provided the basis for the development of four International Plans of Action for the following, which are administered by the FAO (the two primary plans are discussed below):

- IPOA for the conservation and management of sharks (1999).
- IPOA for reducing the incidental catches of seabirds (1999).
- IPOA for the management of fishing capacity (1999).
- IPOA to prevent, deter and eliminate IUU fishing (2001).

International Plan of Action for the Management of Fishing Capacity

Fishing capacity is a measure of the total amount of fishing effort (or fish) that can be produced over a given time period or region for a fully utilized fishing fleet. Currently, the world's fishing fleets have more capacity than is required for most targeted species. This has led to situations where fishing becomes economically and environmentally inefficient. Overcapacity results in lower revenues (and thus wages) for fishers, greater pressures on fish populations (the race to fish) and political pressure on governments to increase catch limits (or provide subsidies) in order to maintain employment in the industry. Overcapacity is a result of open access to fisheries where more fishers invest in technological improvements (e.g. boat size, gear types), regulations are weak or poorly enforced and the growth in fish markets further encourages additional fishing effort (Aranda et al., 2012). It is estimated that the global fishing fleet would need to be reduced by 25 per cent to eliminate overcapacity and that overcapacity (along with overfishing) reduces resource rents (the surplus after costs have been subtracted by revenue) by $50 billion annually (World Bank and FAO, 2009).

The International Plan of Action for the Management of Fishing Capacity (IPOA-Capacity) was adopted 19 February 1999 and applies to all fishers engaged in capture fisheries. The IPOA-Capacity contains the following four strategies:

- Determining fishing capacity at the national, regional and global levels and improving the measurement of capacity.
- Preparing national plans to address overcapacity and immediately attending to coastal fisheries that are threatened by overcapacity.
- Strengthening regional fisheries management organizations to address overcapacity.
- Taking immediate steps to address straddling, highly migratory and high-seas fisheries that are threatened by overcapacity.

The IPOA-Capacity ambitiously set goal for states and regional fisheries management organizations to implement efficient, equitable and transparent programmes to manage overcapacity. Currently, however, only Indonesia, Namibia and the US have national plans of action and the I-ATTC is the only RFMO to have an action plan for overcapacity. The inability to define 'capacity', limitations in the application of vessel registries and impacts of capacity reduction on developing states have been barriers to success. Lack of accurate catch data and lack of a rights-based management regime have also hampered progress (Aranda et al., 2012).

International Plan of Action to Prevent, Deter and Eliminate Illegal, Unregulated and Unregulated Fishing

IUU fishing may comprise up to 30 per cent of the global fisheries catch, with a landed value of between $10 billion and $23 billion (Agnew et al., 2009; Borit and Olsen, 2012; Österblom and Bodin, 2012). IUU fishing may defeat efforts to sustainably manage regional fisheries, manage seafood safety through traceability (chain of custody) and can place fishers in dangerous situations or without adequate compensation (Sumaila et al., 2006). With an annual landed value greater than $5 million, the Patagonian toothfish is perhaps the most heavily exploited IUU species. Moreover, as a result of the use of long-line fishing techniques for toothfish, bycatch impacts on albatross (Diomedidae spp.) and white-chinned petrels (*Procellaria aequinoctialis*) are substantial (Sumaila *et al.*, 2006; Österblom and Bodin, 2012).

The International Plan of Action to Prevent, Deter and Eliminate Illegal, Unreported and Unregulated Fishing (IPOA-IUU) was adopted by the FAO 2 March 2001 (FAO, 2001). The Plan was developed within the framework of the Code of Conduct for Responsible Fisheries and broadly commits signatories to the following actions:

- Ratify, accept or accede to the LOSC, UNFSA and the FAO Compliance Agreement and fully implement the measures in these international agreements (Articles 10–12).
- Establish national legislation to address IUU fishing and ensure that nationals are not participating in IUU fishing (Articles 16–17).
- Ensure that fishing vessels are flagged under states complying with relevant international agreements related to IUU fishing (Article 19).
- Establish sanctions/penalties sufficiently severe to deter IUU fishing and to avoid economic support or subsidies to those participating in IUU fishing (Articles 21 and 23).
- Undertake monitoring, control and surveillance of all aspects of fishing to control IUU fishing (Article 24).
- Cooperate with other states to deter and eliminate IUU fishing as well as publishing details of IUU fishing measures to prevent IUU fishing (Article 28).

The Plan places primary responsibility for the prevention of IUU fishing with the flag state (Articles 34–50), but also tasks coastal (Article 51) and port (Articles 52–64) states with actions to contribute to its overall objectives. Furthermore, the Plan identifies the need to implement market-related measures to reduce the economic incentive to participate in IUU fishing (Articles 65–76). These include: education and awareness; catch certification and documentation; export controls and prohibitions; catch traceability; and publishing information on vessels participating in IUU fishing. Lastly, the Plan outlines the responsibilities of RFMOs (see above) in deterring and eliminating IUU fishing.

Since the Plan's release, the EU has adopted the EU IUU Regulation (1005/2008) and EU Control Regulation (1224/2009) to describe and mandate the development of a new seafood traceability programme. It encompasses all components of the seafood production cycle (Borit and Olsen, 2012). In addition, the CCAMLR has demonstrated to be an effective regional body for addressing IUU fishing of the Patagonian toothfish (Österblom and Bodin, 2012).

Reykjavik Declaration on Responsible Fisheries in the Marine Ecosystem

Following the development of the FAO Code of Conduct, the FAO prepared a background paper in 2001 titled 'Towards Ecosystem-Based Fisheries Management' for the Reykjavik Conference on Responsible Fisheries in the Marine Ecosystem (FAO, 2003). The key messages of this document are captured as the 2001 Reykjavik Declaration on Responsible Fisheries in the Marine Ecosystem; signatories are committed to:

- Advance the scientific basis for developing and implementing management strategies that incorporate ecosystem considerations, and which will ensure sustainable yields while conserving stocks and maintaining the integrity of ecosystems and habitats on which they depend.

- Identify and describe the structure, components and functioning of relevant marine ecosystems, diet composition and food webs, species interactions and predator–prey relationships, the role of habitat and the biological, physical and oceanographic factors affecting ecosystem stability and resilience.
- Build or enhance systematic monitoring of natural variability and its relationship to ecosystem productivity.
- Improve the monitoring of bycatch and discards in all fisheries to obtain better knowledge of the amount of fish actually taken.
- Support research and technological developments for fishing gear and practices. This improves gear selectivity and reduces adverse impacts of fishing practices on habitat and biological diversity.
- Assess adverse human impacts of non-fisheries activities on the marine environment as well as the consequences of these impacts for sustainable use.

Following the Reykjavik Declaration, the FAO published the FAO Operational Guidelines on the ecosystem approach to fisheries (FAO, 2003) that defines the EAF as one that '…strives to balance diverse societal objectives, by taking into account the knowledge and uncertainties about biotic, abiotic and human components of ecosystems and their interactions and applying an integrated approach to fisheries within ecologically meaningful boundaries' (FAO, 2003).

Chapter summary

- Over 1 billion people depend on seafood as their primary source of protein and fishing is estimated to employ 34 million in people in full- or part-time jobs. The first-sale value of the world's fisheries is $84.9 billion.
- Marine fisheries – and fisheries in ABNJ in particular – have been poorly managed or not managed at all.
- Fisheries may be managed under single-species, multi-species or ecosystem-based regimes.
- The FAO is responsible for reporting on the status of the world's fisheries, providing information to decision-makers and supporting RFMOs.
- A number of binding and non-binding agreements have been developed to govern fishing (including whaling).

References

Agnew, D. J. (1997) 'The CCAMLR ecosystem monitoring programme', *Antarctic Science*, 9 (3), 235–242.

Agnew, D. J., Pearce, J., Pramond, G., Peatman, T., Watson, R., Beddington, J. R. and Pitcher, T. J. (2009) 'Estimating the worldwide extent of illegal fishing', *PLoS 1*, 4 (2), 4570.

Aranda, M., Murua, H. and de Bruyn, P. (2012) 'Managing fishing capacity in tuna regional fisheries management organizations (RFMOs): Development and state of the art', *Marine Policy*, 36, 985–992.

Babcock, E. A. and Pikitch, E. K. (2004) 'Can we reach agreement on a standardized approach to ecosystem-based fishery management?', *Bulletin of Marine Science*, 74 (3), 685–692.

Baranov, F. I. (1918) 'On the question of the biological basis of fisheries', *Nauch Issled Ikhtiol Inst Izu*, 1 (1), 81–128.

Borit, M. and Olsen, P. (2012) 'Evaluation framework for regulatory requirements related to data recording and traceability designed to prevent illegal, unreported and unregulated fishing', *Marine Policy*, 36, 96–102.

CEC (Commission of the European Communities) (2008) *Council Facts and Figures on the CFP: Basic data on the Common Fisheries Policy. Edition 2008*, Office for Official Publications of the European Communities, Luxemburg.

Christensen, V., Aiken, K. A. and Villanueva, M. C. (2007) 'Threats to the ocean: On the role of ecosystem approaches to fisheries', *Social Science Information sur les Sciences Sociales*, 46 (1), 67–86.

Christie, P., Fluharty, D. L., White, A. T., Eisma-Osorio, L. and Jatulan, W. (2007) 'Assessing the feasibility of ecosystem-based fisheries management in tropical contexts', *Marine Policy*, 31 (3), 239–250.

CITES (Convention on International Trade in Endangered Species of Wild Fauna and Flora) (1973) *Convention on International Trade in Endangered Species of Wild Fauna and Flora*, www.cites.org/eng/disc/text.php, accessed 3 February 2013.

Cochrane, K. L. (2002) 'A fishery manager's guidebook. Management measures and their application', *FAO Fisheries Technical Paper*, 424, FAO, Rome.

Cochrane, K. L. and Doulman, D. J. (2005) 'The rising tide of fisheries instruments and the struggle to keep afloat', *Philosophical Transactions of the Royal Society of London B*, 360 (1435), 77–94.

Constable, A. J. (2004) 'Managing fisheries effects on marine food webs in Antarctica: Trade-offs among harvest strategies, monitoring, and assessment in achieving conservation objectives', *Bulletin of Marine Science*, 74 (3), 583–605.

Cullis-Suzuki, S. and Pauly, D. (2010) 'Failing the high seas: A global evaluation of regional fisheries management organizations', *Marine Policy*, 34, 1036–1042.

Curry, P. M., Mullon, C., Garcia, S. M. and Shannon, L. J. (2005) 'Viability theory for an ecosystem approach to fisheries', *ICES Journal of Marine Science*, 62 (3), 577–584.

Davies, R. W. D. and Rangeley, R. (2010) 'Banking on cod: Exploring economic incentives for recovering Grand Banks and North Sea cod fisheries', *Marine Policy*, 34, 92–98.

Day, V., Paxinos, R., Emmett, J., Wright, A. and Goecker, M. (2008) 'The Marine Planning Framework for South Australia: A new ecosystem-based zoning policy for marine management', *Marine Policy*, 32 (4), 535–543.

Dew, C. B. and Austring, R. G. (2007) 'Alaska red king crab: A relatively intractable target in a multi-species trawl survey of the eastern Bering Sea', *Fisheries Research*, 85, 265–173.

Drankier, P. (2012) 'Marine protected areas in areas beyond national jurisdiction', *The International Journal of Marine and Coastal Law*, 27, 291–350.

FAO (Food and Agriculture Organization) (1995a) *Agreement to Promote Compliance with International Conservation and Management Measures by Fishing Vessels on the High Seas*, FAO, Rome, www.fao.org/docrep/meeting/003/x3130m/X3130E00.HTM, accessed 3 February 2013.

FAO (1995b) *Code of Conduct for Responsible Fisheries*, FAO, Rome, www.fao.org/docrep/005/v9878e/v9878e00.HTM, accessed 3 February 2013.

FAO (1997) 'Fisheries Management', *FAO Technical Guidelines for Responsible Fisheries*, 4, FAO, Rome, ftp://ftp.fao.org/docrep/fao/003/w4230e/w4230e00.pdf, accessed 26 April 2013.

FAO (1999) *International Plan of Action for the Management of Fishing Capacity*, FAO, Rome, ftp://ftp.fao.org/docrep/fao/006/x3170e/X3170E00.pdf, accessed 3 February 2013.

FAO (2001) *International Plan of Action to Prevent, Deter and Eliminate Illegal, Unreported and Unregulated Fishing*, FAO, Rome, http://www.fao.org/docrep/003/y1224e/y1224e00.htm, accessed 3 February 2013.

FAO (2003) 'The ecosystem approach to fisheries', *FAO Technical Guidelines for Responsible Fisheries*, 4 (2), FAO, Rome, www.fao.org/docrep/011/i0151e/i0151e00.htm, accessed 3 February 2013.

FAO (2009a) *Rome Declaration on the Implementation of the Code of Conduct for Responsible Fisheries*, FAO, Rome, www.fao.org/docrep/005/X2220E/X2220E00.HTM, accessed 3 February 2013.

FAO (2009b) *The State of World Fisheries and Aquaculture 2008*, FAO, Rome, www.fao.org/fishery/publications/sofia/en, accessed 3 February 2013.

FAO (2011) *Review of the State of World Marine Fishery Resources*, FAO Fisheries and Aquaculture

Technical Paper No. 569, FAO, Rome, www.fao.org/docrep/015/i2389e/i2389e.pdf, accessed 3 February 2013.

Freestone, D., Gudmundsdotti, E. and Edeson, W. (2001) *Legislating for Sustainable Fisheries*, World Bank, Washington, DC.

Frid, C. L. J., Paramor, O. A. L. and Scott, C. L. (2005) 'Ecosystem-based fisheries management: Progress in the NE Atlantic', *Marine Policy*, 29 (5), 461–469.

Frid, C. L. J., Paramor, O. A. L. and Scott, C. L. (2006) 'Ecosystem-based management of fisheries: Is science limiting?', *ICES Journal of Marine Science*, 63 (91), 567–1572.

Froese, R., Stern-Pirlot, A., Winker, H. and Gascuel, D. (2008) 'Size matters: How single-species management can contribute to ecosystem-based fisheries management', *Fisheries Research*, 92 (2–3), 231–241.

Goodman, D., Mangel, M., Parkes, G., Quinn, T., Restrepo, V., Smitch, T. and Stokes, K (2002) *Scientific Review of the Harvest Strategy Currently Used in the BSAI and GIA Groundfish Fishery Management Plans*, North Pacific Fishery Management Council, Anchorage, AL.

Graham, M. M. (1939) 'The sigmoid curve and the over-fishing problem', *Rapp. P.-v. Réun. Cons. perm. int. Explor. Mer*, 110 (2), 15–20.

Guerry, A. D. (2005) 'Icarus and Daedalus: Conceptual and tactical lessons for marine ecosystem-based management', *Frontiers in Ecology and the Environment*, 3, 202–211.

Henriksen, T. (2009) 'Revisiting the freedom of fishing and legal obligations on states not party to regional fisheries management organizations', *Ocean Development & International Law*, 40, 80–96.

Hilborn, R. (2004) 'Ecosystem-based fisheries management: The carrot or the stick?', *Marine Ecology Progress Series*, 274, 275–278.

Hirshfield, M. F. (2005) 'Implementing the ecosystem approach: Making ecosystems matter', *Marine Ecology Progress Series*, 300, 253–257.

Hoscha, G., Ferrarob, G. and Faillerc, P. (2011) '1995 FAO Code of Conduct for Responsible Fisheries: Adopting, implementing or scoring results?', *Marine Policy*, 35 (2), 189–200.

ICJ (International Court of Justice) (1975) *Fisheries Jurisdiction Cases: Volume I*, www.icj-cij.org/docket/files/55/9409.pdf, accessed 3 February 2013.

Jennings S. and Revill A. S. (2007) 'The role of gear technologists in supporting an ecosystem approach to fisheries', *ICES Journal of Marine Science*, 64, 1525–1534.

Juda, L. (2002) 'Rio plus ten: The evolution of international marine fisheries governance', *Ocean Development and International Law*, 33, 109–144.

King, D. M. and Sutinen, J. G. (2010) 'Rational noncompliance and the liquidation of Northeast groundfish resources', *Marine Policy*, 34 (1), 7–21.

Lackey, R. T. and Nielsen, L. A. (1980) *Fisheries Management*, John Wiley and Sons, New York.

Larkin, P. A. (1977) 'An epitaph for the concept of maximum sustainable yield', *Transactions of the American Fisheries Society*, 106, 1–11.

Link, J. S. (2005) 'Translating ecosystem indicators into decision criteria', *ICES Journal of Marine Science*, 62 (3), 569–576.

Lotze, H. K. (2004) 'Repetitive history of resource depletion and mismanagement: The need for a shift in perspective', *Marine Ecology Progress Series*, 274, 282–285.

Mace P. M. (2001) 'A new role for MSY in single-species and ecosystem approaches to fisheries stock assessment and management', *Fish and Fisheries*, 2, 2–32.

Mace, P. M. (2004) 'In defence of fisheries scientists, single-species models and other scapegoats: Confronting the real problems', *Marine Ecology Progress Series*, 274, 285–291.

Marasco, R. J., Goodman, D., Grimes, C. B., Lawson, P. W., Punt, A. E. and Quinn, T. J. (2007) 'Ecosystem-based fisheries management: Some practical suggestions', *Canadian Journal of Fisheries and Aquatic Sciences*, 64 (6), 928–939.

McCarthy, M. (2013) 'Is this the end of the bluefin tuna?' www.independent.co.uk/environment/nature/is-this-the-end-of-the-bluefin-tuna-1040246.html, accessed 02 February 2013.

Murawski, S. A. (2007) 'Ten myths concerning ecosystem approaches to marine resource management', *Marine Policy*, 31 (6), 681–690.

Myers, R. A. and Worm, B. (2003) 'Meta-analysis of cod–shrimp interactions reveals top-down control in oceanic food webs', *Ecology*, 84, 162–173.

NAFO (North Atlantic Fisheries Organization) (2008) *Report of the Fisheries Commission Intersessional Meeting, 30 April – 07 May 2008 Montreal, Quebec, Canada*, North Atlantic Fisheries Organization, Dartmouth, Canada.

NMFS (National Marine Fisheries Service) (2009) 'Fisheries of the United States, 2009', www.st.nmfs.noaa.gov/st1/fus/fus09/fus_2009.pdf, accessed 10 September 2010.

New Zealand Ministry of Fisheries (2009) 'Stock status', www.fs.fish.govt.nz/Page.aspx?pk=16, accessed February 2010.

Österblom, H. and Bodin, Ö. (2012) 'Global cooperation among diverse organizations to reduce illegal fishing in the Southern Ocean', *Conservation Biology*, 26 (4), 638–648.

Pauly, D. and Froese, R. (2012) Comments on FAO's State of Fisheries and Aquaculture, or 'SOFIA 2010', *Marine Policy*, 36, 746–752.

Pauly, D., Christensen, V., Dalsgaard, J., Froese, R. and Torres F. (1998) 'Fishing down marine food webs', *Science*, 279 (5352), 860–863.

Pauly, D., Christensen, V., Guenette, S., Pitcher, T. J., Sumaila, U. R., Walters, C. J., Watson, R. and Zeller, D. (2002) 'Towards sustainability in world fisheries', *Nature*, 418 (6898), 689–695.

Pavlikakis, G. E. and Tsihrintzis, V. A. (2000) 'Ecosystem management: A review of a new concept and methodology', *Water Resources Management*, 14 (4), 257–283.

PICES (North Pacific Marine Science Organization) (2004) *Marine Ecosystems of the North Pacific* (edited S. M. McKinnell), PICES Special Publication 1, North Pacific Marine Science Organization, Sidney, British Columbia, Canada.

Pikitch, E. K., Santora, C., Babcock, E. A., Bakun, A., Bonfil, R., Conover, D. O., Dayton, P., Doukakis, P., Fluharty, D., Heneman, B., Houde, E. D., Link, J., Livingston, P. A., Mangel, M., McAllister, M. K., Pope, J. and Sainsbury, K. J. (2004) 'Ecosystem-based fishery management', *Science*, 305 (5682), 346–347.

Pitcher, T. J., Kalikoski, D., Short, K., Varkey, D. and Pramod, G. (2009) 'An evaluation of progress in implementing ecosystem-based management of fisheries in 33 Countries', *Marine Policy*, 33 (2), 223–232.

Rogers, S. I., Tasker, M. L., Earll, R. and Gubbay, S. (2007) 'Ecosystem objectives to support the UK vision for the marine environment', *Marine Pollution Bulletin*, 54 (2), 128–144.

Russell, E. S. (1931) 'Some theoretical considerations on the "overfishing" problem', *Journal de Conseil International pour l'Exploration de la Mer*, 6, 3–20.

Scandol, J. P., Holloway, M. G., Gibbs, P. J. and Astles, K. L. (2005) 'Ecosystem-based fisheries management: An Australian perspective', *Aquatic Living Resources*, 18 (3), 261–273.

Shelton, P. A. (2009) 'Eco-certification of sustainably managed fisheries: Redundancy or synergy?', *Fisheries Research*, 100 (3), 185–190.

Sinclair, M., Arnason, R., Csirke, J., Karnicki, Z., Sigurjonsson, J., Skjoldal, H. R. and Valdimarsson, G. (2002) 'Responsible fisheries in the marine ecosystem', *Fisheries Research*, 58 (3), 255–265.

Sissenwine, M. P. and Murawski. S. (2004) 'Moving beyond "intelligent tinkering": Advancing an ecosystem approach to fisheries', *Marine Ecology Progress Series*, 274, 291–295.

Sumaila, U. R., Alder, J. and Keith, H. (2006) 'Global scope and economics of illegal fishing', *Marine Policy*, 30 (6), 696–703.

Sumaila, U. R., Khan, A., Dyck, A., Watson, R., Munro, G., Tyedmers, P. and Pauly, D. (2010) 'A bottom up reestimation of global fisheries subsidies', *Journal of Bioeconomics*, 12, 201–225.

Sydnes, A. K. (2001) 'Regional fishery organizations: How and why organizational diversity matters', *Ocean Development & International Law*, 32, 349–372.

Symes, D. (2007) 'Fisheries management and institutional reform: A European perspective', *ICES Journal of Marine Science*, 64 (4), 779–785.

Taylor, H. F. (1951) *Survey of Marine Fisheries of North Carolina*, North Carolina University Press, Chapel Hill, NC.

Tyedmers, P. H., Watson, R. and Pauly, D. (2005) 'Fueling global fishing fleets', *Ambio*, 34 (8), 635–638.

UN (United Nations) (1958) *Convention on Fishing and Conservation of the Living Resources of the High Seas*, United Nations, Treaty Series, 559, 285, untreaty.un.org/ilc/texts/instruments/english/conventions/8_1_1958_fishing.pdf, accessed 3 February 2013.

UN (1992) *Convention on Biological Diversity*, www.cbd.int/doc/legal/cbd-en.pdf, accessed 3 February 2013.

UN (1995) *The United Nations Agreement for the Implementation of the Provisions of the United Nations Convention on the Law of the Sea of 10 December 1982 relating to the Conservation and Management of Straddling Fish Stocks and Highly Migratory Fish Stocks*, www.un.org/depts/los/convention_agreements/texts/fish_stocks_agreement/CONF164_37.htm, accessed 3 February 2013.

UNEP (United Nations Environment Programme) (1992) *Protection Of The Oceans, All Kinds Of Seas, Including Enclosed And Semi-Enclosed Seas And Coastal Areas And The Protection Rational Use And Development Of Their Living Resources*, UNEP, Nairobi, Kenya, www.unep.org/documents.multilingual/default.asp?DocumentID=52&ArticleID=65&l=en, accessed 26 April 2013.

Valdermarsen, J. W. and Suuronen, P. (2003) 'Modifying Fishing Gear to Achieve Ecosystem Objectives', in M. Sinclair and G. Valdimarsson (eds) *Responsible Fisheries in the Marine Ecosystem*, FAO, Rome.

Valdimarrson, G., and Metzner, R. (2005) 'Aligning incentives for a successful ecosystem approach to fisheries management', *Marine Ecology Progress Series*, 300, 286–291.

Ware, D. M. and Thomson, R. E. (2005) 'Bottom–up ecosystem trophic dynamics determine fish production in the Northeast Pacific', *Science*, 308, 1280–1284.

Watson, R. A., Cheung, W. W. L., Anticamara, J. A., Sumaila, U. R., Zeller, D. and Pauly, D. (2012) 'Global marine yield halved as fishing intensity redoubles', *Fish and Fisheries*, DOI: 10.1111/j.1467-2979.2012.00483.x.

World Bank and FAO (2009) *Sunken Billions: The Economic Justification for Fisheries Reform*, Washington, DC, http://siteresources.worldbank.org/EXTARD/Resources/336681-1224775570533/SunkenBillionsFinal.pdf, accessed 3 February 2013.

Zacharias, M. A., Gerber, L. R. and Hyrenbach, K. D. (2006) 'External scientific review of the Southern Ocean Sanctuary', *Journal of Cetacean Research and Management*, 8, 1–12.

Chapter 7

Marine transportation and safety policy

The operation and regulation of international shipping

Introduction

Until recently archaeologists believed that humans (*Homo sapiens*) and their recent ancestors (*Homo erectus*) did not cross deep-water straits until approximately 50,000 years ago. However, recent evidence suggests that *Homo erectus* may have crossed the Wallace Line (an imaginary boundary separating Australian and Asian floras and faunas based on marine geographic separation) up to 800,000 years ago and that Australia was colonized more than 60,000 years ago. Evidence from South-east Asia also indicates that humans were frequently travelling between the Moluccas, Borneo and Bali as long ago as 33,000 years ago (Bellwood, 1997; Gibbons, 1998).

The first recorded instances of the use of long rowing boats were documented in Egypt in the time of Sesostris III (3700–4100 years ago). Supporting these written records was the discovery of six boats, averaging 10m in length, next to the pyramid of Sesostris III. Until the appearance of the modern galley in the sixteenth century the Egyptian longboats ruled the seas.

While Sesostris III and subsequent dynasties likely used early galleys to ferry soldiers and supplies to support military endeavours, the first recorded naval battle was the Battle of Artemisium in 480 BC. Xerxes, a Persian King (519–465 BC), is reported to have assembled a fleet comprised 1207 galleys, possibly carrying more than 200,000 men. While the battle against the Greek fleet was indecisive, the clash demonstrated the value of navies to supporting military operations (Mordal, 1959).

The age of navigation (between 1000 and 3000 years ago) saw the advent of the astrolabe (a device that calculates position using sun and star positions), compass and the use of more sophisticated boats based on the Egyptian longboat model. In addition to the Mediterranean nations, Arab, Chinese and Pacific Ocean cultures also exploited celestial navigation during this period.

The first large-scale use of the oceans was to explore and colonize other states. Most likely, the first colonizers to roam beyond the Mediterranean Sea were the Norse explorers of Northern Europe. In the tenth century they settled Greenland and parts of eastern North America. In addition, Polynesian explorers likely landed in the Hawaiian Islands in the tenth century. In the fifteenth century, guided by Papal Bull, Portugal and Spain began a protracted marine campaign to annex the Americas (see Chapter 2). The British and Dutch followed suit in the seventeenth century and the annexation of Africa occurred in the nineteenth and twentieth centuries by a number of European nations.

With colonization arose the need to transport goods and people to and from the colonies. The global shipping trade was born, complete with pirates, competition between shipping companies and the development of naval ships to protect trade routes. The first modern ship designed to carry bulk goods (e.g. coal, ore, grain) was built in 1852 (650 dead weight tonnes – dwt) and followed in 1886 with the 3000 dwt oil tanker named the *Gluckauf* (IMO, 2012). With the advent of modern shipping came the establishment of the modern seaborne trade routes, which currently include the movement of (IMO, 2012):

- Coal from Australia, Southern Africa and North America to Europe and the Far East.
- Grain from North and South America to Asia, Africa and the Far East.
- Iron ore from South America and Australia to Europe and the Far East.
- Oil from the Middle East, West Africa, South America and the Caribbean to Europe, North America and Asia.
- Containerized goods from the People's Republic of China, Japan and South-east Asia.

This chapter presents a short overview of the operation of contemporary shipping. Specifically, it will discuss the configuration of the global shipping fleet, the processes by which shipping between states is conducted and the major international agreements that regulate shipping and the safety of seafarers.

An overview of contemporary shipping

The overall function of shipping has not changed over the millennia: shipping moves goods from a region where they have a low value to a region where they have a high value (Branch, 2007). Today, over 80 per cent of internationally traded goods are moved by shipping and this percentage is higher for trade conducted between developing countries. In 2011, 90 per cent of the EU's foreign trade and 40 per cent of trade between EU member states was conducted by sea (IMO, 2012).

Shipping continues to be the most cost-effective method of moving goods over long distances. The continued globalization of trade combined with more efficient shipping technologies will likely result in the continued dominance of global trade by shipping. As of 2010, the world's merchant shipping fleet consisted of approximately 104,000 ships over 100 GT (gross tonnage – a measure of a ship's internal volume) (Box 7.1), representing approximately 1.4 billion GT with an average ship age of 22 years (UNCTAD, 2012). The world's cargo fleet consists of more than 55,100 ships, representing more than 991 million GT with an average age of 19 years (IMO, 2012) (Table 7.1). In 2010, shipping employed over 1.5 million seafarers from 150 states.

Table 7.1 World fleet size in 2011 by principal types of vessel

Vessel type	Thousands of dwt	Percentage of fleet
Bulk carriers	532,039	38.1
Oil tankers	474,846	34
General cargo ships	108,971	13.2
Other types of ship	103,478	7.4
Liquefied gas carriers	43,339	3.1
Offshore supply	33,227	2.4
Chemical tankers	5,849	0.4
Ferries and passenger ships	6,164	0.4
Total	1,395,743	100

Source: UNCTAD (2011)

Box 7.1 Methods of categorizing ship size

Deadweight tonnage (dwt): The number of tons (2240 lb) of cargo, stores and fuel a vessel can transport.

Cargo tonnage: The weight of cargo carried by the vessel, measured either in American short tons (2000 lb), English long tons (2240 lb) or metric tonnes (1000 kg).

Gross tonnage (GT, G.T or gt): Volume of a vessel's enclosed spaces divided by 100. A GT is 100 cubic feet. Gross tonnage is not a measure of mass (weight) but volume.

Net tonnage (NT, N.T. or nt): Gross tonnage minus all areas of a ship not used to transport cargo. Also referred to as 'earning space'. Net tonnage is not a measure of mass (weight) but volume.

At present, 35 states control 95 per cent of the world's shipping fleet. Of these states, Greece, Japan, Germany, Norway, the US and China control approximately 38 per cent of the world's shipping fleet and, since 2000, China has led the world in the growth of shipping (Branch, 2007). The most popular registries of ships (states in which ships are registered) are Panama, Liberia and the Marshal Islands, which registered 11,636 ships in 2009. This amounts to greater than 39 per cent of the world's shipping capacity (UNCTAD, 2009).

Modern merchant shipping is performed by the following types of ships:

- *Container ships* carry containers that can be onloaded and offloaded onto trucks and trains (referred to as 'intermodal freight transport' or 'fully cellular'). The size of container ships is based on how many twenty-foot equivalent units (TEU) containers the ship can transport. The largest container ships currently have up to a 14,000 TEU capacity. Newer ships coming into operation will have an 18,000 TEU capacity.
- *General cargo ships* include refrigerated cargo, specialized cargo, roll on-roll off (ro-ro – used to transport vehicles) and other types of special purpose ships.
- *Bulk carriers* transport raw materials including grain, ore and coal.
- *Tankers* transport crude oil, chemicals, cooking oils and other liquids. Some specialized tankers carry cargoes such as liquefied natural gas.
- *Ferries and cruise ships* transport passengers, cars and commercial vehicles.
- *Specialist ships* include supply vessels, tugs, ice breakers and research vessels.

In addition to classification of ships by purpose, ships are often categorized by size. Their categorization is according to their ability to transit the world's major canals and straights, as well as the types of ports they are able to service (Table 7.2).

Shipping can be further differentiated into the discrete sectors outlined below, operating under distinct commercial and regulatory regimes.

Passenger shipping services

The IMO defines a passenger ship as any vessel carrying or configured to carry more than 12 non-crew passengers. Passenger ships include ferries, ocean liners and cruise ships, and are normally separated into two types: the *ferry fleet* consists of ships used for short voyages

Table 7.2 Global ship classifications and sizes

Ship type	Applicability	Deadweight tonnes*	Maximum length (m)	Maximum draft (depth m)	Maximum air draft (height m)	Maximum beam (width m)
Handysize	Dry bulk carriers serving smaller ports	<40,000	200	Unlimited	Unlimited	Unlimited
Handymax	Dry bulk carriers serving smaller ports	40,000–50,000	200	Unlimited	Unlimited	Unlimited
Supramax	Dry bulk carriers serving smaller ports	50,000–60,000	200	Unlimited	Unlimited	Unlimited
Suezmax	Maximum ship size to transit the Suez Canal	Unlimited but normally 120,000–150,000	Unlimited	20.1	68	50
Panamax	Maximum ship size to transit the Panama Canal	Unlimited but normally 50,000–80,000	290	12	58	32
Aframax	Oil tankers able to serve most ports. Named after 'average freight rate assessment'	<120,000	Unlimited	Unlimited	Unlimited	44
Chinamax	Bulk carriers able to serve most Chinese ports	<400,000	360	24	Unlimited	65
Capesize	Bulk carriers and oil tankers too large for the Panama and Suez canals. Includes very large ore carriers (VLOC) and very large bulk carriers (VLBC) that service deep water ports	150,000–400,000+	Unlimited	Unlimited	Unlimited	Unlimited
Malaccamax	Ships (bulk, oil or container ships) able to transit the Strait of Malacca	Unlimited	400	25	Unlimited	59
Very large crude carrier (VLCC)	Oil tankers for long-haul crude oil transport	200,000–349,999	Unlimited	Unlimited	Unlimited	60
Ultra large crude carrier (ULCC)	Oil tankers for long-haul crude oil transport	350,000+	Unlimited	Unlimited	Unlimited	Unlimited

Note: *Sum of the weights of cargo, fuel, fresh water, ballast water, provisions, passengers and crew.
Source: UNCTAD (2011); IMO (2012)

without accommodations that may also carry vehicles and cargo. The ferry fleet consists of conventional ferries and fast ferries. The latter includes monohull ships, catamarans, hydrofoils and hovercraft (Branch, 2007). The *cruise fleet* consists of ships specifically designed for longer voyages. Currently, the global passenger fleet consists of 1575 passenger-only vessels (including cruise ships) and 2624 ships suitable for passengers and freight (ISL, 2011). Worldwide, more than 2 billion passengers were transported in 2010 and the total number of cruise ship passengers was almost 21 million (IHS Fairplay, 2011). In 2011, more than 400 million passengers passed through European sea ports (UNCTAD, 2011).

The fastest growing segment of passenger shipping over the past two decades has been the cruise fleet. This segment consists of vessels that carry upwards of 3000 passengers. The cruise market primarily serves the Mediterranean, Caribbean and Baltic Seas, the Atlantic Islands and North American and European rivers.

Passenger shipping is regulated through the international conventions (see below) and domestic law. In addition, the EU regulates passenger shipping through a number of different legislative instruments including EU directives.

Liner cargo services

Liner cargo services transport cargo for several shippers simultaneously and operate on regular schedules between ports ('liner trades'). This contrasts with bulk shipping (see below) where contracts are made with a single shipper on an as-needed basis. Vessels used in the liner trades include general cargo, container, refrigerated (reefer), roll-on/roll-off (ro-ro) and other specialized ships. The liner trade is increasingly multimodal and comprised of large shipping companies that are able to guarantee delivery at specific times.

Currently, container shipping represents over 75 per cent of the liner trade by volume. Annually, half of the world's trade is shipped in over 15 million containers making over 230 million journeys. Over 531 million TEUs were shipped in 2011 with nearly 5000 container ships in operation (IMO, 2012). More than 600 new container ships are on order.

Bulk cargo services

Bulk cargo services exploit a variety of ship configurations to convey their payloads around the planet. They carry petroleum products (e.g. crude oil, condensates and refined products), liquefied gasses (e.g. liquefied natural gas and liquid propane gas), liquid chemicals (e.g. cooking oil), foodstuffs (e.g. grains, sugar) or solids (e.g. ore, coal, sulphur, bauxite/alumina). In 2011, 8687 bulk carriers were in service around the world and 45 per cent of the world's crude oil production was shipped by sea. This represents 1.8 billion tons, or about 70 per cent of the dwt shipped by bulk cargo services (IMO, 2012).

Bulk cargo ships are generally purpose built for specific cargoes and bulk shipping companies often focus on specific cargoes (e.g. oil tankers). Ships participating in the bulk trades generally do not operate on scheduled services but are generally exclusively chartered/contracted by shippers on long-term contracts. This model incents competition between shipping companies but, compared to the liner trades, results in an older fleet. Bulk cargo services are not normally part of the liner conventions (see below). Instead, bulk shipping is regulated by national and international laws. As opposed to the liner trades where anti-competitive practices are a focus of regulation, bulk shipping focuses primarily on ship safety and environmental protection.

Tramp services

Tramp ships, or general traders, are cargo ships, normally with two to six holds, that carry a variety of cargos, including grains, ores, forest products and food stuffs. Tramp services operate throughout the world but are prevalent in East and South Asia. These services are *ad hoc* in nature and do not adhere to fixed schedules. Rather, they anticipate and exploit niches and seasonal markets ('spot markets') that are often too uneconomical for liner and bulk cargo services. Tramp services are generally composed of smaller ships, are often family owned, and respond quickly to market conditions and opportunities as they arise. However, many liner service shipping companies are acquiring tramp services to horizontally and vertically integrate their businesses. While tramp services are not as prevalent as they once were (outside of Asia) they continue to operate in densely populated regions such as the North American Great Lakes, the Baltic, the Mediterranean and other enclosed seas.

Introduction to marine transportation law and policy

The need to regulate shipping

Terrestrial transportation (truck, rail) is primarily focused on moving goods within national boundaries or to and from adjacent states. In contrast, maritime shipping mostly occurs between states and often within ABNJ. Shipping is therefore inherently international in nature and requires cooperation amongst states in order to move more goods efficiently and safely between ports. Shipping regulation is required for a number of reasons, including:

- Ensuring common environmental protection standards.
- Ensuring common safety standards for ship operations.
- Ensuring common crew training standards.
- Holding states and shippers accountable for accidents and negligence.
- Preventing and responding to piracy and maritime terrorism.
- Developing construction and design standards.
- Establishing navigation rules.
- Establishing communications protocols.
- Ensuring cargo liability.
- Ensuring safe and effective ship disposal.
- Meeting greenhouse gas reduction targets and adapting to climate change impacts.

International shipping is a unique economic activity where carriers from multiple states compete with one another to move goods or passengers between ports (often between states) at the lowest cost. States may promote or inhibit competition through subsidies, trade barriers or transportation policies (Box 7.2). There are few economic incentives for shipping companies to invest in safety or environmental protection, especially if their competition fails to do so. Similarly, without shipping regulation or effective insurance regimes (see P&I clubs below), ship operators have no incentive to expend financial resources to clean up accidents, compensate shippers for lost cargoes, or compensate others who may have been negatively affected by the ship operator. Compounding this situation is that most of the world's ships are registered in flag states (under flags of convenience – see Box 7.2) who may not have the desire or financial resources to enforce standards or make ship owners accountable for their actions. Lastly, responding to certain marine accidents (e.g. oil spills), is often out of financial reach for most ship owners. As such, regulation is needed to enable and compel those responsible for costly accidents and damages to be able to fund clean up and financially compensate affected third parties.

Box 7.2 An overview of commercial shipping

Parties involved in commercial shipping

- *Carrier*: The party transporting the goods or cargoes.
- *Coastal state*: Any state with a marine coastline. To protect their interests, coastal states establish laws and regulations with which vessels must comply. The jurisdiction, duties and responsibilities of coastal states are set out in the LOSC.
- *Consignee*: The party (agent, company or person) that receives goods once they reach their destination port (the 'receiver').
- *Consignor*: The party (agent, company or person) sending consignment (the 'shipper' or 'merchant').
- *Classification societies*: Provide technical and surveying services either on behalf of the ship's owner so that the ship can be issued a 'class certificate' (for insurance and operational purposes) or on behalf of flag states as a 'recognized organization'. There are 11 globally recognized classification societies currently in operation.
- *Flag state*: Responsible for verifying that domestic (national) ships comply with national laws and international conventions to which the flag state is a party. These responsibilities are known as flag state control (FSC). Classification societies are often used to implement FSC.
- *International governmental organizations*: Comité Maritime International, International Labour Organization and International Maritime Organization are discussed in the following sections.
- *Port state*: Responsible for verifying compliance (including inspections and boarding) with international conventions for ships in their territorial waters. These responsibilities are known as port state control. Port state control regimes are regional, enabling states to share verification responsibilities. The major port state control regimes include: Paris MoU (Europe and North Atlantic, 27 states), Tokyo MoU (Asia and the Pacific, 18 states), Indian Ocean MoU (Indian Ocean, 16 states), US Coast Guard, Vina del Mar Agreement (Latin America 13 states), Abuja MoU (West and Central Africa, 22 states), Black Sea MoU (Black Sea, 5 states) and Riyadh MoU (Persian Gulf, 6 states).
- *Protection and indemnity (P&I) clubs*: Cooperatives run by ship owners or operators to provide third-party marine insurance to their members. The 13 recognized P&I clubs are non-profit and insure for risks that the conventional insurance market refuses to cover.

Contracts of carriage

- *Bill of lading*: Document that gives title (constructive possession) to the goods and the condition of goods at the time of loading. Usually prepared by the consignor, the bill of lading will state the date(s) goods were loaded and ports of loading and discharge. Since the bill of lading is transferrable, cargo may be sold many times during transit.

- *Charter party*: A formal written contract between the carrier and cargo owner for the exclusive use of a ship for either a period of time ('time charter') or port-to-port voyage ('voyage charter'). In a 'demise' charter party (or bareboat charter), the charterer is responsible for providing the cargo and crew, whereas with a 'non-demise' charter party, the ship owner provides the crew and the charterer provides the cargo.
- *Waybill*: A document outlining the terms and conditions of the contract of carriage as well as a receipt for the goods loaded aboard the carrying vessel. Under a waybill, the shipper owns the cargo until delivery. Unlike a bill of lading, a waybill is non-negotiable and not a document of title. Moreover, under a waybill a cargo cannot be bought or sold during transit under a waybill.

Types of shipping contracts

- *Free on board (fob) contract*: The buyer pays for and arranges carriage. The seller's duty is to load the goods onto the ship.
- *Cost, insurance, freight (cif) contract*: The seller pays for and arranges carriage as well as purchases an insurance policy on the goods, which will be assigned to the buyer.

Other policy-related shipping terminology

- *Cabotage*: Most nations apply some elements of cabotage, where they reserve (protect) a portion of their domestic shipping trades for ships flying their national flag. States may require ships to be crewed or owned by domestic nationals, built at domestic shipyards or registered under the country's national flag.
- *Flags of convenience*: Ships that are registered in states different than the states where the ship is owned/operated. Traditionally, flags of convenience have been used to reduce costs (avoid taxation, utilize lower wage crews) and avoid compliance with domestic regulations that may be more onerous.
- *Short sea shipping*: Movement of cargo between any port on the Baltic, Black or Mediterranean seas.

States also have a vested interest in shipping and in particular are concerned with ensuring:

- the safety of passengers, crews and cargos;
- the orderly movement of goods and people across waterways;
- access to domestic and international shipping routes;
- sufficient competition without compromising the ability for shipping companies, when necessary, to cooperate to provide scheduled, uninterrupted liner services.

Organizations that regulate marine transportation

Global shipping is regulated by over 150 conventions and agreements. In addition, the Organization for Economic Cooperation and Development (OECD) and World Trade Organization (WTO) oversee economic competitiveness in shipping (Table 2.2). While the key conventions are discussed later in this chapter, the following are organizations that oversee the regulation of the international shipping industry.

The LOSC and marine transportation

The LOSC (see Chapter 2) is not an organization per se, nor is it formally administered by the UN General Assembly. However, the Convention is a key regulatory instrument that informs the operations of the IMO. The LOSC regulates marine transportation in two ways. First, the Convention enshrines a number of overarching shipping principles. Among these primary principles are: the rights of innocent passage and transit passage; immunity of warships; freedom of the high seas and use of the high seas for peaceful purposes; rights of all states to participate in shipping; prohibitions on carrying slaves; the necessity of combating piracy; and the right of all states to fish.

Second, the LOSC provides the administrative framework for shipping. Within this framework lies the rights and duties of coastal and flag states by LOSC jurisdiction as well as the rights of states to determine the conditions for granting of their nationality to ships and standards for manning and training crews. The requirement that all flag states ensure that vessels flying their flag meet international shipping regulations also falls within this framework.

International Maritime Organization

Despite the importance of the LOSC, shipping is primarily regulated by the IMO and ILO. The UN established the IMO in 1958. Headquartered in London, England, the IMO is comprised of a 169-member Assembly (composed of states and three associate members), responsible for establishing the rules of international shipping. These rules address maritime safety, security and prevention of marine pollution from ships. The IMO is governed by a Council that is responsible for oversight of the organization. Between Assembly meetings, the Council also performs all the functions of the Assembly except making recommendations to governments (including state parties) on maritime safety and pollution prevention.

The IMO's authority only applies to international shipping. Member states are expected to implement and enforce the 40 IMO conventions and protocols, as well as the codes and recommendations, which now exceed 800 in number. The work of the IMO, under direction from the Council, is undertaken by a number of committees and sub-committees. The current committees are:

- The *Maritime Safety Committee* is the senior IMO committee and responsible for all matters directly affecting marine safety. The Committee is responsible for nine sub-committees on: 1) safety of navigation; 2) radio communications and search and rescue; 3) standards of training and watch keeping; 4) ship design and equipment; 5) fire protection; 6) stability, load lines and fishing vessel safety; 7) flag state implementation; 8) dangerous goods, solid cargoes and containers; and 9) bulk liquids and gases.

- The *Marine Environment Protection Committee* (MEPC) is responsible for the prevention and control of pollution from ships.
- The *Legal Committee* is responsible for all legal matters within the scope of the IMO.
- The *Technical Cooperation Committee* is responsible for providing technical assistance to member states, and has a particular focus on ensuring less developed states are capable of meeting IMO conventions, protocols, codes and recommendations.
- The *Facilitation Committee* is responsible for simplifying the documentation and formalities in international shipping.

The draft conventions produced by these committees are submitted to conferences open to any member of the UN, including UN members that may not be IMO members. All governments, whether coastal, seafaring or landlocked, have the same status and participation in IMO discussions. Similar to other international conventions, the conference adopts a final text that is provided to member governments for ratification. The convention enters force once it is ratified by a set number of member states. Non-IMO members may accede to the convention. Each convention provides the conditions to be met prior to entering into force. For example, the International Convention for the Safety of Life at Sea (see below) required the ratification of at least 25 states whose merchant fleets comprised not less than 50 per cent of the world's tonnage. Conventions are binding on IMO member states while IMO codes and regulations are non-binding. Many states must pass or amend domestic laws to comply with IMO conventions (see Chapter 2).

Comité Maritime International

The CMI, headquartered in Antwerp, Belgium, is a non-governmental organization established in 1897 to oversee the codification of maritime law, commercial practices and the promotion of national associations of maritime law. At present, the CMI is comprised of 51 national maritime law associations, including the IMO and International Oil Pollution Compensation Fund (IOPC Fund – see below).

Early work of the CMI focused on regulations concerning collision and salvage. Among the 22 conventions produced between 1910 and 1969 are the 1952 Arrest of Seagoing Ships and 1968 Hague Rules (Table 2.2). Also, the CMI has assisted on the drafting of many IMO conventions, including the 1994 International Convention on Maritime Liens and Mortgages, the 1997 International Convention Relating to the Arrest of Seagoing Ships and the 2007 Nairobi International Convention on the Removal of Wrecks. In addition, the CMI continues to be the responsible authority for international conventions of bills of lading and salvage (Table 2.2) (CMI, 2012).

International Labour Organization

The ILO is responsible for establishing and overseeing international labour standards. Established in 1919 as part of the Treaty of Versailles (the accord that officially ended World War I), the ILO works to improve working conditions, worker safety and ensure equal pay for equal work between states. The ILO was established under the premise that labour peace is fundamental to national prosperity. At this time the organization currently has four strategic objectives:

- Promote and realize standards and fundamental principles and rights at work.
- Create improved employment opportunities for women and men.
- Enhance the coverage and effectiveness of social protection for all.
- Strengthen tripartism and social dialogue.

Currently, the ILO consists of 185 member states and is headquartered in Geneva, Switzerland. The organization oversees more than 180 conventions and 190 recommendations that cover all aspects of employment. A key responsibility of the ILO is to oversee labour standards of the world's 1.5 million seafarers and 15 million fishers of various nationalities who crew ships that are often owned and flagged in different states. This is required because national labour codes may fail to protect seafarers and, in an effort to maximize profitability, some ships owners compromise worker safety and recruit low-wage crews.

The ILO's primary contribution to maritime law is the 2006 Maritime Labour Convention, which is discussed below.

International law dealing with the ship

Key international agreements and conventions regulating the configuration and operation of ships are discussed below; however, four conventions are frequently referred to as the 'pillars' of maritime shipping law and are as follows:

- International Convention for the Safety of Life at Sea (SOLAS).
- International Convention on Standards of Training, Certification and Watchkeeping for Seafarers (STCW).
- Maritime Labour Convention (MLC).
- International Convention for the Prevention of Pollution from Ships, 1973, as modified by the Protocol of 1978 (MARPOL 73/78).

International Convention for the Safety of Life at Sea

On 15 April 1912 the *Titanic* struck an iceberg and sank, leaving 1503 casualties. The British Wreck Commissioner's Inquiry formed numerous conclusions from the tragedy. While the vessel was in full compliance with the UK Merchant Shipping Acts (1894–1906), they determined that these Acts were established for ships 13,000 tons or less. In the case of the 46,328 ton *Titanic*, this resulted in an insufficient number of lifeboats to accommodate her passengers. Since the iceberg was reported to the bridge, the Inquiry also concluded that the *Titanic's* speed was a factor in the accident. On top of this, the Inquiry noted that passenger lifeboat or safety drills were never conducted, and nor were crews trained in the deployment and use of the safety equipment. Meanwhile, the ship's lack of a public address system and insufficient range of radio communications equipment were contributing factors (Titanic Inquiry, 2012).

Following the *Titanic* calamity, the first Safety of Life at Sea conference was convened in the UK in 1914 (IMO, 1974). Attended by 13 states, the conference produced regulations on emergency equipment (including lifeboats) and procedures and it prescribed new radio communication procedures. The 1914 SOLAS convention was updated in 1929, 1948 and 1960.

In 1974, in an attempt to maintain pace with the considerable number of amendments to the Convention, the Convention was rewritten completely and referred to as SOLAS 1974. The Convention was adopted 1 November 1974 by the International Conference on Safety of Life at Sea and entered into force on 25 May 1980. SOLAS was structured such that all future amendments would be implemented automatically unless the majority of members objected. Additional protocols were adopted in 1978 (International Conference on Tanker Safety and Pollution Prevention – the 1978 SOLAS Protocol – entered into force in 1981) and in 1988 (the International Conference on the Harmonized System of Survey and Certification – the 1988 SOLAS Protocol – entered into force in 2000) (IMO, 2004). The SOLAS 1974 convention has also been amended a further 29 times through resolutions approved by the IMO Maritime Safety Committee or by conferences of SOLAS's contracting governments.

SOLAS 1974 contains 12 chapters. They address:

- Chapter I: Definitions of types of ships and explanations of types of surveys and certificates.
- Chapter II-1: Structure, subdivision and stability, machinery and electrical installations related to ship construction.
- Chapter II-2: Fire protection, fire detection and fire extinction related to ship construction.
- Chapter III: Lifesaving appliances and arrangements.
- Chapter IV: Radio communications.
- Chapter V: Safety of navigation.
- Chapter VI: Carriage of cargoes.
- Chapter VII: Carriage of dangerous goods.
- Chapter VIII: Nuclear ships.
- Chapter IX: Management for the safe operation of ships.
- Chapter X: Safety measures for high-speed craft.
- Chapter XI-1: Special measures to enhance maritime safety.
- Chapter XI-2: Special measures to enhance maritime security.
- Chapter XII: Additional safety measures for bulk carriers.

SOLAS applies to all ships and requires ship construction and equipping to conform to SOLAS on or after the dates specified in various regulations. SOLAS can be retroactive and may require modifications to existing ships to conform with new regulations. The SOLAS Conventions have been managed by the IMO since 1959.

Convention on the International Regulations for Preventing Collisions at Sea

COLREG was adopted 20 October 1972 and entered into force 15 July 1977 (IMO, 1972). The Convention replaced the earlier 1960 International Regulations for Preventing Collisions at Sea, which had been based on a set of British rules formulated in 1862. The COLREG is best conceived as the 'rules of the road' for navigation and navigation safety and contains 38 rules. The rules are divided into sections related to steering and sailing (Part B); lights and shapes (Part C); sound and light signals (Part D); and exemptions (Part D). Amendments to the Convention were adopted in 1981, 1987, 1989, 1993, 2001 and 2007. A short summary of the key rules is as follows:

- Rule 1: The Convention applies to all vessels upon the high seas and in all waters connected to the high seas navigable by seagoing vessels. States reserve the right to retain and implement additional special rules in their inland waters or for specific operations so long as they are consistent with the Convention and do not compromise safety.
- Rule 2: Responsibility for complying with the Convention rests with vessel owners, masters and crew. Vessels may deviate from the Convention to avoid immediate danger.
- Rule 4: Rules apply in any condition of visibility.
- Rule 5: Every vessel shall at all times maintain a proper look-out by sight and hearing as well as by all available means appropriate in the prevailing circumstances and conditions.
- Rule 6: Vessels will be operated at speeds appropriate to the prevailing circumstances and conditions.
- Rule 7: Vessels will use all means to evaluate collision risks.
- Rule 8: Rules to avoid collisions.
- Rule 9: Rules for operation in narrow channels.
- Rule 10: Rules for traffic separation schemes.
- Rule 12: Rules for sailing vessels.
- Rule 13: Rules for overtaking.
- Rule 14: Rules for head-on situations.
- Rule 15: Rules for crossing situations.
- Rule 16: Rules for the 'give-way' vessel.
- Rule 17: Rules for the 'stand-on' vessel.
- Rule 18: Exceptions for Rules 9, 10 and 13.
- Rule 19: Conduct of vessels in restricted visibility.
- Rules 22–31: Lighting requirements in various circumstances.

International Convention on Load Lines

The CLL 66 was adopted 5 April 1966 in London and entered into force 21 July 1968 (IMO, 1966). The purpose of the Convention is to prescribe loading limits on ships in terms of the amount of freeboard (draught) available to prevent overloading and stability issues. The Convention considers geographic location and season when setting freeboards (the height of a ship's deck above the water level) as well as the regulations for any access points through the hull below the water line. The Convention was amended by a protocol in 1988 that entered into force 3 February 2000 and permits the Convention to be amended either by the IMO's Maritime Safety Committee or by a CoP. In 2003, the Convention was revised (entered into force 1 January 2005) to include additional regulations to improve the design and configuration of ships.

International Convention on Arrest of Ships

The ARREST Convention was adopted 12 March 1999 and entered into force 14 September 2011 (IMO, 1999). The Convention updated the 1952 Brussels Convention on the Arrest of Sea-Going Ships. Its purpose is to balance the interests of claimants seeking maritime claims against the owners and operators of ships. Specifically, the Convention outlines the conditions under which a ship may be arrested or released from arrest. Shipping is unique in international law in that a ship – rather than its owner – may be arrested. This is because tracing ownerships of ships can be very difficult.

The Convention applies to any ship within the jurisdiction of a contracting state and covers 22 types of marine claims. Ships may be arrested without proof of their liability, but can only be arrested under the authority of a court of the state party in which the arrest is affected and only as a result of a marine claim (Article 2). Ships may be arrested only when owned or operated by the same individuals that owned or operated the ship at the time of the maritime claim (Article 3). Ships must be released when security (payment) has been received that is satisfactory to the state party in which the arrest was affected (Article 4.1). Courts will determine the amount of security if parties cannot agree but security must not be more than the value of the arrested ship (Article 4.2). Payment of the security does not acknowledge liability on behalf of the ship owner or operator (Article 4.3). Ships may be rearrested if the security previously provided is inadequate or the owner or operator has not provided the security (Article 5). Furthermore, the arrest of multiple ships with the same owner or operator is permitted if the security provided for a marine claim is insufficient (Article 5). However, state courts may award damages for unjustified or wrongful arrest or in situations where the security was set improperly high (Article 6). Lastly, state courts have jurisdiction over these matters unless both parties agree either to submit the case to a court in another jurisdiction or send the case to arbitration (Article 7).

The Athens Convention relating to the Carriage of Passengers and their Luggage by Sea

The Athens Convention relating to the Carriage of Passengers and their Luggage by Sea (abbreviated as 'PAL') was adopted 13 December 1974 and entered into force 28 April 1987. The Convention's purpose is to harmonize regulations relating to passengers and their luggage and establish maximum liability limits for the carrier in the event that an incident was a result of the fault or neglect of the carrier. The Convention places the onus on the carrier to demonstrate it was not at fault.

The Convention is amended by three protocols: the 1976 Protocol (entered into force 30 April 1989) substituted SDRs for the gold franc; the 1990 Protocol (not yet entered into force) increased the compensation limits; and the 2002 Protocol introduced the principles of strict liability and compulsory insurance. The 2002 Protocol requires that the Convention and Protocol be adopted as a single instrument titled the Athens Convention relating to the Carriage of Passengers and their Luggage by Sea, 2002. It entered into force on 23 April 2014. The 2002 Convention increases liability limits from to 250,000 SDRs per passenger, 2250 SDRs for cabin luggage and 12,700 SDRs per vehicle.

International law dealing with the shipping company

International Safety Management Code

The International Safety Management Code (ISMC) is not a stand-alone convention but an amendment to Chapter IX of SOLAS, adopted on 24 May 1994. The Code was enacted after previous voluntary guidelines agreed to in 1989, 1991 and 1993 were deemed to be of sufficient importance to make them mandatory (IMO, 2010). Its objectives are to ensure safety at sea, prevent human injury or loss of life, and avoid damage to property and the environment. The Code became compulsory for all ships in 2002 and requires companies to implement a safety management system with the following components:

- A safety and environmental-protection policy.
- Instructions and procedures to ensure safe operation of ships and protection of the environment in compliance with relevant international and flag state legislation.
- Authority and lines of communication between, and among, shore and shipboard personnel.
- Procedures for reporting accidents and non-conformities with the provisions of the Code.
- Procedures to prepare for and respond to emergency situations.
- Procedures for internal audits and management reviews.

UN Convention on Contracts for the International Carriage of Goods Wholly or Partly by Sea

The United Nations Convention on Contracts for the International Carriage of Goods Wholly or Partly by Sea (the 'Rotterdam Rules') was adopted by the UN General Assembly on 11 December 2008. The Rotterdam Rules replaced a number of previous conventions including the 1924 International Convention for the Unification of Certain Rules of Law relating to Bills of Lading ('the Hague Rules' and the 'the Hague-Visby Rules') and the 1978 United Nations Convention on the Carriage of Goods by Sea ('the Hamburg Rules') (UNCITRAL, 2009). Sixteen states were initial signatories to the Convention while 24 are currently party to the Convention.

The Rotterdam Rules were negotiated over a ten-year period and coordinated by the United Nations Commission for International Trade Law (UNCITRAL). The Rotterdam Rules address modern transport practices, ranging from containerization, door-to-door transport contracts and the use of electronic transport documents (UNCITRAL, 2009). In doing so, the Convention provides greater certainty for cargo owners. These assurances are provided through: increases in the monetary limits on a carrier's liability (for loss, damage or delay), removal of previous exceptions for unforeseen events (e.g. fire, navigation error) and clear articulation of responsibility and liability for all aspects of the transport process.

The Convention applies between the carrier and the consignee in any situation where the port of loading and port of discharge are in different states and the rules apply regardless of the vessel's nationality (Article 5). The general outline of the Convention is as follows:

- Chapter 1: General provisions and definitions.
- Chapter 2: Scope of the Convention and exclusions.
- Chapter 3: Electronic transport records and their application.
- Chapter 4: Obligations of the carrier.
- Chapter 5: Liability of the carrier for loss, damage or delay.
- Chapter 6: Additional provisions relating to particular stages of carriage.
- Chapter 7: Obligations of the shipper to the carrier.
- Chapter 8: Transport documents and electronic transport records.
- Chapter 9: Delivery of the goods.
- Chapter 10: Rights of the controlling party.

- Chapter 11: Transfer of rights.
- Chapter 12: Limits of liability.
- Chapter 13: Time for suit.
- Chapter 14: Jurisdiction.
- Chapter 15: Arbitration.
- Chapter 16: Validity of contractual terms.
- Chapter 17: Matters not governed by the Convention.
- Chapter 18: Final clauses.

UN Convention on a Code of Conduct for Liner Conferences

The liner trades utilize 'liner conferences' (also termed 'shipping conferences') to fashion formal or informal agreements between shipping companies. These conferences enable them to set freight rates or cooperate on the movement of goods between ports while avoiding antitrust scrutiny. Not coincidentally, liner conferences formed shortly after the opening of the Suez Canal (1869) – an engineering feat that halved the shipping time on the Asia to Europe trade route and that led swiftly to massive overcapacity. Since then, the objectives of liner conferences have been to avoid over-competition, to accommodate seasonal fluctuations in shipping demand and to provide economic stability to shippers (Munari, 2012). Liner conferences currently involve up to 40 separate shipping companies and there are currently over 300 liner conferences in operation. While evidence that liner conferences create anticompetitive cartels is mixed (Clyde and Reitzes, 1998), most states view conferences in this light. Until recently, conferences were tolerated or exempted from most antitrust laws as they seemed to provide shippers with a stable access to their markets. This stability was perceived to have lucrative ramifications. Indeed, as far back as 1909 a British Royal Commission on Shipping Rings concluded that liner conferences were anticompetitive but, nonetheless, overall beneficial to the British economy (Liu, 2009).

The Convention on a Code of Conduct for Liner Conferences ('UN Liner Code') was signed in Geneva, 6 April 1974 and entered into force on 6 October 1983. The purpose of the Convention is to:

- facilitate the orderly expansion of seaborne trade;
- stimulate the development of regular and efficient liner services;
- balance the interests between suppliers and users of liner shipping services;
- foster communication and collaboration between shippers and liner services.

Many states, however, began to use the Convention to reserve a portion of the shipping trade for national shipping lines. Liner conferences thus became a means by which national shipping lines negotiated their involvement in international trade. The Convention was also used by less developed states – formerly excluded from participating in international shipping – to become involved in shipping. The Convention developed the 40:40:20 rule where 80 per cent of the shipping trade was reserved for national shipping lines of the source (40 per cent) and destination (40 per cent) states and 20 per cent made available for third-party shipping lines (Munari, 2012). Currently, there are 21 signatories and 76 parties to the Convention.

Since the adoption of the UN Liner Code, the influence of liner conventions and monopolistic shipping behaviour has waned. New entrants into the shipping business, relaxation of price fixing by shipping companies, growth of intermodal transportation and changes to competition laws in various nations have freed up the market. The Code has also enabled increased information on supply and demand trends in order to anticipate shipping needs. On top of this, monopolistic practices ebbed when the EU, in 2006, repealed the exemption from the European Commission Treaty ban on restrictive business practices (Article 81) for liner conferences on routes to and from the EU. The repealed exemption came into force in October 2008 (Hummels *et al.*, 2009).

International law dealing with the seafarer

On 25 October 1859, 195 ships were wrecked and 685 souls perished in a single night in North-west Europe. Casualties from the passenger ship *Royal Charter* were in excess of 450 and the storm is referred to eponymously as the *Royal Charter* Storm. By the end of the series of storms on 9 November 1859, 325 ships had been lost with 748 passengers and crew perishing. These losses triggered a global dialogue on how to improve marine safety and avoid such events in the future. Marine safety has improved considerably since this time; however in 2010, 250 seafarers lost their lives to ship-borne accidents (IHS Fairplay, 2011).

Maritime Labour Convention

The ILO is responsible for the 2006 MLC, which consolidates more than 68 maritime labour standards adopted over 80 years (ILO, 2006). The MLC was adopted on 23 February 2006 in Geneva and entered into force on 20 August 2013. The purpose of the Convention, also called the 'seafarer's bill of rights', is to establish working conditions for seafarers and ensure fair competition for ship owners. The MLC applies to all public and private ships engaged in commercial activities except fishing vessels, traditional vessels (e.g. dhows and junks) and vessels used solely within a state's internal waters (Article 2). Like other international legal instruments, the MLC does not apply directly to seafarers or ship owners/operators. Instead, it requires state parties to the Convention to pass domestic legislation or enact regulations consistent with the Convention. As such, the MLC provides considerable flexibility to states to develop new or use existing standards that are equivalent to the requirements in the Convention.

The Convention consists of articles and regulations that set out the basic principles and responsibilities of state parties to the Convention. Under each regulation, the Code contains both mandatory standards (Part A) and voluntary (Part B) guidelines. Key articles embodied within the MLC are:

- Article III: Parties to the Convention will: permit seafarers the right to collective bargaining and union association; eliminate forced labour; abolish child labour and eliminate discrimination.
- Article IV: Seafarers have the right to a safe work environment, medical services, decent living conditions and fair terms of employment.
- Article V: States must ensure that, for ships flying their flag, they carry evidence that they are in compliance with MLC. Flag states must also undertake inspections and oversee seafarer recruitment in their jurisdictions. Port states may inspect ships of another flag when they are in port of a port state.

The regulations are separated into the following five titles, under which Parts A and B of the Code are supplied:

- Title 1: Minimum requirements for seafarers to work on a ship
  - Regulation 1.1: Minimum age
  - Regulation 1.2: Medical certificate
- Title 2: Conditions of employment
  - Regulation 2.1: Seafarers' employment agreements
  - Regulation 2.2: Wages
  - Regulation 2.3: Hours of work and hours of rest
  - Regulation 2.4: Entitlement to leave
  - Regulation 2.5: Repatriation
  - Regulation 2.6: Seafarer compensation for the ship's loss or foundering
  - Regulation 2.7: Manning levels
  - Regulation 2.8: Career and skill development and opportunities for seafarers' employment
- Title 3: Accommodation, recreational facilities, food and catering
  - Regulation 3.1: Accommodation and recreational facilities
  - Regulation 3.2: Food and catering
- Title 4: Health protection, medical care, welfare and social security protection
  - Regulation 4.1: Medical care on board ship and ashore
  - Regulation 4.2: Ship owners' liability
  - Regulation 4.3: Health and safety protection and accident prevention
  - Regulation 4.4: Access to shore-based welfare facilities
  - Regulation 4.5: Social security
- Title 5: Compliance and enforcement
  - Regulation 5.1: Flag State responsibilities
  - Regulation 5.2: Port State responsibilities
  - Regulation 5.3: Labour-supplying responsibilities

International Convention on Standards of Training, Certification and Watchkeeping for Seafarers

The STCW was adopted on 7 July 1978 and entered force on 28 April 1984. The Convention responded to many state-specific regulations on training and certification that caused problems when ships transited foreign waters or put into foreign ports. The STCW establishes the basic requirements and standards for masters (captains), officers and watch personnel on merchant ships (IMO, 1978a). The overall format of the convention is as follows:

- Chapter I: General provisions.
- Chapter II: Master and deck department.
- Chapter III: Engine department.
- Chapter IV: Radiocommunication and radio personnel.
- Chapter V: Special training requirements for personnel on certain types of ships.
- Chapter VI: Emergency, occupational safety, medical care and survival functions.
- Chapter VII: Alternative certification.
- Chapter VIII: Watchkeeping.

The Convention was amended in 1995 to: provide clarification on such phrases as '... to the satisfaction of the Administration'; move the technical regulations into a new STCW Code that can be amended without convening a full conference; and require parties to the Convention to provide evidence of complaints – the first incidence of IMO involvement in compliance. There are currently 155 parties to the Convention.

The STCW Code has mandatory (Part A) and optional (Part B) sections where Part B provides advice to assist parties with complying with the Convention. Both the STCW Convention and Code were further amended on 25 June 2010 and entered force on 1 January 2012. Referred to as the 'Manila Amendments' they updated the standards necessary to comply with the Convention.

The STCW does not apply to ships less than 200 GT or ships operating only in internal waters. The Convention does apply to ships registered in non-party states when they visit port states that have ratified the Convention.

Maritime security

Maritime security encompasses piracy, terrorism and criminal activities related to the safety of ships, crews, passengers and cargoes when underway as well as in port. Shipping is a unique endeavour with respect to safety as illegal acts are often perpetrated in international waters (ABNJ). These ships are often registered in flag states different from their ownership state and crewed by citizens of neither state. As such, maritime security regulation requires a degree of international collaboration far in excess of most other modes of transportation. Furthermore, modern terrorism utilizes not just the ship as a means to transport weapons, for the ship itself can be used as a weapon, particularly if the ship is carrying dangerous maritime cargoes (DMCs), which include hydrocarbons and chemicals (Nincic, 2005).

Addressing maritime security is important: research has shown that, for example, without international action against Somali pirates operating in the Gulf of Aden, shipping along the Far East–Europe route would decline by 30 per cent (Lu et al., 2010). However, the costs of maritime security are prohibitively high and leave the industry vulnerable. As an example, currently less than 1 per cent of the 230 million container voyages are inspected (Broder, 2004).

International Ship and Port Facility Security Code

The International Ship and Port Facility Security Code (ISPS) was adopted on 13 December 2002, partly in response to the 11 September 2001 terrorist attacks on the US (IMO, 2003). The ISPS applies to all passenger ships, cargo ships greater than 500 GT, mobile offshore drilling units and ports serving such ships engaged on international voyages. Broadly, the Code:

- Establishes cooperative mechanisms between governments, shipping and port industries to detect security threats and take preventative measures to protect ships and ports involved in international trade.
- Establishes roles and responsibilities between governments, shipping and port industries with respect to maritime security.
- Ensures the exchange of maritime security information.

- Provides a methodology for security assessments in order to develop ship and port security plans able to react to changing security levels.

The Code entered into force 1 July 2004 in the form of the following amendments to SOLAS:

- Chapter V Safety of Navigation: Regulation 19 was amended to require upgraded navigation equipment and to ensure that ships equipped with automated information systems use the system at all times.
- Chapter XI-I Special Measures to Enhance Maritime Safety: Regulation 3 was amended to specify how a ship's identification number is to be marked on the vessel.
- Chapter XI-I Special Measures to Enhance Maritime Safety: Regulation 5 was amended to require ships to carry a continuous synopsis record, which is an onboard document record of the history of the vessel with respect to information including the vessel's previous names, ownership, classification societies and administrations/organizations issuing the ISM Code Safety Management Certificate.
- Chapter XI-II Special Measures to Enhance Maritime Safety: Regulation 1 provides updated definitions.
- Chapter XI-II Special Measures to Enhance Maritime Safety: Regulation 2 provides the types of ships and ports (discussed above) to which the Code applies, and exempts warships or ships owned by a contracting government in non-commercial service.
- Chapter XI-II Special Measures to Enhance Maritime Safety: Regulation 3 commits states to convey security status information to ships flying their flag and their ports.
- Chapter XI-II Special Measures to Enhance Maritime Safety: Regulations 4 and 10 commit states to comply with part A of the Code and to take into account part B.
- Chapter XI-II Special Measures to Enhance Maritime Safety: Regulation 5 requires a ship's master to be able to identify who is responsible for hiring crew and determining the business of the ship.
- Chapter XI-II Special Measures to Enhance Maritime Safety: Regulation 6 requires ships to carry a security alert system.
- Chapter XI-II Special Measures to Enhance Maritime Safety: Regulation 7 commits states to communicate security threat levels to vessels operating in their waters.
- Chapter XI-II Special Measures to Enhance Maritime Safety: Regulation 8 provides a ship's master authority over a ship's security and allows the master, in the interests of security, to override decisions/policies of the ship owners.
- Chapter XI-II Special Measures to Enhance Maritime Safety: Regulation 9 specifies the information that port states may require of vessels intending to enter its ports. If information is not produced the port state may inspect the ship or deny entry into its ports.
- Chapter XI-II Special Measures to Enhance Maritime Safety: Regulations 11 and 12 permit contracting states to enter into bilateral/multilateral security agreements and develop their own security standards so long as they are equivalent to the Code.

The Convention for the Suppression of Unlawful Acts against the Safety of Maritime Navigation and Protocol for the Suppression of Unlawful Acts against the Safety of Fixed Platforms Located on the Continental Shelf

The Convention for the Suppression of Unlawful Acts against the Safety of Maritime Navigation (SUA) and Protocol for the Suppression of Unlawful Acts against the Safety of

Fixed Platforms Located on the Continental Shelf was adopted on 10 March 1988 and entered into force 1 March 1992 (IMO, 1988a, 1988b). The Convention is currently ratified by 156 states. It was partially a result of the growing awareness of maritime security issues, but ultimately convened after the Italian cruise ship *Achille Lauro* was high-jacked sailing from Alexandria to Port Said, Egypt. While piracy and hot pursuit are addressed in the LOSC, the Convention is silent on the international response to maritime terrorism.

The IMO requested that the Maritime Safety Committee develop measures to address unlawful acts against civilian shipping. The primary purpose of the Convention is to require state parties to the Convention to extradite or prosecute persons committing unlawful acts against ships. Key articles of the Convention include:

- Article 2: The Convention does not apply to warships and government ships operated for non-commercial purposes, and nor does it apply to ships withdrawn from navigation or ships removed from water for repair.
- Article 4: The Convention applies to ships operating or planning to operate outside of their flag state's territorial waters, or when an alleged offender is found in the territory of a state party.
- Article 6–12: Responsibilities of state parties (including ship's masters) in the apprehension, extradition, conviction and punishment of those committing unlawful acts against ships and fixed platforms.
- Articles 13: Commits state parties to share information on marine security and cooperate to prevent marine terrorism.

The Convention was amended on 14 October 2005 via a protocol (2005 SUA Convention and 2005 SUA Fixed Platforms Protocol) (IMO, 2005a, 2005b, 2005c). The Protocol was prompted by increased attention to maritime security following the 11 September 2001 terrorist attacks on the US as well as the increased security threats as a result of the growth in containerized shipping (cargo ownership becomes more difficult to trace when shippers use liner trades to move cargo in containers). Furthermore, the rise of suicide bombers was never contemplated in the original SUA. Also required in the amendments was the need to expand the categories of 'unlawful acts' and to authorize boarding of foreign flagged ships when on the high seas in a manner that respected freedom of navigation and right of innocent passage. The 2005 SUA Convention is significantly amended from the original Convention and fundamentally expands the definition of unlawful acts. It permits, with the permission of the flag state, a state party to the Convention to board a ship on the high seas if there is reasonable suspicion that individual(s) aboard a ships have been or are expected to contravene the Convention (IMO, 2003).

Piracy

In 2011, 439 incidents of piracy were recorded worldwide with 45 captured vessels and 802 seafarers taken hostage (IMB, 2013). Currently, Somalia continues to be a prime centre of piracy and, as of 2012, there were 237 reported attacks resulting in the capture of 11 ships and 222 crew members. Besides Somalia, South-east Asia, the Horn of Africa and Gulf of Guinea also experience high levels of piracy. Piracy, however, may be underreported by up to 50 per cent since ship owners may want to avoid dealings with the authorities, avoid increases to insurance rates or conceal actions or inactions by their crews (Chalk, 2008).

Moreover, the globalization of the shipping industry can make it difficult to identify the victims of piracy. For example, on 4 April 2010 Somali pirates captured the MV *Sambo Dream*, which was registered in the Marshal Islands, owned by a South Korean company and crewed predominantly by Filipino nationals (Coggins, 2012).

The LOSC addresses piracy in Articles 100–107 and 110, and provides the overall legal foundation for addressing piracy and related incidents such as armed robbery. With respect to piracy, perhaps the most important aspect of the LOSC is that the Convention defines piracy in Article 101 as an action that can only occur on the high seas or areas outside of national jurisdiction and encompasses '... any acts of illegal violence or detention, or act of depredation, committed for private ends by the crew or the passengers of a private ship or a private aircraft'. Acts of piracy in territorial waters are considered domestic crimes under the LOSC. Furthermore, under the LOSC piracy requires one ship (or aircraft) to seize another and therefore the actions of a ship's crew or passengers against their ship cannot be considered piracy. Another key component of the LOSC regarding piracy includes the ability of states to seize pirate ships and ships under the control of pirates on the high seas, arrest persons on board and seize cargo (Article 105). A ship of another state may only be seized by a warship or government ship (Article 107) while a state may board a ship believed to be engaged in piracy (Article 110).

The problem with the LOSC's definition of piracy is that acts of piracy often occur in territorial waters of jurisdictions without effective control of their seas (e.g. Somalia). As such, a more modern definition of piracy is '... an act of boarding or attempting to board any ship with the apparent intent to commit theft or any other crime and with the apparent intent or capability to use force in the furtherance of that act' (IMB, 2009). Piracy is often differentiated from terrorism by the fact that piracy is generally undertaken for financial gain, while terrorism is conducted to further political or ideological causes (Nincic, 2005).

Since 2000, over 30 reports on piracy and related activities have been submitted to the UN General Assembly, resulting in multiple resolutions on piracy (Table 2.2). In addition, between 2008 and 2012 the UN Security Council passed 12 specific resolutions on piracy adjacent to the coast of Somalia and the Gulf of Guinea. Lastly, the IMO is the lead international agency on combating piracy and passed the Code of Practice for the Investigation of Crimes of Piracy and Armed Robbery Against Ships on 2 December 2009 (A.1025(26)) in order to facilitate the investigation of piracy and robbery against ships (IMO, 2009a). The Code urges states to adopt national legislation to prosecute offences of piracy and armed robbery (Article 3.1) and to implement the provisions of the LOSC, the Convention for the Suppression of Unlawful Acts Against the Safety of Navigation and the Protocol for the Suppression of Unlawful Acts Against the Safety of Fixed Platforms Located on the Continental Shelf (Article 3.2). The Code implores masters to report piracy and armed robbery attempts and encourages coastal/port states to ensure investigations do not unduly delay ships (Article 3.3). States are also requested to cooperate with each other on investigations (Article 3.4), train investigators (Article 4) and follow established procedures regarding investigation (Article 5) and initial response (Article 6) (IMO, 2009a).

In addition to the IMO Code of Practice, the IMO adopted the Code of Conduct concerning the Repression of Piracy and Armed Robbery against Ships in the Western Indian Ocean and the Gulf of Aden on 29 January 2009 (the 'Djibouti Code of Conduct'). Currently, 20 of 21 eligible states have signed the Code and agree to cooperate on reducing piracy in the Western Indian Ocean.

Other maritime security measures

Proliferation Security Initiative

The US launched the Proliferation Security Initiative (PSI) with the purpose of halting the transport of weapons of mass destruction, their constituent materials or their delivery systems. The PSI allows states to develop bilateral agreements to board each other's vessels in ABNJ if believed to be involved in fostering terrorism or illegal activities. The seizure of weapons of mass destruction and related cargo in ABNJ if found is also authorized by the PSI (Nincic, 2005). The principles of the PSI include:

- Interdicting the transfer of weapons of mass destruction or their components to states and non-state actors believed to be seeking weapons of mass destruction.
- Exchanging information between states.
- Strengthening domestic laws to facilitate interdiction.
- Taking specific interdiction actions to halt the proliferation of weapons of mass destruction.

The PSI is not an international agreement or convention. Rather than becoming signatories, states either endorse or participate in the PSI. The PSI has no secretariat, no statistics on success or compliance and, as of 2012, 101 states were participating in the PSI (United States Department of State, 2003).

Container Security Initiative

The Container Security Initiative (CSI) was established by the US in 2002. The CSI ensures that high-risk containers are identified and inspected prior to loading on US-bound ships. The programme stations US nationals at 58 ports in North America, Europe, Asia, Africa, the Middle East and Latin and Central America. Since the establishment of the initiative, Canada and Japan have signed bilateral agreements to share information and station their nationals at US ports to monitor exports to their nations (Haveman et al., 2007). The EU is examining the CSI for use in European trade.

Stowaways

Stowaways are individuals or groups of people who hide (or are hidden) on ships (or in their cargos) to flee one state for another for economic, political or religious reasons. Addressing the issue of stowaways encompasses aspects of basic human rights, immigration and refugee policies and efforts to combat organized crime. Stowaways are relevant to this section because international agreements and conventions related to them generally focus on preventing access to ports and ships rather than addressing the underlying causes (e.g. inequity and oppression) that cause people to flee their jurisdiction. In 2010, the IMO reported 253 stowaway cases involving 721 stowaways found on all types of commercial ships (IMO, 2011a). Currently nearly 50 per cent of the world's stowaways are from West Africa. Stowaways interfere with shipping as they must be processed by the relevant authorities in port and cared for by a vessel's crew while underway.

The overarching international law related to stowaways is the United Nations Declaration of Human Rights. The 1948 Declaration confers upon stowaways the right to life and freedom from torture, slavery, discrimination and degradation. The 1957 International Convention Relating to Stowaways (the 'Brussels Convention'), adopted 10 October 1957, attempted to establish common procedures to address stowaways; however, the Convention has not yet entered into force (UNHCR, 1957).

The next official attempt to address stowaways was the Convention on Facilitation of International Maritime Traffic (the 'FAL'), adopted 9 April 1965 and entered into force 5 March 1967 (IMO, 1965). The Convention set the formalities, documentary requirements and procedures to be used on the arrival, stay and departure of a ship. It was amended in 2002 (adopted 10 January 2002, entered into force 1 May 2003) to add new standards and recommended practices to grapple with issues surrounding stowaways.

Other efforts to address stowaways include:

- The 1997 IMO Guidelines on the Allocation of Responsibilities to seek the Successful Resolution of Stowaway Cases (A.871(20)) (IMO, 1997).
- The 2002 SOLAS amendments and adoption of the ISPS Code (see above) to prevent unauthorized access to ships and port facilities (IMO, 2003).
- The 2011 IMO Revised Guidelines on the Prevention of Access by Stowaways and the Allocation of Responsibilities to Seek the Successful Resolution of Stowaway Cases (FAL.11(37)) (IMO, 2011b).

Shipping and the environment

Shipping has many potential negative consequences to the marine environment, to adjacent terrestrial environments and to the atmosphere. Nearly all commercial ships in operation today are powered by hydrocarbons, which contribute to atmospheric pollution and greenhouse gases. Ships may intentionally (dumping) or unintentionally (accidents, antifouling hull paint) pollute (e.g. oil, garbage, cargoes, emissions) the ocean, land or atmosphere. Besides suffering detrimental effects of pollution, marine mammals are affected by shipping noise and ship strikes. Light pollution from ships may alter navigational patterns of marine birds. Ships are vectors for invasive or alien species that may reside in ballast water, on a ship's hull or within a ship's piping (see Chapter 5). Recycling or disposal of retired ships often releases toxic substances into the environment.

This section outlines the primary global institutions and agreements that govern shipping's impacts on the marine environment. This is not a comprehensive examination but, instead, provides the reader with an understanding of how environmental impacts from shipping are addressed, both in terms of preventing or minimizing environmental impacts and ensuring adequate compensation to those affected by the environmental externalities of shipping. This section begins with the International Convention for the Prevention of Pollution from Ships (MARPOL 73/78). It continues with the introduction of a number of other marine protection-related conventions.

Note that Chapter 5 addresses the non-shipping related impacts to marine environments from human activities.

International Convention for the Prevention of Pollution from Ships

Early conventions to protect the marine environment included the 1921 Conference on the Protection of the Area from Pollution by Oil and the 1954 International Convention for the Prevention of Pollution of the Sea by Oil. Termed the 'Oil Pollution Convention' (OPC), it prohibited the deliberate discharge of oil or oily mixtures from all seagoing vessels. The Conventions were amended in 1969 to prohibit the discharge of oil cargo within 50nmi of the nearest land and to regulate the proportion of oil and ballast water discharges. The OPC was amended again in 1971 to provide additional protection to Australia's Great Barrier Reef.

The OPC was eventually replaced by the International Convention for the Prevention of Pollution from Ships, which is now the primary convention that governs the protection of the marine environment from shipping. The Convention addresses both the operational and accidental causes of pollution in solid, liquid and airborne forms. The pollutants covered by the Convention include oil, chemicals, harmful substances in packaged form, sewage and garbage. The Convention, often referred to as 'MARPOL 73/78', is a combination of the 1973 and 1978 Treaties and a 1997 Protocol.

The Convention contains six annexes. Annexes I and II are mandatory for those states that have ratified the Convention while Annexes II, IV, V and VI are optional:

- Annex I: Regulations for the Prevention of Pollution by Oil (1983). Reinforces the 1954 Oil Pollution Convention with additional provisions for minimizing oil discharges when loading/unloading, prohibiting the use of cargo oil tanks for carrying ballast water, and structural requirements to prevent spills in the event of collisions or stranding.
- Annex II: Regulations for the Control of Pollution by Noxious Liquid Substances in Bulk (1967). Identifies four categories (X, Y, Z, other) of liquid substances transported in bulk ranked by relative hazard to marine resources or human health. Ships constructed after 1 July 1986 must comply with different safety codes depending on which category of liquid substances they are constructed to transport.
- Annex III: Regulations for the Prevention of Pollution by Harmful Substances Carried at Sea in Packaged Form (1992). Outlines eight regulations related to the labelling, packing, stowing, documenting and transport of harmful substances. Revised in 2010.
- Annex IV: Regulations for the Prevention of Pollution by Sewage from Ships (2003). Regulates the discharge of sewage into the sea from ships and specifies ship- and land-based sewage management facilities. Does not apply to the high seas.
- Annex V: Regulations for the Control of Pollution by Garbage from Ships (1998). Prohibits the disposal of plastics into the sea and limits other types of garbage that can be introduced into the marine environment including food, domestic and operational waste. The Annex also requires port states to provide reception facilities to accept and process ship-borne garbage.
- Annex VI: Regulations for the Prevention of Air Pollution from Ships (1997). Regulates the harmful emissions from ships and, in particular, diesel emissions (sulphur oxide and nitrogen oxide). Establishes emissions thresholds in sulphur emission control areas (see below). Regulations on reducing the release of greenhouse gases from shipping are currently under development.

MARPOL 73/78 introduced the concept of 'special areas' for Annexes I, II, IV, V and VI. For certain environments of concern ('special areas'), more stringent regulations were applied. Special areas in Annex I prevent oil discharges of any kind and currently include the Mediterranean Sea (1973), the Black Sea (1973), the Baltic Sea (1973), the Red Sea (1973), the Persian Gulf area (1973, 1987, 2004), the Antarctic (1992), North-west

European (1997) and Southern South African (2006) waters. Annex II prohibits the release of noxious liquid substances in Antarctic areas (1992). Annex IV outlines special provisions for the release of sewage into the Baltic Sea (1992) and Annex V further restricts the dumping of garbage from ships into the Baltic (1973), Black (1973), Mediterranean (1973), North (1989) and Red (1973) seas as well as the Antarctic (1992), the 'Gulfs' area (1973) and wider Caribbean (1991). Annex VI establishes Emission Control Areas in the Baltic (1997) and North (2005) Seas, North America (2010) and US Caribbean Sea (2011).

MARPOL 73/78 has no direct enforcement provisions but does permit any flag state or party to the Convention to enforce its provisions. Under both the Convention and the LOSC, primary responsibility for enforcing the conventions rests with the flag state. Generally, the flag state's duties are to issue international certificates (e.g. International Oil Pollution Prevention Certificate) and to carry out periodic inspections of vessels to ensure that they comply with international laws. Coastal and port states have no direct obligation under the LOSC and MARPOL 73/78 to enforce environmental protection aspects of the conventions. Yet many coastal and port states are active in ensuring the compliance and enforcement of international and domestic laws. Furthermore, port states maintain the right to inspect and detain vessels that either fail to produce valid international certificates or are believed to be in contravention of MARPOL 73/78 and the LOSC.

It has long been recognized in international law that a ship's captain (master) can be criminally liable for contravention of international laws (including MARPOL 73/78) and can also be liable for the actions of those under his/her command (Karim, 2010).

Regulation of marine oil spills and oil spill response and countermeasures

Approximately half of the world's oil production (approximately 1.6 billion tonnes) is transported by sea. Much of this transport is over long distances in VLCC and ULCC class tankers (Table 7.2) that pass through straits and transit along coastlines. This means of transport has resulted in many serious accidents and spills over the past several decades (Table 7.3). In addition, a significant amount of oil is transported short distances by seabed pipelines from production wells to offshore or onshore facilities. While the average number of major oil spills per year has dropped from 25 in the 1970s to 3 today, spills continue to impact fisheries, tourism and coastal economic activities (UNCTAD, 2012).

While ship owners, ship registries (flag states), classification societies, port states and cargo owners all have responsibilities to avoid oil spills into the marine environment, the IMO is the primary international agency responsible for establishing the regulatory framework related to the prevention, clean up and compensation for oil spills from ships. Although MARPOL 73/78 is the primary international instrument used to prevent the introduction of pollution into the marine environment from shipping, the mandated standards for the construction, equipment and operation of vessels under SOLAS may reduce the risk of oil pollution. The IMO also administers a number of oil pollution-specific conventions, which are discussed below.

International Convention relating to Intervention on the High Seas in Cases of Oil Pollution Casualties

The International Convention relating to Intervention on the High Seas in Cases of Oil Pollution Casualties ('Intervention Convention') was adopted on 29 November 1969 and

Table 7.3 Largest marine oil spills from ships, pipelines and oil production platforms

Ship/platform	Estimated spilled (maximum tonnes)	Year	Location of spill
Gulf War oil spill	820,000	1991	Persian Gulf
Deepwater Horizon	627,000	2010	Gulf of Mexico
Ixtoc I	480,000	1980	Gulf of Mexico
Atlantic Empress	287,000	1979	Trinidad and Tobago
Nowruz Field Platform	260,000	1983	Persian Gulf
ABT Summer	260,000	1991	Angola
Castillo de Bellver	252,000	1983	South Africa
Amoco Cadiz	227,000	1983	France
MT Haven	144,000	1991	Italy
Odyssey	132,000	1988	Eastern Canada
Torrey Canyon	119,000	1967	United Kingdom
Sea Star	115,000	1972	Gulf of Oman
Exxon Valdez	104,000	1989	Alaska

Source: UNCTAD (2012)

entered into force on 6 May 1975 (IMO, 1969a). The Convention currently has 87 contracting states representing 75.1 per cent of the world tonnage (UNCTAD, 2012). It was developed following the March 1967 loss of the tanker *Torrey Canyon*, which ran aground off the western coast of Cornwall, England, en route from Kuwait to Milford Haven in Wales. The tanker spilled 119,000 tonnes of oil and extensively damaged 190km of the western coast of southern England.

The Convention enables coastal states to take necessary actions in their territorial waters or the high seas to prevent, mitigate or eliminate oil pollution that would otherwise be illegal under international law (Article I). The Convention applies to all ships except warships, non-commercial government vessels and coastal installations and defines 'oil' as crude oil, fuel oil, diesel oil and lubricating oil (Article II). However, the coastal state must first consult with the flag state prior to initiating intervention (Article III).

In 1973 the Intervention Convention was modified by the Protocol relating to Intervention on the High Seas in Cases of Marine Pollution by Substances other than Oil. The Protocol entered force 30 March 1983 and amends the Convention by adding a number of non-oil substances listed in the Annex to the Protocol. Among non-oil substances are noxious substances, liquefied gasses and radioactive substances. Currently the Protocol has been ratified by 54 contracting states representing 50.4 per cent of the world's tonnage. Amendments to the Protocol by the IMO's MEPC to add further substances to the Annex occurred in 1991 (MEPC.49(31)), 1996 (MEPC.72(38)), 2002 (MEPC.100(48)), and 2007 (MEPC.165(56)).

International Convention on Civil Liability for Oil Pollution Damage

The International Convention on Civil Liability for Oil Pollution Damage (also referred to as the 'Civil Liability Convention' or CLC 1969) was adopted 29 November 1969 and entered into force 19 June 1975 (IMO, 1969b). The Convention was amended by Protocols in 1976 (CLC 1976) and 1992 (CLC 1992) with the CLC 1992 consolidating the 1969 text and 1976 Protocol. The CLC 1992 has effectively replaced CLC 1969 and was adopted 17

November 1992 and entered into force 30 May 1996 (IMO, 1992). As of 2013, the CLC 1992 has 130 contracting parties.

Maritime law has a long tradition of operating under the principle of limitation of liability (financial caps). The purpose of the CLC 1992 is to ensure that adequate compensation is available to persons who suffer damage caused by pollution resulting from the escape or intentional discharge of oil from ships. Claimants may be governments, private bodies (e.g. tourism operators) or individuals so long as the incident occurs within the jurisdiction of a state party to the Convention. The Convention also sets out the procedures for determining liability and appropriate compensation and creates a system of compulsory liability insurance. Key articles of the Convention include (IPIECA, 2005):

- Article I: The Convention only applies to tankers carrying oil as a bulk cargo (fuel and lubricants are excluded).
- Article II: The Convention only applies to the territory (i.e. coastal land areas), territorial sea and EEZ of a contracting state.
- Article III: The ship owner is not liable if a spill is a result of hostilities, exceptional natural phenomena, intentional acts by third parties or failure of (or incorrect) navigation aids. Only the ship owner may be held liable under CLC 1992.
- Article V: Liability limits are set at 4,510,000 SDR (as defined by the International Monetary Fund, approximately $6.8 million in 2012) for ships not exceeding 5000 units of tonnage. Each additional unit of tonnage adds 631 units of account (approximately $1000) up to a maximum of 89,770,000 units of account (approximately $136 million).
- Article VII: Owners of ships registered in a contracting state carrying more than 2000 tons of oil as bulk cargo must have insurance or guarantees to the liability limits outlined in Article V.
- Article XI: The Convention does not apply to warships or government ships used for non-commercial purposes.

International Convention on the Establishment of an International Fund for Compensation for Oil Pollution Damage

The International Convention on the Establishment of an International Fund for Compensation for Oil Pollution Damage ('1971 Fund Convention') was adopted 18 December 1971 and entered into force 16 October 1978. The original Convention was superseded by the 1992 Protocol to amend the 1971 International Convention on the Establishment of an International Fund for Compensation for Oil Pollution Damage ('1992 Fund Protocol') adopted 27 November 1992 and entered into force 30 May 1996 (IMO, 1992). The 1992 Fund Protocol has 111 parties as of November 2012 (Figure 7.1).

The purpose of the Fund Conventions is to provide compensation for pollution damage in instances where the CLC 1969/1992 is inadequate. This transpires when the ship owner can demonstrate an exemption under Article III of the CLC 1992 or the ship owner lacks the financial resources (or insurance) to pay the full compensation required under the CLC 1992. The 1971 Fund Convention created the IOPC Funds through a levy on both shippers and cargo owners.

The IOPC Funds, based in London, are used to pay damages over and above the limits of the CLC 1992. Three funds – which have different member states and amounts of compensation – are managed by the IOPC Funds: the 1971 Fund, the 1992 Fund and the

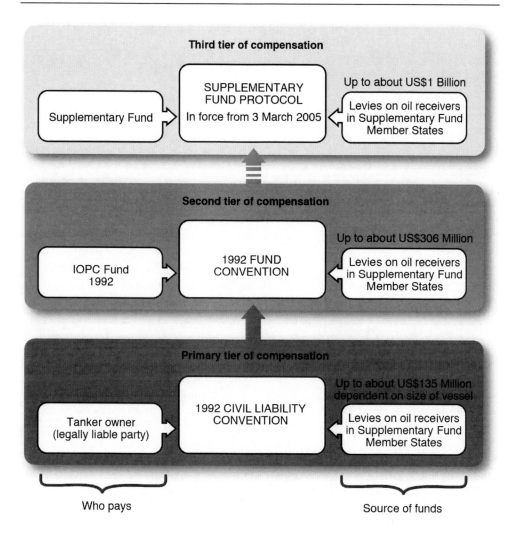

Figure 7.1 Compensation levels established by international conventions to fund oil spill clean up
Source: reproduced with permission from IPIECA (www.cid.csic.es/ampera/images/reports/IPIECA%20AAOP.pdf)

Supplementary Fund. Compensation maximums in the 1971 Fund were deemed by many members to be insufficient and, as such, the 1992 Fund and Supplementary Fund were established. The 1971 Fund currently has no members and ceased to be in force on 24 May 2002. Nevertheless, it continues to exist to address compensation claims for older incidents.

As of November, 2003 the liability limits of the 1992 Fund were 203 million SDR ($312 million). The 1992 Fund is financed by a levy on any person (e.g. business, government) operating within a state party to the Convention who has received more than 150,000 tonnes of crude oil and heavy fuel oil in one calendar year (IOPC, 2012).

In 2005, the Supplementary Fund was established with a maximum liability limit of 750 million SDR ($1.15 billion). It is financed in a similar fashion to the 1992 Fund with the exception that, for contribution purposes, each state will be levied for a minimum of 1 million tonnes of contributing oil.

International Convention on Oil Pollution Preparedness, Response, and Co-operation

The International Convention on Oil Pollution Preparedness, Response, and Co-operation (OPRC) was adopted 30 November 1990 and entered into force 13 May 1995 (IMO, 1990). Article 3 requires ships to carry onboard an oil pollution emergency plan that may be inspected when a ship is in the port of a state party to the Convention. Ships must report the discharge of oil to the nearest coastal state while oil handing facilities must report to the appropriate national authority (Article 4). Each state party to the Convention must establish a national organization with responsibility for oil spill preparedness and response. This authority must develop national contingency plans for oil spills, maintain a minimum level of prepositioned oil spill clean up equipment, coordinate oil spill response exercises and share information amongst various parties and governments (Article 6). Parties must also cultivate research and development related to oil spill preparedness and response (Article 8). In addition they must cooperate with other jurisdictions to provide technical support, training and technology transfer (Article 9).

The Protocol on Preparedness, Response and Co-operation to Pollution Incidents by Hazardous and Noxious Substances (OPRC-HNS) was adopted 15 March 2000 and entered into force 14 June 2007. The Protocol mirrors the OPRC Convention but is augmented by regulations concerning discharge, release/emission of hazardous or noxious substances that threaten the marine environment, coastline or other interest of a state (Article 2).

Emissions and climate change

While shipping is an efficient means of moving goods over short and long distances, the burning of fossil fuels for propulsion results in emissions of sulfur dioxide (SO_2), nitrogen oxides (NOx), volatile organic compounds (VOCs), particular matter, carbon dioxide (CO_2) and other greenhouse gases (Miola et al., 2011). Passenger and cruise ships must also produce power when in ports that results in increases in air pollution adjacent to the port.

The global shipping industry emits significant amounts of greenhouse gases. The IMO estimates that shipping was responsible for 1.046 million tonnes of CO_2 emissions, or about 3.3 per cent of global emissions, in 2007 (IMO, 2009b). If the global shipping industry were a country, it would be the sixth largest emitter of CO_2 after the US, China, Russia, India and Japan. Without additional regulations on CO_2 emissions or the adoption of more efficient shipping technologies, shipping emissions are forecast to increase by a factor of two to three by 2050. However, if technical and operational improvements are implemented, they could reduce CO_2 emissions by 25 to 75 per cent below current levels.

The 1992 UNFCCC commits state parties to the Convention to develop national policies for limiting greenhouse gas emissions (see Chapter 5). The 1997 Kyoto Protocol requires 37 developed states to reduce their greenhouse gas emissions to 1990 levels. However, the Kyoto Protocol acknowledges that emissions from international shipping are difficult to ascribe to particular states and, as such, tasks the IMO with addressing greenhouse gas reductions for the global shipping industry.

In July 2011, the IMO's MEPC adopted mandatory energy efficiency measures for international shipping that were added as a new Chapter 4 to MARPOL 73/78 Annex VI. The measures, which entered into force on 1 January 2013, apply to the entire global international shipping industry and are the first example of global industry-wide CO_2 reduction measures. The Energy Efficiency Design Index (EEDI) for new ships mandates energy efficiency targets of 10 per cent improvement for ships built between 2015 and 2019, 15 per cent or 20 per cent for between 2020 and 2024 (depending on the type of ship) and 30 per cent for ships delivered after 2024. The Ship Energy Efficiency Management Plan (SEEMP), required for all new and existing ships, establishes best practices for fuel efficiency and voluntary guidelines for measuring CO_2 emissions.

While the IMO MEPC has considered a number of market-based mechanisms (incentives) for reducing ship emissions, no such market mechanisms are yet in place. However, market mechanisms will likely be incorporated into international shipping in the future, and, if the IMO is unable to broker a global consensus, the EU is likely to bring into force an emissions trading programme for shipping (Miola et al., 2011).

Green shipping practices

Green shipping practices (GSPs) is a broad term that captures efforts by shipping companies to reduce waste and to conserve resources. Opportunities to 'green' shipping are numerous: reducing atmospheric pollution (e.g. NOx, SOx, CO_2) from ship propulsion systems, more efficient routing and reducing shipboard waste are important measures. Two other practices to help make shipping greener are the reduction of 'vapour emissions' or the minimization of cargo loss (e.g. hydrocarbons) to the atmosphere while in transit and tracking improvements related to the environmental impacts of shipping (Mair, 1995). Green shipping is driven by regulation (including the Emissions Control Areas outlined in MARPOL 73/78 Annex VI and the Kyoto Protocol), consumer demand for products shipped in a sustainable manner and the need to improve efficiencies (e.g. reduce fuel costs).

Industry-led green shipping efforts include the International Chamber of Shipping and International Shipping Federation (www.shippingandco2.org) and Green Award programme, which certifies ships according to a number of environmental management and

---

Box 7.3 Der Blaue Engel (Blue Angel) eco-label and shipping

In 2002, Germany included ships in their Der Blaue Engel environmental labelling ('eco-labelling') programme. The programme sets 10 mandatory and 20 optional standards to reduce air emissions and harmful discharges into the marine environment. Standards include environmental protection management, staff management, ship design and equipment, collision protection and leakage stability, redundant systems, hull stress monitoring, emergency towing systems, SOx emissions, particle emissions, emissions from cooling and refrigerating devices, waste disposal, incineration of waste, waste water, bilge water, antifouling, ballast water and firefighting foams. Ships of any flag may be certified, but the programme excludes warships, high-speed passenger vessels and fishing vessels. Currently, seven ships have been certified by the programme.

Source: Der Blaue Engel (2012)

safety criteria. Green Award certified ships received reduced port dues at more than 20 ports worldwide (www.greenaward.org).

Ship recycling

Ship recycling, also termed 'ship breaking' is the process of disassembling and disposing of ships at the end of their life. It is estimated that approximately 1000 ships per year are recycled, while more than 40,000 ships over the next 30 years will need to be recycled (Mikelis, 2007). Currently, 98 per cent of the world's recycled ships (by tonnage) are recycled in Bangladesh, China, India, Pakistan and Turkey. Around the globe, the industry provides considerable local employment and economic benefits. Nearly every part of a recycled ship can be reused. On the downside, recycling practices are often hazardous to both workers and the environment since ships – and particularly older ships – often contain hazardous materials including asbestos, poly-chlorinated biphenols (PCBs), ozone depleting substances, antifouling compounds, heavy metals and radioactive substances.

To minimize the impact of these hazards, the 1989 Basel Convention on the Control of Transboundary Movements of Hazardous Wastes and their Disposal (Chapter 5) prohibits the transboundary dumping or transfer of hazardous wastes. Not until 2004 did the Convention contemplate the ship recycling industry as a vector for the movement of hazardous waste. It was in this year that the CoP invited the IMO to establish regulations to ensure ship recycling conformed to the Basel Convention. Exploiting their experience with their 2003 voluntary Guidelines on Ship Recycling, the IMO agreed to lead the development of a new convention for ship recycling in 2005.

The Convention on the Safe and Environmentally Sound Recycling of Ships (the 'Hong Kong Convention') was adopted 19 May 2009 (IMO, 2009). Its purpose is to ensure that the disposal and recycling of ships is executed in a way that protects human health and the environment. In addition, the Convention includes provisions that apply to the construction and operation of ships in order to improve public and environmental health throughout a ship's lifecycle. The Convention builds upon existing conventions related to hazardous substances, worker safety and past work of the IMO on ship recycling. It includes 25 regulations grouped into the following categories: requirements for ships (regulations 4–14); requirements for ship recycling facilities (regulations 15–23); and reporting requirements (regulations 24–25).

Chapter summary

- Shipping moves goods from a region where they have a low value to a region where they have a high value.
- The world's merchant shipping fleet consists of approximately 104,000 ships over 100 GT representing approximately 1.4 billion GT with an average ship age of 22 years.
- Thirty-five states control 95 per cent of the world's shipping fleet. Of these states, Greece, Japan, Germany, Norway, the US and China control approximately 38 per cent of the world's shipping fleet.
- The LOSC enshrines the right of innocent passage and transit passage; immunity of warships; freedom of the high seas and use of the high seas for peaceful purposes; rights of all states to participate in shipping; prohibitions on carrying slaves; the necessity of combating piracy; and the right of all states to fish.

- The LOSC also articulates the rights and duties of coastal and flag states as well as the rights of states to determine the conditions for granting of their nationality to ships and standards for manning and training crews.
- Shipping is primarily regulated by the IMO and the ILO.
- The IMO's authority only applies to international shipping. Member states are expected to implement and enforce the 40 IMO conventions, protocols, as well as the 800+ codes and recommendations.
- The CMI is a non-governmental organization established to oversee the codification of maritime law, commercial practices and the promotion of national associations of maritime law.
- The major marine shipping conventions are SOLAS, STCW, MLO and MARPOL 73/78.

References

Bellwood, P. (1997) 'Ancient seafarers', *Archaeology*, 50 (2).

Branch, A. E. (2007) *Elements of shipping*, Routledge, Milton Park.

Broder, J. M. (2004) 'At ports cargo backlog raises security questions', *New York Times*, 27 July.

Chalk, P. (2008) *The Maritime Dimension of International Security: Terrorism, Piracy and Challenges for the United States*, RAND Corporation, Santa Monica, CA.

Clyde, P. and Reitzes, J. D. (1998) 'Market power and collusion in the ocean shipping industry: Is a bigger cartel a better cartel?', *Economic Inquiry*, 36 (2), 292–304.

CMI (Comité Maritime International) (2012), www.comitemaritime.org, accessed 01 December 2012.

Coggins, B. L. (2012) 'Global patterns of maritime piracy, 2000–2009: Introducing a new dataset, *Journal of Peace Research*, 49 (4), 605–617.

Der Blaue Engel (2012) 'Environment-conscious ship operation RAL-UZ 110', www.blauer-engel.de/en/products_brands/search_products/produkttyp.php?id=506, accessed 1 December 2012.

Gibbons, A. (1998) 'Ancient island tools suggest *homo erectus* was a seafarer', *Science*, 279 (5357), 1635–1637.

Haveman, J. D., Shatz, H. J., Jennings, E. M. and Wright, G. C. (2007) 'The container security initiative and ocean container threats', *Journal of Homeland Security and Emergency Management*, 4 (1), 1–19.

Hummels, D., Lugovskyy, V. and Skiba, A. (2009) 'The trade reducing effects of market power in international shipping', *Journal of Development Economics*, 89 (1), 84–97.

IHS Fairplay (2011) *2011 World Fleet Statistics*, IHS Fairplay, Englewood, CO.

ILO (2006) *Maritime Labour Convention*, Geneva, Switzerland, www.ilo.org/wcmsp5/groups/public/@ed_norm/@normes/documents/normativeinstrument/wcms_090250.pdf, accessed 1 December 2012.

IMB (International Maritime Bureau) (2013) 'Piracy falls in 2012, but seas off East and West Africa remain dangerous, says IMB', Piracy Reporting Centre, London, www.icc-ccs.org/news/836-piracy-falls-in-2012-but-seas-off-east-and-west-africa-remain-dangerous-says-imb, accessed 1 May 2013.

IMO (International Maritime Organization) (1965) *Convention on Facilitation of International Maritime Traffic*, IMO, London, treaties.un.org/doc/Publication/UNTS/Volume%20591/volume-591-I-8564-English.pdf, accessed 1 December 2012.

IMO (1966), *International Convention on Load Lines*, IMO, London seafarers.msa.gov.cn/InternationalPact/InternationalFile/LOAD%20LINE/LL%201966/%E8%8B%B1LL%201966.pdf.

IMO (1969a) *International Convention relating to intervention on the high seas in cases of oil pollution casualties*, IMO, London, treaties.un.org/doc/Publication/UNTS/Volume%20970/volume-970-I-14049-English.pdf, accessed 1 December 2012.

IMO (1969b) *International Convention on Civil Liability for Oil Pollution Damage*, IMO, London, treaties.un.org/doc/Publication/UNTS/Volume%20970/volume-973-I-1497-English.pdf, accessed 1 December 2012.

IMO (1972) *Convention on the International Regulations for Preventing Collisions at Sea*, IMO, London, treaties.un.org/doc/Publication/UNTS/Volume%201050/volume-1050-I-15824-English.pdf, accessed 1 December 2012.

IMO (1974) *International Convention for the Safety of Life at Sea*, IMO, London, treaties.un.org/doc/Publication/UNTS/Volume%201184/volume-1184-I-18961-English.pdf, accessed 1 December 2012.

IMO (1978a) *International Convention on Standards of Training, Certification and Watchkeeping for Seafarers*, IMO, London, www.cosco.com/en/pic/research/11058200875274005.pdf, accessed 1 December 2012.

IMO (1978b) *1978 Protocol Relating to the 1973 International Convention for the Prevention of Pollution from Ships*, IMO, London, treaties.un.org/doc/Publication/UNTS/Volume%201340/volume-1340-A-22484-English.pdf, accessed 1 December 2012.

IMO (1988a) *Convention for the Suppression of Unlawful Acts against the Safety of Maritime Navigation*, IMO, London, treaties.un.org/doc/db/Terrorism/Conv8-english.pdf, accessed 1 December 2012.

IMO (1988b) *Protocol for the Suppression of Unlawful Acts against the Safety of Fixed Platforms Located on the Continental Shelf*, IMO, London, www.un.org/en/sc/ctc/docs/conventions/Conv9.pdf, accessed 1 December 2012.

IMO (1990) *International Convention on Oil Pollution Preparedness, Response and Co-operation*, IMO, London, treaties.un.org/doc/Publication/UNTS/Volume%201891/volume-1891-I-32194-English.pdf, accessed 1 December 2012.

IMO (1992) *1992 Protocol to Amend the 1969 International Convention on Civil Liability for Oil Pollution Damage*, IMO, London, www.transportrecht.org/dokumente/HaftungsUe_engl.pdf, accessed 1 December 2012.

IMO (1997) *Resolution A.871(20), Guidelines on the Allocation of Responsibilities to Seek the Successful Resolution of Stowaway Cases*, FAL.2/Circ.43, Annex; Resolution A.871(20), IMO, London, www.unhcr.org/refworld/docid/3ae6b31db.html, accessed 1 December 2012.

IMO (1999) *International Convention on Arrest of Ships*, IMO, London, treaties.un.org/doc/Publication/MTDSG/Volume%20II/Chapter%20XII/XII-8.en.pdf, accessed 1 December 2012.

IMO (2003) *International Ship and Port Facility Security Code*, IMO, London, www.portofantwerp.com/sites/portofantwerp/files/ISPS_code_en.pdf, accessed 1 December 2012.

IMO (2004) *SOLAS – Consolidated Edition*, IMO, London.

IMO (2005a) *Convention for the Suppression of Unlawful Acts Against the Safety of Maritime Navigation*, IMO, London, www.unodc.org/tldb/pdf/Convention&Protocol%20Maritime%20Navigation%20EN.pdf, accessed 1 December 2012.

IMO (2005b) *2005 Protocol to the 1988 Protocol for the Suppression of Unlawful Acts against the Safety of Fixed Platforms Located on the Continental Shelf*, IMO, London, www.unodc.org/tldb/en/2005_Protocol2Protocol_Fixed%20Platforms.html, accessed 1 December 2012.

IMO (2005c) *2005 Protocol to the 1988 Convention for the Suppression of Unlawful Acts against the Safety of Maritime Navigation*, IMO, Loondon, www.unhcr.org/cgi-bin/texis/vtx/refworld/rwmain?docid=49f58c8a2&page=search, accessed 1 December 2012.

IMO (2009a) *Code of Practice for the Investigation of Crimes of Piracy and Armed Robbery Against Ships*, IMO, London, www.imo.org/OurWork/Security/PiracyArmedRobbery/Guidance/Documents/A.1025.pdf, accessed 1 December 2012.

IMO (2009b) *Second IMO GHG Study 2009*, IMO, London www.imo.org/blast/blastDataHelper.asp?data_id=27795&filename=GHGStudyFINAL.pdf, accessed 1 December 2012.

IMO (2009c) *The Hong Kong International Convention for the Safe and Environmentally Sound Recycling of Ships*, IMO, London, ec.europa.eu/environment/waste/ships/pdf/Convention.pdf, accessed 1 December 2012.

IMO (2010) *International Safety Management Code*, IMO, London, www.imo.org/ourwork/human element/safetymanagement/documents/ismcode_4march2010_.pdf, accessed 1 December 2012.

IMO (2011a) *Reports on Stowaway Incidents: Annual statistics for the year 2010*, FAL.2/Circ.121, IMO, London, www.imo.org/blast/blastDataHelper.asp?data_id=30513&filename=121.pdf, accessed 1 December 2012.

IMO (2011b) *Revised Guidelines on the Prevention of Access by Stowaways and the Allocation of Responsibilities to Seek the Successful Resolution of Stowaway Cases*, FAL.11(37), IMO, London, www.imo.org/ourwork/facilitation/stowaways/documents/resolution%2011(37)_revised%20guideli nes%20on%20the%20prevention%20of%20access%20by%20stowaways%20and%20the%20alloc ation%20of%20responsibilities.pdf, accessed 1 December 2012.

IMO (2012) *International Shipping Facts and Figures Information Resources on Trade, Safety, Security, Environment*, IMO, London, www.imo.org/KnowledgeCentre/ShipsAndShippingFactsAndFigures/ TheRoleandImportanceofInternationalShipping/Documents/International%20Shipping%20-%20Facts%20and%20Figures.pdf, accessed 1 December 2012.

IOPC (International Oil Pollution Compensation Funds) (2012) *The International Regime for Compensation for Oil Pollution Damage: Explanatory note*, www.iopcfund.org/npdf/genE.pdf, accessed 1 December 2012.

IPIECA (International Petroleum Industry Environmental Conservation Association) (2005) 'Action against oil pollution', www.cid.csic.es/ampera/images/reports/IPIECA%20AAOP.pdf, accessed 1 December 2012.

ISL (Institute of Shipping Economics and Logistics) (2011) *Shipping Statistics and Market Review*, 55 (8), www.infoline.isl.org/index.php%3Fmodule%3DDownloads%26func%3Dprep_hand_out %26lid%3D695&sa=U&ei=v2iMUKrCAsbligLH94CoDg&ved=0CBgQFjAA&usg=AFQjCNGl xupdci4YNxQ5lnm5v9sNbJ-n6Q, accessed 1 December 2012.

Karim, S. (2010) 'Implementation of the MARPOL convention in developing countries', *Nordic Journal of International Law*, 79 (2), 303–337.

Liu, H. (2009) *Liner Conferences in Competition Law: A Comparative Analysis of European and Chinese Law*, Springer, Berlin.

Lu, C.-S., Chang, C-C., Hsu, Y.-H. and Prakash, M. (2010) 'Introduction to the special issue on maritime security', *Maritime Policy & Management*, 37 (7), 663–665.

Mair, H. (1995) 'Green shipping', *Marine Pollution Bulletin*, 30 (6), 360.

Mikelis, N. E. (2007) 'A statistical overview of ship recycling', International Symposium on Maritime Safety, Security & Environmental Protection, Athens, September 2007.

Miola, A., Marra, M. and Ciuffo, B. (2011) 'Designing a climate change policy for the international maritime transport sector: Market-based measures and technological options for global and regional policy actions', *Energy Policy*, 39, 5490–5498.

Mordal J. (1959) *Twenty-five Centuries of Sea Warfare*, Abbey Library, London.

Munari, F (2012) 'Competition in liner shipping', in J. Basedow, M. Jürgen and R. Wolfrum Rüdigeret (eds) *The Hamburg Lectures on Maritime Affairs 2009 & 2010*, Springer-Verlag, Berlin.

Nincic, D. J. (2005) 'The challenge of maritime terrorism: Threat identification, WMD and regime response', *Journal of Strategic Studies*, 28 (4), 619–644.

UNCITRAL (United Nations Commission on International Trade Law) (2009) *United Nations Convention on Contracts for the International Carriage of Goods Wholly or Partly by Sea*, London, www.uncitral.org/pdf/english/texts/transport/rotterdam_rules/09-85608_Ebook.pdf, accessed 1 December 2012.

UNCTAD (United Nations Conference on Trade and Development) (2009) *Review of Maritime Transport*, UNCTAD, New York, unctad.org/en/Docs/rmt2009_en.pdf, accessed 25 October 2013.

UNCTAD (2011) *Review of Maritime Transport 2011*, New York, unctad.org/en/docs/rmt2011ch1_en.pdf, accessed 1 December 2012.

UNCTAD (2012) *Liability and Compensation for Ship-source Oil Pollution: An Overview of the International Legal Framework for Oil Pollution Damage from Tankers*, New York, unctad.org/en/PublicationsLibrary/dtltlb20114_en.pdf, accessed 1 December 2012.

UNHCR (United Nations High Commission for Refugees) (1957) *International Convention Relating to Stowaways*, www.unhcr.org/refworld/docid/3ae6b3a80.html, accessed 1 December 2012.

United States Department of State (2003) *Proliferation Security Initiative*, www.state.gov/t/isn/c10390.htm, accessed 28 February 2013.

Chapter 8

International law and policy of the polar oceans

Governing the Southern and Arctic oceans

Introduction

The polar regions – comprising the Antarctic and Arctic – are characterized by their remoteness, inhospitable climate, distinctive biological systems and, until recently, perceived limited economic utility. Yet, while similar in climate, the Antarctic and Arctic regions differ from each other in almost all other respects, including governance, patterns of human use and oceanographic structure.

The Antarctic Ocean (termed the 'Southern Ocean') is singular in that no sovereignty claims are permitted on the Antarctic continental shelf. Most human activities are attentively regulated in the region under international agreements. As such, the Southern Ocean is the world's largest global 'commons' and available for use by any states or individual with the means to operate in its inhospitable environs.

The Arctic Ocean, in contrast, is surrounded by continents on which five states have near-exclusive jurisdiction. This is a result of the extent of the Arctic continental shelves, and, as such, the Arctic has few elements of a global commons. Furthermore, with climate change engendering the retreat of sea ice, no ocean region is changing faster. These transformations are focusing global attention on the potential for resource development (particularly hydrocarbons) and the use of the Arctic for shipping.

This chapter introduces how the Antarctic and Arctic environments are governed and also explores some of the challenges of managing the polar regions. A more detailed summary on Antarctic governance can be found in Mitchell and Sandbrook (1980), Joyner (1998) and Stokke and Vidas (1996). Emmerson (2010), Byers (2009) and Osherenko and Young (2005) are excellent resources on the management of the Arctic.

Introduction to the Antarctic and Southern Ocean

The Southern Ocean comprises approximately 35–36 million square kilometres, about 15 per cent of the world's ocean surface. The northern boundary of the Southern Ocean is oceanographically defined by the Antarctic Convergence, which is a 30–50km wide zone where the warmer, oligotrophic (nutrient poor – see Chapter 1) waters of the Atlantic, Indian and Pacific oceans merge with nutrient and biologically rich Antarctic waters. This convergence creates a zone of high biological productivity (Figure 8.1). The Antarctic Convergence is between 50 and 60 degrees south latitude except in the Western Indian Ocean where the Convergence extends to 45 degrees south latitude.

For the purposes of Antarctic administration, two boundaries are commonly employed.

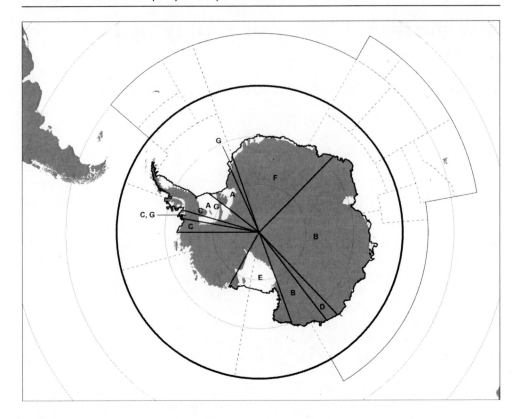

Figure 8.1 Overview of the Antarctic maritime area (Southern Ocean)

Note: The Antarctic maritime area is defined as areas south of 60 degrees south latitude (thick line). The Antarctic Convergence is found between 50 and 60 degrees south latitude except in the Western Indian Ocean where the Convergence extends to 45 degrees south latitude (thin line). The Antarctic Convergence is used to define the northern boundary of the Commission for the Conservation of Antarctic Marine Living Resources (sub-areas shown as dashed lines). Sovereignty claims are as follows: A) Argentina, B) Australia, C) Chile, D) France, E) New Zealand, F) Norway, G) UK.

The Antarctic Treaty System (discussed below) is generally applicable to all areas south of 60 degrees south latitude (the 'Antarctic maritime area'). This latitude, therefore, is the boundary that separates Antarctic management from the management of oceans elsewhere (Vidas, 2000). However, the 1980 Convention on the Conservation of Antarctic Marine Living Resources (see below) is defined by the Antarctic Convergence.

The Antarctic continent is 14 million square kilometres (about twice the size of Australia). It was first discovered in the 1820s by British, American and Russian fleets (Mitchell and Sandbrook, 1980). By the end of the century, Belgium, Germany, the UK and Sweden were staging overwintering expeditions. Norway's Roald Amundsen reached the South Pole in December 1911 and exploration activities continued throughout the early part of the twentieth century. Between 1980 and 1943, seven nations – Argentina, Australia, Chile, France, New Zealand, Norway and the UK – made territorial claims to Antarctica. Despite these declarations, Antarctica is the only continent to date that is free from permanent human non-research-related settlement and use.

Throughout much of human history, the Southern Ocean was too inhospitable to support human activities. In the eighteenth century, however, humans began hunting in the Southern Ocean. Populations of fur seals (>2 million harvested), whales (>500,000 harvested), elephant seals (~1 million harvested) and penguins (unknown harvest) were all driven to near commercial extinction (Kock, 2007). In the twentieth century, commercial fishing interests moved into the area. First, cod icefishes (nothothens) were targeted using trawling methods; then, in the 1980s, a longline fishery for Patagonian toothfish (*Dissostichus eleginoides*) emerged. In the process, bycatches caused declines of albatrosses and petrels. In the 1970s, a fishery developed for krill (*Euphausia superba*). Krill is a food staple for many fish, birds and marine mammals; this raised concerns that a large, unregulated low trophic level fishery may have ecological repercussions throughout the Southern Ocean.

The threats of sovereignty claims, commercial interests and environmental damage led to a suite of international agreements negotiated over a period of several decades beginning in the 1940s. The most important international agreements governing the Southern Ocean are (the year of entry into force in brackets):

- International Whaling Convention (see Chapter 6) (1946).
- The Antarctic Treaty (AT) (see below) (1961).
- The LOSC (see Chapter 2) (1994).
- The Protocol on Environmental Protection to the Antarctic Treaty (1998).

In addition, many other treaties and conventions (e.g. marine pollution, fishing, climate change) apply to the Southern Ocean area and are discussed in other chapters. Specific conventions that apply only to the Southern Ocean are as follows (the year of entry into force in brackets):

- CCAS (1978).
- CCAMLR (1982).
- The Agreement on the Conservation of Albatrosses and Petrels (ACAP) (2004) under the Convention on the Protection of Migratory Species of Wild Animals (CMS).

The Antarctic Treaty and Antarctic Treaty System

The AT regulates the management and governance of all land and ice shelf areas south of 60 degrees south latitude. The Treaty and its related agreements are generally termed the Antarctic Treaty System (ATS). Negotiations were concluded over a six-week period and the Treaty was signed on 1 December 1959. It entered into force on 23 June 1961 after ratification by the 12 parties (Argentina, Australia, Belgium, Chile, France, New Zealand, Norway, South Africa, UK, US, USSR) (ATS, 1959). The primary intent of the Treaty is to ensure 'in the interests of all mankind that Antarctica shall continue forever to be used exclusively for peaceful purposes and shall not become the scene or object of international discord'. The Treaty currently has 50 signatories representing 65 per cent of the world's population and is administered by the Antarctic Treaty Secretariat in Buenos Aires, Argentina.

The specific Treaty articles are outlined in Box 8.1; however, the primary focus of the Treaty is to prohibit military activity, ban nuclear explosions and the disposal of nuclear waste, promote scientific cooperation and prevent additional territorial claims.

Box 8.1 Articles of the Antarctic Treaty

- Article I: Antarctica is to be used for peaceful purposes and no state may establish or present a military presence.
- Article II: Antarctica is free to all for scientific investigation.
- Article III: States will cooperate on scientific investigation.
- Article IV: No state may claim new or additional sovereignty in the Antarctic region.
- Article V: Nuclear tests and the disposal of nuclear wastes are prohibited.
- Article VI: The Treaty applies to terrestrial and ice shelf areas south of 60 degrees south latitude.
- Article VII: Contracting Parties to the Treaty may designate inspectors who may inspect any part of Antarctica at any time.
- Article VIII: Anywhere within Antarctica, observers discussed in Article VII shall be subject to the jurisdiction of the Contracting Party of which they are nationals.
- Article IX: Contracting Parties agree to meet to discuss the Treaty after the Treaty enters into force.
- Article X: Contracting Parties may not defeat the purpose and intent of the Treaty.
- Article XI: Disputes will be resolved through negotiation, inquiry, mediation, conciliation, arbitration, judicial settlement or other peaceful means of their own choice. If a settlement cannot be reached the dispute may be referred to the ICJ.
- Article XII: The Treaty may be modified by unanimous agreement of the Contracting Parties and enter into force when ratified by the national governments of all the Contracting Parties. The Treaty expires 30 years from the date of entry into force.
- Article XII: Non-party states may enter the Treaty by accession and are bound by the Treaty once it is received by the depositary government.
- Article XIV: The Treaty will be deposited in the archives of the government of the US.

With respect to the Southern Ocean, only Argentina, Australia and Chile have the ability under the LOSC to claim marine jurisdiction south of 60 degrees south latitude. To date, however, these states have not done so and continue to work within the ATS. Japan is prevented from claiming Antarctic areas by its post-war Peace Treaty.

The ATS is administered annually through the Antarctic Treaty Consultative Meeting (ATCM) where all parties (50 to date) to the Treaty have the right to attend (Table 8.1). However, only the 28 Consultative Parties (CPs) have the right to participate in decision-making. CPs include the 12 original signatories to the Treaty as well as 16 additional states that have carried out significant scientific investigations in the Antarctic region. The Netherlands is the only CP without a scientific base in Antarctica. Conversely, Pakistan established a summer station on the continent but was not granted CP status. The remaining parties, termed Non-Consultative Parties, are those states who have acceded to the AT but without sufficient scientific activities on the continent to warrant CP status.

Although unable to participate in resolutions, observers and invited experts are permitted to attend ATCM meetings. Currently, the three observers are the Scientific Committee

Table 8.1 Parties to the Antarctic Treaty and their status

State	Entry into force	Consultative status	Environment Protocol*	CCAS	CCAMLR
Argentina	23 Jun 1961	23 Jun 1961	14 Jan 1998	X	X
Australia	23 Jun 1961	23 Jun 1961	14 Jan 1998	X	X
Austria	25 Aug 1987				
Belarus	27 Dec 2006		15 Aug 2008		
Belgium	23 Jun 1961	23 Jun 1961	14 Jan 1998	X	X
Brazil	16 May 1975	27 Sep 1983	14 Jan 1998	X	X
Bulgaria	11 Sep 1978	5 Jun 1998	21 May 1998		X
Canada	4 May 1988		13 Dec 2003	X	X
Chile	23 Jun 1961	23 Jun 1961	14 Jan 1998	X	X
China	8 Jun 1983	7 Oct 1985	14 Jan 1998		X
Colombia	31 Jan 1989				
Cuba	16 Aug 1984				
Czech Republic	14 Jun 1962		24 Sep 2004		
Denmark	20 May 1965				
Ecuador	15 Sep 1987	19 Nov 1990	14 Jan 1998		
Estonia	17 May 2001				
Finland	15 May 1984	20 Oct 1989	14 Jan 1998		X
France	23 Jun 1961	23 Jun 1961	14 Jan 1998	X	X
Germany	5 Feb 1979	3 Mar 1981	14 Jan 1998	X	X
Greece	8 Jan 1987		14 Jan 1998		X
Guatemala	31 Jul 1991				
Hungary	27 Jan 1984				
India	19 Aug 1983	12 Sep 1983	14 Jan 1998		X
Italy	18 Mar 1981	5 Oct 1987	14 Jan 1998	X	X
Japan	23 Jun 1961	23 Jun 1961	14 Jan 1998	X	X
Korea (Democratic Peoples Republic of)	21 Jan 1987				
Korea (Republic of)	28 Nov 1986	9 Oct 1989	14 Jan 1998		X
Malaysia	31 Oct 2011				
Monaco	30 May 2008		31 Jul 2009		
Netherlands	30 Mar 1967	19 Nov 1990	14 Jan 1998		X
New Zealand	23 Jun 1961	23 Jun 1961	14 Jan 1998		X
Norway	23 Jun 1961	23 Jun 1961	14 Jan 1998	X	X
Pakistan	1 Mar 2012		31 Mar 2012		X
Papua New Guinea	16 Mar 1981				
Peru	10 Apr 1981	9 Oct 1989	14 Jan 1998		X
Poland	23 Jun 1961	29 Jul 1977	14 Jan 1998	X	X
Portugal	29 Jan 2010				
Romania	15 Sep 1971		5 Mar 2003		
Russian Federation	23 Jun 1961	23 Jun 1961	14 Jan 1998	X	X
Slovak Republic	1 Jan 1993				
South Africa	23 Jun 1961	23 Jun 1961	14 Jan 1998	X	X
Spain	31 Mar 1982	21 Sep 1988	14 Jan 1998		X
Sweden	24 Apr 1984	21 Sep 1988	14 Jan 1998		X
Switzerland	15 Nov 1990				
Turkey	24 Jan 1996				
Ukraine	28 Oct 1992	4 Jun 2004	24 Jun 2001		X
UK	23 Jun 1961	23 Jun 1961	14 Jan 1998	X	X
US	23 Jun 1961	23 Jun 1961	14 Jan 1998	X	X
Uruguay	11 Jan 1980	7 Oct 1985	14 Jan 1998		X
Venezuela	24 Mar 1999				

Note: *Protocol on Environmental Protection to the Antarctic Treaty (1991).
Source: www.ats.aq/devAS/ats_parties.aspx?lang=e

on Antarctic Research (SCAR), the CCAMLR and the Council of Managers of National Antarctic Programs (COMNAP) (Box 8.2) Recurrent invited experts include the Antarctic and Southern Ocean Coalition (ASOC), the International Association of Antarctic Tour Operators (IAATO) and Coalition of Legal Toothfish Operators (COLTO) (Dodds, 2010; Dudeney and Walton, 2012).

Decision-making at the ATCM is by consensus. Discussion topics are presented in two forms: working papers, which require debate and action, and information papers, which do not. Working papers may be brought forward by CCAMLR, SCAR, COMNAP and the Committee for Environmental Protection (see below). Other invited experts/groups may only introduce information papers. During each ATCM, there is also a meeting of the Committee of Environmental Protection.

The AT/ATS has been lauded as one of the most successful international treaties. The AT has managed over 10 per cent of the earth's surface both peacefully and in the interests of scientific research. Not only has the AT avoided disputes over sovereignty between the seven claimant states, but it has adroitly balanced the interests of the non-claimant states. Despite continued pressure to allow exploitation and development, the AT has continually prohibited the quest for mineral resources. All of this has been accomplished during five decades of profound political, technological and economic change (Triggs, 2009).

Box 8.2 Organizations formed to advise the Antarctic Treaty System

SCAR is an independent, interdisciplinary scientific body established to provide scientific advice to the ATCMs and other organizations on issues of science and conservation affecting the management of Antarctica and the Southern Ocean. It was established in 1958 and currently has 37 member states. SCAR establishes standing scientific groups and committees who address a wide range of scientific topics related to the life sciences, geosciences and physical sciences.

COMNAP was established in 1988 to foster the coordination of scientific research by the 28 CPs. COMNAP's purpose is to 'develop and promote best practice in managing the support of scientific research in Antarctica'. The organization's goals are to:

- serve as a forum to develop practices that improve effectiveness of activities in an environmentally responsible manner;
- facilitate and promote international partnerships;
- provide opportunities and systems for information exchange;
- provide the ATS with objective, technical and non-political advice.

The *Committee for Environmental Protection* was convened in 1998 to provide advice with respect to the implementation of the Protocol on Environmental Protection to the Antarctic Treaty. The Committee meets annually in conjunction with the ATCM and provides specific advice on environmental impact assessment, environmental emergencies, the establishment and operation of protected areas, scientific information management and environmental monitoring.

Source: www.comnap.aq; www.cep.aq; www.scar.org

Nonetheless, prickly issues in the Southern Ocean continue: the threat of overfishing (both legal and illegal), whaling, ozone depletion, pollution from inside and outside the region, climate change (regional warming, ocean acidification, changes in sea ice distribution), tourism, bioprospecting and human security (Chown et al., 2012). While the CLCS (see Chapter 2) has not yet been asked to consider the limits of continental shelf claims for claimant states, there is a future risk of these states requesting the CLCS to make delineations. On top of this, 30 states currently operate research bases on the continent, and there are concerns that some may use the area to advance their military objectives (Anonymous, 2012). With 200 organizations, representing 27 states, complications are inevitable undertaking commercial-based research in the AT region.

Other serious challenges are emerging. Disputes between claimant and non-claiming states over access to offshore hydrocarbon resources, ownership of intellectual property from bioprospecting and conflicts between whaling and non-whaling nations all endanger the fragile coexistence in the Southern Ocean (Joyner, 2009).

Many blame the AT itself for the myriad conflicts. In contrast to other international treaties, the AT has been criticized for its ambiguous and vague language as well as weak compliance and enforcement mechanisms. Others argue the polar opposite, contending that the absence of definitive provisions in the Treaty has, perhaps inadvertently, created a political climate where claimant states feel comfortable entering into agreements without jeopardizing their claims.

The Protocol on Environmental Protection to the Antarctic Treaty

The Protocol on Environmental Protection to the Antarctic Treaty (the 'Environmental Protocol' or 'Madrid Protocol') was signed on 4 October 1991 (ATS, 1991). It entered force on 14 January 1998. The Protocol has been ratified by 38 states and signed by a further 16 states. The Protocol designates Antarctica as a 'natural reserve, devoted to peace and science' (Article 2). Article 3 commits to the protection of Antarctica's environmental values and its vital role for scientific pursuits. The Article then sets out the principles to avoid or mitigate human impacts on Antarctic systems.

The Protocol is clear that it neither modifies nor amends the AT and that nothing in the Protocol shall relieve any party from their rights and obligations under the ATS. The Protocol also commits signatories to cooperate on the planning and conduct of activities in the Antarctic, including completing environmental assessments and evaluating cumulative impacts. Importantly, except for scientific research, the Protocol prohibits any activity related to mineral resources. Lastly, the Protocol establishes the Committee for Environment Protection to provide advice with respect to the implementation of the Protocol.

The Environmental Protocol contains the following six annexes:

- Annex I: Guidelines and processes for conducting environmental impact assessments in the Antarctic region.
- Annex II: Instructions regarding permitting for the removal or use of Antarctic flora and fauna for scientific purposes and rules on the introduction or use of non-native species in Antarctica.
- Annex III: Guidelines on the storage, removal and disposal of wastes.
- Annex IV: Guidelines on the prevention of marine pollution, including emergency preparedness and the disposal of garbage and sewage.

- Annex V: Instructions for the establishment of Antarctic Specially Protected Areas and Antarctic Specially Managed Areas.
- Annex VI: Prevention of environmental emergencies and setting liability limits.

Convention for the Conservation of Antarctic Seals

For almost three centuries, hunting for pelts has decimated various marine mammal species. By the 1820s, populations of southern fur seals (*Arctocephalus* sp.) were believed to be commercially extinct. Close to 95 per cent of the fur seal population breeds on South Georgia Island, making them particularly vulnerable to overexploitation. The decline of the fur seal combined with interest in the commercial harvest of other Antarctic seal species resulted in early prohibitions on sealing on South Georgia Island. These were followed by the 1964 Agreed Measures for the Conservation of Antarctic Fauna and Flora, which were endorsed to at the third ATCM.

The Convention for the Conservation of Antarctic Seals (the 'Antarctic Seals agreement') was signed on 1 June 1972 and entered into force on 11 March 1978 (BAS, 1972). There are 16 parties to the Convention. These include Argentina, Australia, Belgium, Brazil, Canada, Chile, France, Germany, Italy, Japan, New Zealand (signed but not ratified), Norway, Poland, Russia, South Africa, UK and the US. The Convention applies to all marine waters south of 60 degrees south latitude and to the following species:

- Southern elephant seal *Mirounga leonine*.
- Leopard seal *Hydrurga leptonyx*.
- Weddell seal *Leptonychotes weddelli*.
- Crabeater seal *Lobodon carcinophagus*.
- Ross seal *Ommatophoca rossi*.
- Southern fur seals *Arctocephalus* sp.

The Convention commits signatories to establish a permitting system for the harvest of species covered by the Convention, to share scientific and harvest information with other contracting parties, and to conform to harvest regulations outlined in the Annex. These regulations address protected species, harvest levels by species, closed seasons, seal reserves and sealing methods.

Conference on the Conservation of Antarctic Marine Living Resource

The 1977 ATCM invoked Article IX of the ATS and held a number of meetings that, on 1 August 1980, led to the adoption of the CCAMLR (ATS, 1980). With 31 signatories, the Convention entered into force on 7 April 1982. The overall objective is to conserve the marine life of the Southern Ocean while allowing 'rational use' of its resources. In particular, and unique to international marine law in 1982, CCAMLR used Article II of the Convention as a basis to manage the Southern Ocean using both 'precautionary' and 'ecosystem' approaches (discussed in Chapter 10). It has now successfully exploited these approaches for three decades (Constable, 2011). CCAMLR specifically excludes whales and seals, which are managed through the ICRW (Chapter 6) and Convention for the Conservation of Antarctic Seals (see above). CCAMLR is complimentary to and cooperates with the Annex II (Conservation of Antarctic Fauna and Flora) of the Protocol on Environmental Protection to the Antarctic Treaty.

Key Articles of CCAMLR include:

- Article I: Outlines the CCAMLR area (Figure 8.1) and specifies that the Convention applies to apply to all living marine organisms in the CCAMLR area, including birds.
- Article II: Defines the principle of 'conservation' for the Convention and outlines how harvesting and associated activities are to be conducted in accordance to the definition of 'conservation'.
- Article III: Commits parties to the Convention to also respect the AT whether or not they are signatories.
- Article IV: Asserts that no state may claim new or additional sovereignty in the Antarctic region.
- Article V: Parties to the Convention recognize their obligation to the living resources provisions in the AT.
- Article VI: The Convention will not override the ICRW and the Convention for the Conservation of Antarctic Seas.
- Article VII: A Commission for the Conservation of Antarctic Marine Living Resources will be established.
- Article VIII: The Commission discussed in Article VII will have the authority to operate in all CCAMLR areas.
- Article XIV: The Scientific Committee for the Conservation of Antarctic Marine Living Resources will be established.

A Commission, headquartered in Hobart, Tasmania, Australia oversees the implementation of the Convention. The Commission is comprised of 25 members who are involved in fishing and/or scientific research in the Southern Ocean. The primary duties of the Commission are to establish catch levels for harvested species and minimize impacts to non-target species. Compliance and enforcement of harvest levels are the responsibility of Commission members. The Commission meets annually to review reports on the members' activities, analyse compliance with Commission decisions, study and adopt regulatory measures and discuss the financing and administration of the Commission.

In addition to the Commission, sub-committees on science (Scientific Committee), ecosystem monitoring (Ecosystem Monitoring and Management), fish stock assessment (Fish Stock Assessment), finance and administration (Standing Committee on Administration and Finance) and implementation and compliance (Standing Committee on Implementation and Compliance) were established.

CCAMLR is widely viewed as successful in its efforts to balance environmental protection with sustainable use. At the request of the Commission, CCAMLR underwent a performance review in 2008. The review suggested that CCAMLR devote attention to invasive species issues, vessel safety standards, marine pollution, krill management and bycatch issues. The review also recommended that CCAMLR be more proactive with respect to the designation of marine protected areas (see Chapter 10), controlling fishing effort and revisiting mechanisms to resolve disputes.

Introduction to the Arctic

The Arctic comprises 30 million square kilometres or one sixth of the earth's surface. The region is generally defined as areas north of the Arctic Circle (66.5 degrees north latitude), which is the approximate limit of the midnight sun and the polar night. The Arctic is also defined as the areas where the average temperature in the warmest month (July) is below 10°C (Figure 8.2). There are eight Arctic states (Canada, US, Russia, Norway, Sweden, Finland, Denmark including Greenland and Faroe Islands, and Iceland), five of which border the Arctic Ocean.

The Arctic Ocean is a relatively shallow entrained sea, primarily covered by ice, and exhibits the lowest salinities of any ocean as a result of freshwater inputs. Due to the remote location and harsh environment, there is still much unknown about the region. Recent technological advances have allowed a growth in knowledge of the region's physiographic, oceanographic and ecological characteristics. In addition, traditional knowledge stemming from the regions' vast indigenous populations has emerged as an important source of information to better understand Arctic processes.

Humans have been active in Arctic areas for tens of thousands of years. Approximately 4 million people living in eight states inhabit the Arctic, of which indigenous populations

Figure 8.2 Overview of Arctic circumpolar area

Note: The Arctic Circle (66° 33′N) is marked by the solid gray line; the 10°C July isotherm is the dashed line; and the historical summer extent of permanent sea ice is the shaded area.

number approximately 300,000 (ACIA, 2005). Indigenous peoples have domesticated rein-deer and continue to rely on the ocean for whales, walrus, seals, fish and polar bears.

With some notable differences, the Arctic Ocean experiences similar environmental impacts from human activities as other ocean areas. Oceanic and atmospheric advection of pollutants from more southern areas is common in the Arctic. In particular, PCBs, persist-ent organic pollutants, heavy metals and radionucleides accumulate through the Arctic food web (Kallenborn *et al.*, 2011). As part of this web, indigenous Arctic peoples are some of the most contaminated peoples on earth.

Arctic sovereignty and politics

Historically the Arctic has been occupied by indigenous peoples for tens of thousands of years, followed by sojourns by western explorers, missionaries and scientists. With the estab-lishment of the state of Alaska in 1959 and the advent of the cold war, nuclear submarines – able to transit under sea ice – made the Arctic an important military interest and led to decades of submarine activity. Similarly, the threat of missiles and bombers using Great Circle routes to reach North America and the former USSR resulted in the establishment of distant early warning radar installations that, to this day, continue to be a source of pollu-tion. More recently, melting sea ice has provoked an interest in resource development and the use of the northern sea passages for trade. This has caused both Canada and Russia to declare the Northwest Passage and Northern Sea Route respectively as internal waters under the LOSC. As early as the fifteenth century, traders pondered the use of the Arctic for shipping between Europe and Asia. A trip between London and Yokohama through the Northwest Passage is 15,700 km and 13,841 through the Northern Sea Route (Northeast Passage). These distances are significantly shorter than routes through the Suez (21,200 km) or Panama (23,300 km) (Lasserre and Pelletier, 2011) (Figure 8.3).

Figure 8.3 Northern Sea Route (A) and Northwest Passage (B)

Indigenous populations are governed by capitals much further south. As such, many Arctic residents feel removed and isolated from decisions made in their territories. The converse can be said of their fellow countrymen to the south. While they rarely visit their northern climes, they often exhibit nationalist sentiments towards the Arctic and feel a strong kinship with Arctic regions. This sense of ownership towards Arctic environments has been growing in recent years as offshore hydrocarbons are discovered and the potential for the commercialization of northern shipping routes is realized.

Arctic governance

The interests of the five Arctic coastal states are reasonably well known (Table 8.2). With the exception of the 1973 Agreement on the Conservation of Polar Bears, no specific international treaty regime governs the Arctic. Within the Arctic region, continental shelves within the EEZ, as per the LOSC, are governed by domestic law. All other marine regions are considered high seas areas, so the LOSC is the de facto primary international instrument governing the Arctic. The Arctic is not specifically addressed in the LOSC. The only reference within the Convention related to the Arctic is Article 234, which states that coastal states have the right to adopt and enforce non-discriminatory laws and regulations for the prevention, reduction and control of marine pollution from vessels in ice-covered areas.

Article 76 of the LOSC permits states to claim sovereignty over areas beyond 200nmi provided applications are approved by the CLCS (see Chapter 2). Arctic states that have

Table 8.2 Summary of national Arctic interests

Nation	Arctic interests
Canada	Sovereignty and responsible development of the Northwest Passage
	Defining Arctic waters as 'internal waters' under the LOSC to better regulate shipping
	Oil and gas exploration and development
	Tourism
US	Does not acknowledge Canadian sovereignty claims
	Sovereignty and responsible development of the Northwest Passage
	Oil and gas exploration and development
Russia	Development of the Northern Sea Route for transit
	Oil and gas exploration and development
Sweden/Denmark/ Norway	Strive for regulation of Arctic environmental pollution while maintaining protection for Arctic communities
Indigenous peoples	Concern that changes in ecosystem structure due to sea ice melting and increased coastal development could have detrimental effects on community structure and traditional lifestyles
	Want responsible economic development and social progress
	Fear that increased oil pollution due to expansion of Arctic shipping, oil and gas development could lead to ecological destruction and health risks for local populations

Source: Modified from Pietri et al. (2008)

applied to the CLCS to increase their territorial waters beyond 200nmi include Russia (2001), Norway (2006), Iceland and Denmark (2009) and Canada (2010). If all five Arctic states are successful in their claims, approximately half of the Arctic will be managed as territorial seas and therefore will be under national jurisdiction.

Despite the absence of a formalized Arctic treaty system, the Arctic states have cooperated among themselves for many decades to manage fisheries, undertake joint search and rescue exercises, establish scientific programmes and collaborate on boundary identification. In 1987, the president of the Soviet Union, Mikhail Gorbachev, called for Arctic states to cooperate on arms control and resource development. This demand soon prompted the International Arctic Science Committee in 1990, the Arctic Environmental Protection Strategy in 1991 and the Arctic Council in 1996 (Young, 2011).

Despite these beneficial actions, certain government and non-government organizations maintain that existing cooperative and jurisdictional arrangements are insufficient to address the scope, scale and pace of climate change impacts in the region and the increase in economic activity. In particular, warming is expected to shift the distribution and composition of biological communities. This may contribute additional management challenges for commercially important fish stocks, including the movement of stocks across national borders and calls to increase fishery quotas. Currently, there is no pan-Arctic RFMO to manage rapidly changing fish stocks, and there have been a number of calls to establish more binding and comprehensive international governance mechanisms to address Arctic issues (Pietri et al., 2008).

The absence of an Arctic legal framework has led the European Parliament (2008) and WWF (2010) to propose the development of an Arctic Treaty (European Parliament, 2008; Koivurova and Molenaar, 2010). However, to date the coastal state Arctic nations oppose the development of an Arctic treaty and have expressed their position in the Ilulisaat Declaration (Ilulisaat Declaration, 2008).

Arctic Council

Signed on 19 September 1996 in Ottawa and referred to as the Ottawa Declaration, the Arctic Council emerged from earlier efforts by the eight Arctic nations (Canada, Denmark – including Greenland and the Faroe Islands, Finland, Iceland, Norway, Russian Federation, Sweden and the US) to develop the 1989 Arctic Environmental Protection Strategy (AEPS) that was formally adopted on 14 June 1991 (Arctic Council, 1996). The overall purpose of the Council (Article 1.a) is to provide a means for promoting cooperation, coordination and interaction among member states and indigenous communities on common Arctic issues including sustainable development and environmental protection. Other duties of the Council are to oversee AEPS programmes (Article 1.b), supervise a sustainable development programme (Article 1.c), disseminate information, encourage education and promote interest in Arctic-related issues (Article 1.d). The Council has no independent legal power, avoids military matters and is a forum for discussion only (Jeffers, 2010). The chairmanship of the Arctic Council rotates between Arctic nations every two years. Since its inception each Arctic state has chaired the Council once. Canada is now chair from May 2013–2015, followed by the US, 2015–2017.

The following six indigenous organizations are permanent participants on the Arctic Council:

- Arctic Athabaskan Council (AAC).
- Aleut International Association (AIA).
- Gwich'in Council International (GGI).
- Inuit Circumpolar Council (ICC).
- Russian Arctic Indigenous Peoples of the North (RAIPON).
- Saami Council (SC).

Observers are permitted on the Council if they are endorsed by the Council. Observer status is open to non-Arctic states, inter-governmental organizations and non-governmental organizations. In May 2013, the Arctic Council added six new non-Arctic state observers including China, India, Italy, Japan, Singapore and South Korea. These are in addition to France, Germany, The Netherlands, Poland, Spain and the UK. Nine inter-governmental organizations (e.g. the International Union for the Conservation of Nature – IUCN) and 11 non-governmental organizations (e.g. WWF) have received observer status. The EU has requested observer status on the Council but the Council has not yet approved this request as of May 2013.

The Council directs the activities of the following working groups comprised of government representatives and researchers:

- Arctic Contaminants Action Program.
- Arctic Monitoring and Assessment Programme (AMAP).
- Conservation of Arctic Flora and Fauna (CAFF).
- Emergency Prevention, Preparedness and Response (EPPR).
- Protection of the Arctic Marine Environment (PAME).
- Sustainable Development Working Group (SDWG).

Furthermore, the Council has established a number of task forces that address time-limited issues. Examples of task forces include the Task Force on Ecosystem-based Management, Circumpolar Business Forum, Black Carbon, Arctic Marine Oil Pollution Preparedness and Response, and Search and Rescue.

Since the Ottawa Declaration, the Council has met eight times. A summary of the key issues and provisions is provided in Table 8.3.

Hydrocarbons and transport have galvanized interest in the area. Almost half of the Arctic marine area is comprised of continental shelves, the largest percentage of any ocean. These shelves are estimated to contain 40–83 billion barrels of oil (13 per cent of the world's undiscovered reserves) and 770–1547 trillion cubic feet of natural gas (30 per cent of the world's undiscovered reserves) at 95 and 50 per cent probabilities respectively. In addition, up to 44 billion barrels of natural gas liquids are predicted (Bird et al., 2008; Gautier et al., 2009). These resources will become more accessible to development as sea ice continues to recede. Shipping, too, is poised to exploit this recession.

Despite economic interests in the Arctic, rapid industrial expansion in the region is unlikely. For offshore hydrocarbon exploration and development to occur, many remaining boundary disputes will need to be resolved and advances in technology to address ice risks (for platforms, ships and pipelines) will need to be developed. Additionally, a social license will need to be obtained from indigenous Arctic peoples and citizens further south. For shipping, the northern routes may be decades away from significant commercial use. Arctic shipping requires lower average vessel speeds in order to circumvent ice, thus reducing the

Table 8.3 Declarations of the Arctic Council and key provisions

Declaration	Signed	Key provisions
Iqaluit Declaration	17 September 1998	Approval of new Permanent and Observer Participants in the Arctic Council Establish the Sustainable Development Program and Working Group Announce the University of the Arctic
Barrow Declaration	13 October 2000	Approval of new Permanent and Observer Participants in the Arctic Council Approval of new proposals to conduct a Survey of Living Conditions in the Arctic Endorse and adopt the Arctic Climate Impact Assessment (ACIA)
Inari Declaration	10 October 2002	Approval of new Permanent and Observer Participants in the Arctic Council Expansion of the work of the Council Continued focus on human conditions in the Arctic Partnering with other organizations
Reykjavik Declaration	24 November 2004	Approval of new Permanent and Observer Participants in the Arctic Council Request for Council working groups to review Arctic contaminants, conduct a marine shipping assessment and assess oil and gas and acidification issues
Salekhard Declaration	26 October 2006	Approval of new Permanent and Observer Participants in the Arctic Council Request to continue climate research Request working groups to research the behaviour of oil and other hazardous substances in Arctic waters
Tromso Declaration	29 April 2009	Approval of new Permanent and Observer Participants in the Arctic Council Urge early implementation actions on methane and other short-lived climate forcers Urge Council members to strengthen their work on adaptation Encourage cooperation with the IMO on measures to reduce shipping impacts Decide whether to hold deputy minister-level Council meetings
Nuuk Declaration	12 May 2011	Announce the Agreement on Cooperation in Aeronautical and Maritime Search and Rescue in the Arctic as the first binding agreement under the Arctic Council Establishment of the Arctic Council Secretariat Establish Ecosystem-based Management Task Force
Kiruna Declaration	15 May 2013	Announce the second binding Agreement on Oil Spill Preparedness and Response Establish Circumpolar Business Forum Task Force, Black Carbon Task Force and Scientific Cooperation Task Force

Source: www.arctic-council.org

competitive advantage over the Panama and Suez routes. In addition, shipping can only operate in the near ice-free summer (at least for the immediate future). There are other disincentives as well: vessels will require specialized navigation equipment and training, revised cargo management practices and technologies to prevent cargo freezing will be required, and the northern routes will most likely require higher insurance rates (Stokke, 2011).

Overall, however, the Arctic is a politically stable and peaceful region. Sovereignty claims are mature and generally accepted, states work well together and towards common purposes, and the remaining boundary disputes between states are – relative to states in other parts of the world – proceeding towards resolution. Yet, some challenges remain. The Arctic, as a global commons continues to shrink as seabed mapping technologies lengthen the boundaries of the Arctic states' extended continental shelves (as allowed under the LOSC). Indeed, this expansion has developed to a point where Russia and Canada are expected to require a maritime boundary treaty (Baker and Byers, 2012). With the exception of a few high-latitude areas, it is quite conceivable that in the future most of the accessible Arctic will be within an EEZ of one or more Arctic coastal states.

China, while currently lacking an Arctic policy, considers itself an Arctic stakeholder and has lobbied the international community to prevent Arctic coastal states from securing claims to their extended continental shelves. The state has expressed a strong interest in several aspects of the Arctic: the effects of changes in Arctic climate on China, securing access to the Northern Sea Route to move goods to Europe and beyond, and participating in Arctic fisheries and natural resource development. With these interests in mind, China became a permanent observer at the Arctic Council in 2013 (Jakobsen and Peng, 2012).

Climate change and political implications in the Arctic

Relative to other areas of the planet, the effects of climate change have a disproportionate impact on Arctic environments. While the reasons for the Arctic's accelerated rate of warming are not fully understood, increases in its average temperature are almost twice that of the rest of the world. This can be attributed in part to sea ice melt, reducing regional albedo and increasing ocean heat absorption. Annual temperatures have increased by an average of 1°C over northwest Russia, east Greenland and Scandinavia over the past 50 years; yet they have cooled by 1°C over Iceland and the North Atlantic Ocean over the same period. In addition, average atmospheric temperatures in the Arctic have increased up to 3°C over land and are projected to increase between 6 and 10°C over the Arctic Ocean by 2090 (ACIA, 2005).

As Arctic environments warm, sea ice and glaciers melt and thin. This adverse circumstance contributes to sea-level rise and decreases in the earth's albedo: the amount of incoming solar radiation reflected back into space. Since the Arctic is an important regulator of the entire earth's temperature, malignant changes on a global scale are also on the horizon. Over the past 30 years, Arctic sea ice has decreased in area by approximately 30 per cent and modelling suggests that the Arctic may be ice free by 2030 (Jones, 2011). Other Arctic-specific impacts of climate change include increased methane emissions with the melting of permafrost, reductions in atmospheric ozone in the Arctic and human settlements being jeopardized with escalating coastal erosion.

Climate change is expected to have significant economic, social and environmental impacts. Along with coastal erosion, the continued retreat of sea ice and increased ocean temperatures will affect the distribution and abundance of commercially and socially important fisheries and marine mammal species (Huntington, 2009). In particular, polar bear populations are expected be imperiled by climate-induced habitat change (Laidre *et al.*, 2008). Increasing resource development – and oil and gas exploration and development in particular – while creating economic opportunities will increase risks of accidents and spills (Meek, 2011).

Chapter summary

- The Southern Ocean is defined by the Antarctic Convergence, which is found between 50 and 60 degrees south latitude except in the Western Indian Ocean where the Convergence extends to 45 degrees south latitude.
- The boundary that separates Antarctic management from the management of oceans elsewhere is 60 degrees south latitude (the 'Antarctic maritime area').
- The primary conventions governing the Antarctic are the International Whaling Convention, AT, the LOSC and the Protocol on Environmental Protection to the Antarctic Treaty.
- The primary conventions that apply only to the Southern Ocean are as follows: CCAS, CCAMLR, ACAP under CMS.
- The Arctic is defined as areas north of the Arctic Circle (66.5 degrees north latitude) or areas where the average temperature in the warmest month (July) is below 10°C.
- The Arctic is bordered by five coastal states (Canada, Denmark, Norway, Russia) and three non-coastal states (Finland, Iceland and Sweden). In addition, the 'near-Arctic' states (China, EU, Japan, Korea) have interests in the Arctic but no territory above the Arctic circle.
- No specific international treaty regime governs the Arctic and the Arctic is not specifically referenced in the LOSC.
- The Arctic Council (founded by the eight Arctic coastal and non-coastal states) promotes cooperation, coordination and interaction among member states and indigenous communities on common Arctic issues including sustainable development and environmental protection.
- The Arctic Council has no independent legal powers.

References

ACIA (Arctic Climate Impact Assessment) (2005) *Arctic Climate Impact Assessment*, Cambridge University Press, Cambridge.

Anonymous (2012) 'Antarctic Treaty is cold comfort', *Nature*, 481, 237.

ATS (Antarctic Treaty Secretariat) (1959) *Antarctic Treaty*, Buenos Aires, www.ats.aq/documents/ats/treaty_original.pdf, accessed 6 December 2012.

ATS (1980) *Convention on the Conservation of Antarctic Marine Living Resources*, Buenos Aires, www.ats.aq/documents/ats/ccamlr_e.pdf, accessed 6 December 2012.

ATS (1991) *Protocol on Environmental Protection to the Antarctic Treaty*, Buenos Aires, www.ats.aq/documents/recatt/Att006_e.pdf, accessed 6 December 2012.

Arctic Council (1996) *Declaration on the Establishment of the Arctic Council*, www.arctic-council.org/index.php/about/documents/category/5-declarations?download=13:ottawa-declaration, accessed 3 December 2012.

Baker, J. S. and Byers, M. (2012) 'Crossed lines: The curious case of the Beaufort Sea maritime boundary dispute', *Ocean Development and International Law*, 43 (1), 70–95.

BAS (British Antarctic Survey) (1972) 'Convention for the Conservation of Antarctic Seals', Cambridge, www.antarctica.ac.uk/about_antarctica/geopolitical/treaty/update_1972.php, accessed 2 December 2012.

Bird, K. J., Charpentier, R. R., Gautier, D. L., Houseknecht, D. W., Klett, T. R., Pitman, J. K., Moore, T. E., Schenk, C. J., Tennyson, M. E. and Wandrey, C. J. (2008) 'Circum-Arctic resource appraisal: Estimates of undiscovered oil and gas north of the Arctic Circle', U.S. Geological Survey Fact Sheet 2008-3049, pubs.usgs.gov/fs/2008/3049/, accessed 2 December 2012.

Byers, M. (2009) Who Owns the Arctic?, Douglas & McIntyre, Toronto, ON.

Chown, S. L., Lee, J. E., Hughes, K. A., Barnes, J., Barrett, P. J., Bergstrom, D. M., Convey, P., Cowan, D. A., Crosbie, K., Dyer, G., Frenot, Y., Grant, S. M., Herr, D., Kennicutt, M. C., Lamers, M., Murray, A., Possingham, H. P., Reid, K., Riddle, M. J., Ryan, P. G., Sanson, L., Shaw, J. D., Sparrow, M. D., Summerhayes, C., Terauds, A. and Wall, D. H. (2012) 'Challenges to the future conservation of the Antarctic', Science, 337, 158–159.

Constable, A. (2011) 'Lessons from CCAMLR on the implementation of the ecosystem approach to managing fisheries lessons from CCAMLR on EBFM', Fish and Fisheries, 12 (2), 138–151.

Dodds, K. (2010) 'Governing Antarctica: Contemporary challenges and the enduring legacy of the 1959 Antarctic Treaty', Global Policy, 1 (1), 108–115.

Dudeney, J. R. and Walton, W. H. (2012) 'Leadership in politics and science within the Antarctic Treaty', Polar Research, 31, 1–9.

Emmerson, C. (2010) The Future History of the Arctic, Perseus Books, Jackson, TN.

European Parliament (2008) 'Resolution of 9 October 2008 on Arctic governance', www.arcticgoverance.org, accessed 3 December 2012.

Gautier, D. L., Bird, K. J., Charpentier, R. R., Grantz, A., Houseknecht, D. W., Klett, T. R., Moore, T. E., Pitman, J. K., Schenk, C. J., Schuenemeyer, J. H., Sørensen, K., Tennyson, M. E., Valin, Z. C. and Wandrey, C. J. (2009) 'Assessment of undiscovered oil and gas in the Arctic', Science, 324 (5931), 1175–1179.

Huntington, H. (2009) 'A preliminary assessment of threats to arctic marine mammals and their conservation in the coming decades', Marine Policy, 33 (1), 77–82.

Ilulisaat Declaration (2008) 'Declaration of 28 May 2008 adopted by the representatives of Canada, Denmark, Norway, the Russian Federation, and the United States', www.arcticgoverance.org, accessed 3 December 2012.

Jakobsen, L. and Peng, J. (2012) 'China's Arctic aspirations', SIPRI Policy Paper, 34, Stockholm International Peace Research Institute, Stocklolm.

Jeffers, J. (2010) 'Climate change and the Arctic: Adapting to changes in fisheries stocks and governance regimes', Ecology Law Quarterly, 37 (3), 917–977.

Jones, N. (2011) 'Towards an ice-free Arctic', Nature Climate Change, 1, 381.

Joyner. C. C. (1998) Governing the Frozen Commons: The Antarctic Regime and Environmental Protection, University of South Carolina Press, Columbia, SC.

Joyner, C. C. (2009) 'Potential challenges to the Antarctic Treaty', in P. A. Berkman, M. A. Lang, D. W. H. Walton and O. R. Young (eds), Science Diplomacy: Antarctica, Science and the Governance of International Spaces, Smithsonian Institution Scholarly Press, Washington, DC, http://www.atsummit50.org/media/book-15.pdf, accessed 10 July 2013.

Kallenborn, R., Borgå, K., Christensen, J. H., Dowdall, M., Evenset, A., Odland, J. Ø. Ruus, A. Aspmo Pfaffhuber, K., Pawlak, J. and Reiersen, L.-O. (2011) Combined Effects of Selected Pollutants and Climate Change in the Arctic Environment, Arctic Monitoring and Assessment Programme, Oslo.

Kock, K.-H. (2007) 'Antarctic marine living resources: Exploitation and its management in the Southern Ocean', Antarctic Science, 19 (2), 231–238.

Koivurova, R. and Molenaar, E. J. (2010) 'International governance and the regulation of the marine Arctic: A proposal for a legally binding agreement', WWF International Arctic Programme, Oslo.

Laidre, K. L., Stirling, I., Lowry, L. F., Wiig, Ø., Heide-Jørgensen, M. P. and Ferguson, S. H. (2008) 'Quantifying the sensitivity of Arctic marine mammals to climate-induced habitat change',

Ecological Applications, 18 (Supplement), S97–S125.

Lasserre, F. and Pelletier, S. (2011) 'Polar super seaways? Maritime transport in the Arctic: An analysis of shipowners' intentions, *Journal of Transport Geography*, 19, 1464–1473.

Meek, C. L. (2011) 'Adaptive governance and the human dimensions of marine mammal management: Implications for policy in a changing north', *Marine Policy*, 35 (4), 466–476.

Mitchell, B. and Sandbrook, R. (1980) *The Management of the Southern Ocean*, International Institute for Environment and Development, London.

Osherenko, G. and Young, O. R. (2005) *The Age of the Arctic: Hot Conflicts and Cold Realities*, Cambridge University Press, Cambridge.

Pietri, D., Soule, A. B., Kershnera, J., Solesa, P. and Sullivana, M. (2008) 'The Arctic shipping and environmental management agreement: A regime for marine pollution' *Coastal Management*, 36 (5), 508–523.

Stokke, O. S. (2011) 'Environmental security in the Arctic: The case for multilevel governance', *International Journal*, 66, 835–848.

Stokke, O. S. and Vidas, D. (1996) *Governing the Antarctic: The Effectiveness and Legitimacy of the Antarctic Treaty System*, Cambridge University Press, Cambridge.

Triggs, G. (2009) 'The Antarctic Treaty System: A model of legal creativity and cooperation', in P. A. Berkman, M. A. Lang, D. W. H. Walton and O. R. Young (eds), *Science Diplomacy: Antarctica, Science and the Governance of International Spaces*, Smithsonian Institution Scholarly Press, Washington, DC, http://www.atsummit50.org/media/book-8.pdf, accessed 10 July 2013.

Vidas, D. (2000) 'Emerging law of the sea issues in the Antarctic maritime area: A heritage for the new century?', *Ocean Development and International Law*, 31 (1–2), 197–222.

Young, O. R. (2011) 'If an Arctic Ocean treaty is not the solution, what is the alternative?', *Polar Record*, 47 (4), 327–334.

Chapter 9

International law and policy related to offshore energy and mining

Renewable and non-renewable resource development on continental shelves and in the Area

Introduction

The nearshore marine environment has long been utilized as a source for materials. For centuries, beaches and nearshore areas were mined for gold (Au), tin (Sn), diamonds, sand and gravel (Scott, 2007). As mining and shipping technologies improved, exploration and development radiated seaward onto the continental shelves. Offshore oil drilling was pioneered in the 1930s in the Gulf of Mexico while the first offshore windfarm was established in Denmark in the 1990s. Currently, one third of the world's petroleum is produced from continental shelves. This output is likely to continue to increase as sea ice retreats in the polar regions, opening up new areas for exploration and development.

It is estimated that deep sea areas (outside of continental shelves) contain two thirds of the world's mineral wealth. For the past 60 years, anticipated global mineral shortages have created periods of interest in mining the mineral rich deep sea. However, these prognosticated shortages failed to transpire, due to combination of increased metal recycling, refinements in industrial processes and new terrestrial discoveries. This has left commodity prices insufficiently high to make deep sea mining economically attractive. The rapid industrialization of China and India has recently altered this dynamic. Mineral prices have been driven to a point where commercial deep sea mineral production may be profitable. Long-term demand – based on forecasted development of land-based deposits – suggests that marine deposits high in zinc (Zn) and enriched in silver (Ag) are expected to become commercially viable at some point in the future. The first deep water offshore mineral exploration licenses were issued for two 2500 square kilometre claims by the government of Papua New Guinea in 1997 (Jones and Morgan, 2003). In contrast, terrestrial supplies of copper (Cu), nickel (Ni) and cobalt (Co) are anticipated to meet global needs for the next several decades (Antrim, 2005).

This chapter explores the different non-biological resources that are extracted – or anticipated to be extracted – from continental shelf and deep sea (ABNJ) areas. These realms are treated differently given that most continental shelves are either under the sovereignty of one or more states or are in the process of being annexed via applications to the Commission on the Limits of the Continental Shelf (CLCS).

Differences between continental shelves and deep ocean areas with respect to energy and mineral development

Marine energy and mineral development can be separated into activities that occur on the continental shelf versus activities that occur in the deep ocean. From a resource perspective, traditional hydrocarbons such as oil, natural gas and coal are only found on, or adjacent to, continental shelves. Formed by the decomposition of organic matter produced in terrestrial environments, these resources are not found in the deep ocean. Similarly, continental shelves contain aggregate resources (e.g. sand and gravel) transported from terrestrial environments that may contain metals (e.g. gold, tin) or diamonds that can be mined using placer operations. Unless interrupted by volcanic processes, continental shelves contain little to no mineralization beyond minerals transported from terrestrial environments as a result of erosion. The shallow depths of continental shelves also provide two other opportunities to harvest energy: anchoring wind energy projects to the seabed and generating electricity by harnessing water movement amplified by shallow depths and constrained passages.

Under the LOSC, continental shelves – and their corresponding mineral and energy resources – can be appropriated (owned) by states. These areas are therefore managed under domestic legal and regulatory regimes and treated similarly to their terrestrial counterparts under international law. Continental shelves, however, tend to be subject to overlapping jurisdictional claims more so than terrestrial environments.

In contrast to the continental shelves, the deep sea environments contain significant mineral resources. Deep sea deposits include high demand minerals including manganese (Mn), cobalt, copper (Cu), yttrium (Y) as well as rare earth elements used in electronics, batteries, solar panels and materials required for the production or processing of other composite materials. While traditional hydrocarbons (i.e. oil and gas) and coal are not found in the deep sea, methane hydrates (a type of hydrocarbon) are present if temperature and pressure regimes are favorable. Unlike the continental shelves, the deep seas are not currently used for energy production. However, there have been a number of proposals to generate electricity using the temperature gradient between surface and deep ocean water.

The LOSC and energy and mineral development

Under the LOSC, the non-biological resources on or under the seabed outside of the continental shelf areas are termed 'the Area' (see Chapter 2) and comprise approximately 260 million square kilometres, or about three times the area of all marine areas under state control.

In many respects, the LOSC owes its existence to uncertainties regarding ownership of deep sea resources. Deep sea environments are truly examples of *res nullis* (belonging to no one – see Chapter 2). Interest in seabed mining in the 1960s raised concerns that deep sea areas would be developed in the absence of a regulatory framework and without consideration of the economic and environmental concerns of states adjacent to potential mineral development. In 1968, the UNGA called for the exploitation of the deep seabed to be carried out for the 'benefit of mankind as a whole' (UN, 1968). This was followed by further resolutions placing a moratorium on deep sea mining and prohibiting the recognition of any claims to the seabed until an international regulatory regime was in place. As such, the original intent of the LOSC was to regulate seabed mining, which was furthered in 1974 by the 'Group of 77' (120+ less developed states) who demanded that less developed states be allowed and enabled to participate in seabed mining.

The LOSC confers on the coastal state the rights to explore, exploit, conserve and manage the resources in the water column on and under the seabed within the state's territorial waters and EEZ (Article 56). Within continental shelves extending beyond a coastal state's EEZ, states have rights to the non-biological resources on and within continental shelves as well as their living sedentary species (Article 77). This includes the exclusive right to authorize and regulate drilling on continental shelves.

Coastal states are required to make payments to the ISA (see Chapter 2) based on profits produced from the exploitation of the extended continental shelf. This revenue is redistributed to other states based on equitable sharing and also by taking account of the needs of less developed states (Article 82).

The Area is formally defined under the LOSC as those non-biological resources on or under the seabed outside of continental shelf areas. The LOSC defines these non-biological resources as 'minerals' (Article 133) regardless of whether they are metallic in nature. The Area does not apply to the water column or airspace above the seabed outside of continental shelves (Article 135).

Under the LOSC, the Area is open to all for the benefit of mankind and is to be used exclusively for peaceful purposes. States are required to cooperate to achieve these ends (Articles 136 and 138). The LOSC provides for the economic development of the Area under the condition that its development contributes to the world economy and growth in international trade (Article 150). Specifically, all states, regardless of geographic location (e.g. landlocked) or state of economic development have equal rights to access and gainful participation (Article 141). Economic benefits derived from the Area will be shared among all states (Article 139).

No state is permitted to claim territory or the natural resources within the Area. However, mineral resources obtained from the Area may be appropriated (owned) (Article 137). In addition, archaeological artefacts or discoveries of historical significance must be managed for the benefit of humanity, recognizing the importance of the state or culture of origin (Article 149). States bear the responsibility for actions of their nationals (individuals, corporations or government) and are liable for their actions (Article 139).

Management of marine energy and mineral resources on continental shelves

Although the LOSC confers on coastal states the exclusive right to develop their seabed areas within their territorial waters, EEZs and continental shelves, it does not provide guidance on mineral and energy development in these areas. Nevertheless, many states have entered bilateral or multilateral agreements on the development of these areas. The following sections briefly outline the various types of mining and energy activities that take place on continental shelf areas and broadly explain how they are managed.

Aggregate sand and gravel deposits

Aggregate deposits are non-mineral (mostly sand and gravel) resources formed by the erosion of terrestrial areas and their subsequent deposition in the marine environment. By tonnage, aggregates are the world's most mined substance. They are most commonly used for construction purposes, although marine sands are often used for beach nourishment. They are mined in shallow marine areas in depths up to 50m. Many states, including Japan and the US and some in Europe, currently mine marine aggregates. Nearshore aggregate mining is expected to increase as existing terrestrial sand and gravel resources are depleted and new deposits become more difficult to mine due to public opposition (Scott, 2011).

There are very few international agreements that address the management of marine aggregate development. The Convention for the Protection of the Marine Environment of the North East Atlantic 1992 (OSPAR Convention) was convened to eliminate pollution and protect the marine environment. In 2003, an Agreement on Sand and Gravel Extraction was adopted that required OSPAR signatories to adhere to the ICES *Guidelines for the Management of Marine Sediment Extraction*, which established a set of procedures for environmentally responsible marine aggregate development (Radzevičius *et al.*, 2010). In addition, under the Barcelona Convention (see Chapter 5), the Protocol for the Protection of the Mediterranean Sea against Pollution Resulting from Exploration and Exploitation of the Continental Shelf and the Seabed and its Subsoil (adopted in 1994 but not yet in force) provides guidance on marine aggregate extraction. Lastly, the European Code of Conduct for Coastal Zones (www.eucc.net) was developed by the Coastal Union and adopted by the Council of Europe Ministers in 1991. The Code provides detailed recommendations to minimize the effects of marine aggregate extraction on coastal processes, seasonal biological events and water quality (Radzevičius *et al.*, 2010).

Marine placer deposits

Marine placer deposits are formed by the deposition and transport of eroded terrestrial materials into marine environments. Certain placer deposits may concentrate heavy minerals (e.g. gold, silver, tin) or diamonds in certain locations. These deposits, termed 'secondary occurrences', are not produced at the mining site but transported there by fluvial (river) processes. They may be further sorted or concentrated by tidal or wave action.

Marine diamonds have been mined adjacent to the Namibian coast since 1989. Diamonds are transported by fluvial processes into shallow nearshore sediments where they can be excavated in waters up to 200m deep. Marine diamond production in Namibia now exceeds terrestrial production (Garnett, 2002).

Conventional offshore oil and gas

Marine environments currently produce one third of the world's hydrocarbons and have the potential for significant expansion (Sandrea and Sandrea, 2007). The shallow marine (<400m) and deep marine (>400m) environments are estimated to contain 36 per cent and 11 per cent of the global distribution of oil-bearing reservoir rocks respectively. Hydrocarbon-bearing rocks are found in the deep marine as a result of avalanches that move sediments into the deep ocean. Early efforts to develop offshore hydrocarbons were undertaken in the 1920s and 1930s in Azerbaijan, the Gulf of Mexico and Venezuela. However, it was not until the 1950s that offshore oil and gas extraction became economically and technically feasible in the Gulf of Mexico. Globally, 500 billion barrels of oil have been discovered in offshore areas, of which 200 billion barrels have been extracted. It is estimated that another further 300 billion barrels may yet be discovered in offshore areas (Sandrea and Sandrea, 2007).

In terms of the global production of energy, the oceans have become invaluable. In 2010, offshore facilities produced 30 per cent of the world's oil and 27 per cent of the world's gas. Current investment in offshore exploration and development is approximately $100 billion, or about 20 per cent of all oil and gas investment. Offshore areas contain approximately 20 per cent and 25 per cent of the world's remaining oil and gas reserves respectively (www.ihs.com) and between 1999 and 2009 over half of new oil and gas discoveries were found in offshore areas.

Offshore production, in order of total production, is concentrated in the Persian Gulf/Middle East, North Sea, West Africa, the Gulf of Mexico (US and Mexico), Asia/Australasia, Brazil, China, Caspian Sea and Russia/Russian Arctic. Currently, 17,000 platforms are operating and another 400 production facilities are constructed annually. Of the new offshore exploration, over half is centred in Southeast Asia (www.ifpenergiesnouvelles.com).

In the 1970s, the first deepwater (400–1500m) wells were completed. Ultra deepwater (>1500m) wells are currently producing in Brazil (1500m) and in the Gulf of Mexico (2700m) (Sandrea and Sandrea, 2007).

With respect to the management and regulation of offshore oil and gas resources, individual states issue exploration and development leases on their seabed areas and prescribe the operational framework under which activities are conducted. The exception is the EU that – through its Hydrocarbons Licensing Directive (94/22/EC) – has a common approach for the prospecting, exploration and production of offshore activities. Moreover, the European Commission is contemplating establishing common safety standards for all aspects of offshore hydrocarbon exploration and development.

Methane hydrates

Methane hydrates, produced by the decomposition of organic materials, consist of molecules of natural gas (predominately methane) trapped in ocean sediments by a combination of low temperature and high pressure. They are generally found in water depths greater than 400–500m and up to 1100m below the sea floor. One cubic metre of hydrate is equivalent to 164 cubic metres of methane at atmospheric pressure and temperature and the energy potential of the world's reserves is estimated at between 10^5 and 10^8 trillion standard cubic feet (Antrim, 2005; Scott, 2011). These astounding numbers make the energy potential of methane hydrates greater than all the world's known coal, oil and natural gas reserves (Morgan, 2012).

Methane hydrates are recovered by increasing the temperature or releasing the pressure within the deposit. This dissociates the gas from its bonds with water, whereupon the gas is piped to the surface. Given the physical and technological obstacles, there are no commercial methane hydrate recovery operations currently functioning; however, India and Japan are expected to begin commercialization of their hydrate resources within the next several decades (Antrim, 2005).

Marine renewable energy production

Marine renewable energy ('marine renewables') is a broad term encompassing all types of electricity production generated from the ocean. Marine renewables include the use of wind, waves, currents or temperature gradients between bottom and surface waters to generate electricity.

Offshore energy production using wind-driven turbines ('wind farming') to generate electricity is the most mature of the marine renewable industries. While wind has been used for millennia to drive pumps and produce mechanical energy for other uses, the earliest commercial wind turbine installation for the purposes of generating electricity was installed in 1980. The first offshore turbine was installed in 1991 in Denmark. Although it adds another level of complexity to construction and maintenance, locating wind turbines in offshore environments reduces public opposition and capitalizes on more consistent and favourable wind conditions (Lozano-Minguez, 2011).

Total offshore wind production at the end of 2011 was almost 4 gigawatts (GW), or sufficient energy to power 4 million homes. The UK is the largest producer (2GW), followed by Denmark (830 megawatts (MW) and the Netherlands (250MW). Globally, approximately 100GW of offshore wind infrastructure is under development, predominantly centred in Europe and Asia. The average annual growth rate of the industry is expected to be greater than 80 per cent until 2016 and approximately 15.6 per cent thereafter (BTM Consult, 2013).

Wave energy is harvested by harnessing the kinetic energy produced by floating structures as waves move them up and down on the ocean surface. The first wave energy demonstration projects were initiated in the 1970s, and while no large-scale wave energy installations are in operation, many states, including Australia, Chile and the UK have significant opportunities to develop wave energy projects (Bahaj, 2011). Relative to offshore wind energy production, wave energy has fewer environmental impacts (e.g. bird strikes), is less visible (as buoys have a low visibility profile) and waves can be predicted accurately up to a week in advance, which allows for less reliance on alternative power sources (Dunnett and Wallace, 2009).

Tidal energy production uses water movement generated by tidal streams/marine currents to turn impellers to produce electricity. While tidal energy production is in its infancy, the approach is appealing as, unlike wind and wave energy, tidal stream velocity is predictable to the hour, years in advance. Furthermore, generation sites can be located in constrained topography that increases the speed of water movement (Bahaj, 2011).

Management of marine energy and mineral resources in deep sea areas

Hydrothermal sulfide deposits

Hydrothermal sulfide deposits are found where cold ocean waters come into contact with heated areas of the earth's crust. This results in the precipitation and deposition of minerals. The most well-known sulphide deposits are the 'black smokers' (chimneys) surrounding hydrothermal vents. Sulphide depositions are found anywhere with geothermal activity, and include volcanic arcs, mid-oceanic ridges and hydrothermal vents. These areas commonly support concentrations of sulphur (S), copper, zinc, gold and iron (Fe) far above those found in terrestrial mines.

The first commercial mining of these deposits is expected to commence in Papua New Guinea's EEZs, which contain up to 40,000 hydrothermal vents. This form of mining is not without contention. There is considerable concern over the impacts of sulphide deposit mining, ranging from benthic disturbances, sediment plumes and toxic effects on the water column (Halfar and Fujita, 2007; Morgan, 2012).

Ferromanganese oxide deposits (polymetallic nodules)

Ferromanganese oxide deposits, also termed polymetallic nodules, first received attention in the 1960s as a potential source of manganese, nickel, copper, cobalt and iron. Deposits are normally found between 4000 and 6000m. They are formed as a result of the precipitation of metals from seawater into nodules in soft sediments and crusts on hard sediments. The nodules are readily accessible as they rest on the seabed, thus making them commercially important.

More recently, these deposits have been found to contain concentrations of the 17 rare earth elements at much higher levels than terrestrial rare earth mines (Zhang et al., 2012). The estimated global reserve of these nodules is more than 3×10^{12} metric tons. Recent estimates in the tropical north-east Pacific suggest that they contain significant deposits of manganese (2000×10^8 metric tons), nickel (90×10^8 metric tons), copper (50×10^8 metric tons) and cobalt (30×10^8 metric tons) (Morgan, 2012; Zhang et al., 2012).

Ferromanganese oxide deposits (polymetallic crusts)

Polymetallic (ferromanganese) crusts are similar in formation and composition to polymetallic nodules but formed when cold ocean bottom waters precipitate minerals onto hard-bottom substrates such as rocky outcrops or seamounts. Crusts may be up to 25 centimetres thick, are formed between depths of 400 and 4000 meters, and are most frequently found in the volcanic island arcs of the Pacific Ocean and Indian Ocean seamounts. Less than 2 per cent of the seabed is believed to support ferromanganese crusts.

Ferromanganese crusts concentrate elements above normal concentrations in the earth's crust (e.g. up to 50,000 times greater for tellurium (Te)), thus making these areas potentially commercially important. However, these crusts are not as important as polymetallic nodules given the difficulties in mining polymetallic crusts. Commercially important minerals found in crusts include manganese, iron, cobalt, nickel and copper. In particular, crusts are potential sources for a number of commercially important rare earth elements include cerium (Ce), europium (Eu), lanthanum (La), tellurium and yttrium (Y) (Hein et al., 2009).

Phosphate deposits

Phosphorous (P) is a necessary element to support life and is a limiting nutrient in most biological systems (including marine environments). It is also a key ingredient in modern agriculture and, as such, is a commercially valuable resource that cannot be substituted by any other element or compound. Of mined phosphate, 80 per cent – predominantly from China, Morocco and the US – is used in the production of fertilizer and the remainder in other uses such as detergents.

Phosphorus accumulates in the marine environment from a number of sources, but primarily from terrestrial erosion and the deposition of material by fluvial processes. Phosphorous transported into the ocean is consumed by biota and later deposited and buried in ocean sediments. Bacteria and enzymes then dissolve the organic matter, which releases the element into ocean sediments (Delaney, 1998). The process is anything but rapid: phosphate deposits can take 10–15 million years to accumulate in concentrations sufficient for commercial exploitation. It can therefore be considered a non-renewable resource. As global phosphate deposits are depleted, new sources will need to be found to maintain current agricultural production rates. It is estimated that, at current phosphorous usage rates, the world will run out of over the next several hundred years (Cordell et al., 2009).

Given the looming phosphate scarcity, the search for phosphorous in marine environments has become paramount. Phosphates are normally found in waters less than 1000m deep and are currently commercially mined 40–60km off the coast of Namibia in waters 180–300m deep. A new phosphate mining operation is expected to commence operations in 2013 at Chatham Rise, a seamount rising to 400m below the ocean surface east of New Zealand's South Island.

Chapter summary

- One third of the world's petroleum is produced from deposits on continental shelves. It is estimated that the deep sea areas (outside of continental shelves) contain two thirds of the world's mineral wealth.
- Under the LOSC, continental shelves and their corresponding mineral and energy resources can be appropriated (owned) by states and therefore managed under domestic legal and regulatory regimes.
- The Area is defined by the LOSC as those non-biological resources on or under the seabed outside of the continental shelf areas. The Area is open to all for the benefit of mankind and to be used exclusively for peaceful purposes.
- All states, regardless of geographic location (e.g. land locked) and state of economic development, have equal rights to participation in the Area. Economic benefits derived from the Area will be shared among all states.

References

Antrim, C. (2005) 'What was old is new again: Economic potential of deep ocean minerals the second time around', *Oceans*, 2, 1311–1318.

Bahaj, A. S. (2011) 'Generating electricity from the oceans', *Renewable and Sustainable Energy Reviews*, 15 (7), 3399–3416.

BTM Consult (2013) 'Offshore Report 2013', btm.dk/news/offshore+report+2013/?s=9&p=1&n=50&p_id=2&year=2012, accessed 23 June 2013.

Cordell, D., Drangert, J.-O. and White, S. (2009) 'The story of phosphorus: Global food security and food for thought', *Global Environmental Change*, 19, 292–305.

Delaney, M. L. (1998) 'Phosphorus accumulation in marine sediments and oceanic phosphorus cycle', *Biogeochemical Cycles*, 12 (4), 563–572.

Dunnett, D. and Wallace, J. S. (2009) 'Electricity generation from wave power in Canada', *Renewable Energy*, 31 (1), 179–195.

Garnett, R. H. T. (2002) 'Recent developments in marine diamond mining', *Marine Georesources and Geotechnology*, 20, 137–159.

Halfar, J. and Fujita, R. M. (2007) 'Danger of deep-sea mining', *Science*, 316 (5827), 987–997.

Hein, J. R., Conrad, T. A. and Dunham, R. E. (2009) 'Seamount characteristics and mine-site model applied to exploration- and mining-lease-block selection for cobalt-rich ferromanganese crusts', *Marine Georesources & Geotechnology*, 27 (2), 160–176.

ITLOS (International Tribunal for the Law of the Sea) (2011) *Responsibilities and Obligations of States Sponsoring Persons and Entities with Respect to Activities in the Area*, Advisory Opinion of the Seabed Disputes Chamber, www.itlos.org/fileadmin/itlos/documents/cases/case_no_17/adv_op_010211.pdf, accessed 9 January 2013.

Jones, A. and Morgan, C. L. (2003) 'Code of practice for ocean mining: An international effort to develop a code for environmental management of marine mining', *Marine Georesources & Geotechnology*, 21 (2), 105–114.

Lozano-Minguez, E. (2011) 'Multi-criteria assessment of offshore wind turbine support structures', *Renewable Energy*, 36 (11), 2831–2837.

Morgan, C. L. (2012) 'Deep-seabed mining of manganese nodules comes around again', *Sea Technology Magazine*, April, www.sea-technology.com/news/archives/2012/soapbox/soapbox0412.php, accessed 23 June 2013.

Plakokefalos, I. (2012) 'Seabed Disputes Chamber of the International Tribunal for the Law of the Sea: Responsibilities and obligations of states sponsoring persons and entities with respect to activities in the Area. Advisory opinion', *Journal of Environmental Law*, 24 (1), 133–143.

Radzevičius, R., Velegrakis, A.F., Bonne, W., Kortekaas, S., Garel, E., Blažauskas N. and Asariotis, R. (2010) 'Marine aggregate extraction: Regulation and management in EU Member States', *Journal of Coastal Research*, 51, 151–164.

Sandrea, I. and Sandrea, R. (2007) 'Global offshore oil: Geological setting of producing provinces, E&P trends, URR, and medium term supply outlook', *Oil and Gas Journal*, 105 (10), 34–40.

Scott, S. D. (2007) 'The dawning of deep sea mining of metallic sulfides: The geologic perspective', *Proceedings of the Seventh (2007) ISOPE Ocean Mining Symposium*, International Society of Offshore and Polar Engineers, Danvers, MA, 6–11.

Scott, S. D. (2011) 'Marine minerals: Their occurrences, exploration and exploitation', *Oceans 2011*, IEEE Publications, New York, 1–8.

UN (United Nations) (1968) *United Nations General Assembly Resolution 2467 (XXIII)*, www.un.org/ga/search/view_doc.asp?symbol=A/RES/2467(XXIII)&Lang=E&Area=RESOLUTION, accessed 26 June 2013.

UN (1994) *Agreement relating to the implementation of Part XI of the United Nations Convention on the Law of the Sea of 10 December 1982*, www.un.org/depts/los/convention_agreements/convention_overview_part_xi.htm, accessed 26 June 2013.

Vidas, D. (2000) *Implementing the Environmental Protection Regime for the Antarctic*, Kluwer Academic Publishers, London.

Zhang, Z., Du, Y., Gai, L., Zhang, Y., Shi, G., Liu, C., Zhang, P. and Duan, X. (2012) 'Enrichment of REEs in polymetallic nodules and crusts and its potential for exploitation', *Journal of Rare Earths*, 30 (6), 621–627.

Chapter 10

Integrated approaches to ocean management

Putting policy into practice and managing across sectors

Introduction

The application of policy has been an iterative process, predominantly evolving through trial and error and periodic refinements over time. Improvements to marine management are evolutionary rather than revolutionary, borrow extensively from experiences in terrestrial environments, and are inconsistently applied. If proven to be successful, they are often enshrined in domestic policy or international agreements. Prior to the 1970s, marine issues (e.g. oil spills, overfishing) were addressed on a sector-by-sector basis as problems arose. There was little incentive or political appetite to address anticipated future problems or find solutions that solved more than one issue. Furthermore, the LOSC and other marine conventions that committed states to develop integrated approaches to address socioeconomic and environmental marine issues were not completed until the final two decades of the twentieth century.

Integrated approaches to ocean management initially borrowed from the terrestrial experience. Early coastal management programmes in the 1970s were modelled on land use planning (also termed regional planning) that incorporated stakeholder engagement, scenario modelling and the identification of socioeconomic and environment objectives to develop zoning plans to regulate future growth. These early plans attempted to link disparate sectors (e.g. housing, transport, employment, environmental quality) that were formerly managed in isolation. They were applied to large landscapes (e.g. maintaining ecological function in watersheds), political areas (e.g. growth plans for local and regional governments), environmental assessments (e.g. proposed mine developments) and protecting property and public safety (e.g. floodplain planning). Notably, they were geospatial (mapped), involved multiple levels of government participation, attempted to maximize multiple objectives, engaged a range of user groups and the public and identified trade-offs (impacts to objectives). Additionally, these early terrestrial planning exercises were frequently designed to be updated (refreshed) as needed.

Integrated policy approaches versus integrated ocean management

It is important to distinguish integrated policy approaches from integrated approaches to ocean management. Integrated policy approaches are attempts to realize efficiencies from aligning current sector-based policies (e.g. pollution prevention, fishing) towards more overarching societal goals (see Box 10.1). In contrast, integrated ocean management uses specific approaches and tools to assist with marine decision-making. These approaches may also be used to analyse existing and formulate new policies. They are characteristically

---

Box 10.1 An example of an integrated policy approach: the European Commission
Blue Growth Strategy

In 2007, the European Commission established the Integrated Maritime Policy to
promote sustainable development and to protect European marine environments. In
2012, the Commission endorsed the development of a Blue Growth Strategy to
increase jobs and growth in the EU's marine sectors. A particular focus is to ensure
sustainable growth and to address climate change. The Union defines its 'blue economy' as those activities that relate to shipping, fishing, aquaculture, energy
development, coastal tourism and research and innovation. The blue economy represents 5.4 million full-time jobs with a gross added value of EU€500 million annually
(European Commission, 2012). The Blue Economy Strategy includes the following key
areas:

- Blue energy: Focus on offshore wind development, tidal energy and thermal energy
  conversion.
- Aquaculture: Focus on culturing new species, moving operations offshore and
  continuing to open up new markets for EU aquaculture products.
- Maritime, coastal and cruise tourism: Ensuring environmental quality, infrastructure improvements, improved training and promoting off-season coastal tourism
  opportunities.
- Marine mineral resources: Promoting EU participation in seabed mineral exploration and development and extracting dissolved minerals from seawater.
- Blue biotechnology: Promoting industry development, commercialization of
  research and reducing barriers to attract investors.

---

multi-sector, forward-looking, inclusive, transparent and attempt to inform durable
decisions.

The integrated approaches discussed below are ordered chronologically to illustrate the
linkages between approaches and how they have evolved. The following integrated
approaches are therefore not mutually exclusive. Rather, they are complementary
approaches that have been refined over time and adapted to different environments and
spatial and temporal scales (see also Craig, 2012).

The application of integrated ocean management approaches should contextualize the
problems under consideration. Technological and methodological improvements to data
collection, analysis and modelling combined with continued improvements in the amount
and quality of marine information will continue to improve marine decision-making and
further expand capabilities of the approaches discussed below.

Coastal zone management

The coastal zone is the part of the ocean with which we are most familiar. It is the most
biologically and physiologically complex of all marine environments (see Chapter 1) and
supports the greatest human use. Worldwide, coastal environments are characterized by
high population growth rates, degraded fisheries, conflicts between users both within and

between sectors, multiple levels of government with overlapping jurisdiction and people that take coastal environments for granted. As such, coastal environments are a particular focus of marine policy. While this book is primarily concerned with the international aspects of ocean governance, coastal policy has informed the development of ocean governance beyond the territorial seas and EEZs. For this reason it warrants treatment here.

A variety of terms and definitions have been used to describe the areas of the coastal realm or coastal zone. Some of these terms are illustrated in Figure 10.1. While considerable disagreement exists on where the coastal zone begins and ends, a useful description is provided by the *Encyclopedia of the Earth* (www.eoearth.org/):

> The coastal ocean is the portion of the global ocean where physical, biological and biogeochemical processes are directly affected by land. It is either defined as the part of the global ocean covering the continental shelf or the continental margin. The coastal zone usually includes the coastal ocean as well as the portion of the land adjacent to the coast that influences coastal waters. It can readily be appreciated that none of these concepts has a clear operational definition.

Regardless of how the coastal zone is defined, approximately 40 per cent of the world's population and 60 per cent of the world's economic production are concentrated in a 100km swath along the world's coasts (MA, 2005; Martínez *et al.*, 2007). Of the world's 33 megacities, 21 are coastal, and it is estimated that by 2020 up to 75 per cent of the world's 7.5 billion people may be living within 60km of the coastal zone. The confluence of such a

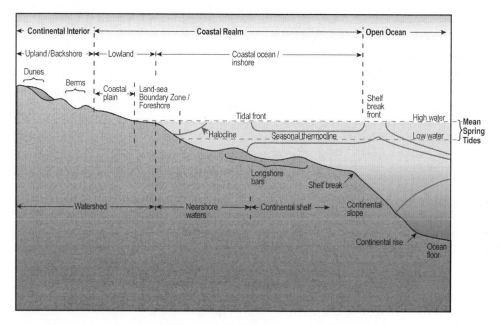

Figure 10.1 Various definitions and extents of the coastal realm and coastal zone

Source: Adapted from Ray and McCormick-Ray (2004) and University of Liverpool *Encyclopedia of Earth* www.eoearth.org/

significant proportion of the world's population on the narrow fringe separating the marine and terrestrial realms has significant ecological and socioeconomic implications. Ultimately, this concentration of people inevitably culminates in a reduction of the ecological goods and services provided by coastal environments. Corresponding socioeconomic ramifications (e.g. reductions in fish catches) and political conflicts between resource users are inescapable.

Coastal zone management (CZM) emerged in the 1960s in response to the failure of sector-based management (e.g. fisheries management, pollution prevention) to address the environmental and socioeconomic effects of multiple coastal and ocean uses. For example, the effects of overfishing on the health of fisheries were compounded by the degradation or loss of nearshore spawning or rearing habitat. Further, competition from introduced species and pollution from land- and ocean-based sources only exasperated the situation. While the notion of managing for cumulative effects (see below) was not yet fully understood as a management principle, early CZM efforts fundamentally attempted to manage the cumulative impact of human activities on coastal and marine cultural, economic and environmental values. Marine resource users, states and international organizations realized that an integrated approach was required to regulate human activities in order to maintain the vitality of ecological processes and socioeconomic systems. Coastal management also became increasingly necessary as states began to exert their jurisdiction over their coastal resources through expansions of territorial waters and declarations of EEZs and extended continental shelves. This expansion also resulted in the proliferation of laws at the local, regional and national levels to manage activities in these areas. These laws often conflict with one another as well as with the laws of adjacent states.

The need for an integrated approach to the management of coastal areas became apparent in debates surrounding the early Law of the Sea Conventions. However, coastal management as a discipline was formally launched by the introduction of the US Coastal Zone Management Act in 1972 (NOAA, 1972). The Act defined the objectives, principles, concepts and guidelines for CZM in the US. Ten years later, the LOSC created a formal legal structure for regulating the use of the sea and mandated that CZM be employed to manage coastal areas. Under the LOSC, the basic obligation of every nation is to 'protect and preserve the marine environment'. The agreement follows the basic principles of international environmental law and dictates that activities within a state's jurisdiction should be conducted in a manner that does not cause damage to other states. The LOSC also extends these principles to ABNJ (i.e. the high seas and deep seabed).

The LOSC (see Chapter 2) clarified several important jurisdictional issues that resulted in additional impetus for states to implement CZM. Most important, the LOSC extended the sovereignty of the territorial seas for coastal states to 12nmi and bestowed sovereignty and jurisdiction over state's EEZs. In return, the state must perform a number of obligations, including the conservation and management of living resources in the zone. This includes determining the allowable catch of the living resources in its EEZ and ensuring that the maintenance of the living resource within the EEZ is not endangered by overexploitation (Chapters 2 and 6). The LOSC therefore supersedes the former doctrine of freedom of fishing, which now applies only to the high seas. Lastly, the LOSC granted coastal states sovereignty over the natural resources of the continental shelf and gave coastal states exclusive rights to the sedentary species of the seabed.

With their newly affirmed sovereignty under the LOSC, many nations developed CZM programmes in the 1980s. Also contributing to the rise in CZM programmes was the 1987 publication of *Our Common Future* by the WCED. It outlined the following 'sustainable development principles':

- Integration of conservation and development.
- Maintenance of ecological integrity.
- Economic efficiency.
- Satisfaction of basic human needs.
- Opportunities to fill other non-material human needs.
- Progress towards equity (present/future, cultural/economic) and social justice.
- Respect and support for cultural diversity.
- Social self-determination.

The final international agreement that set the stage for CZM was UNCED (see Chapter 5). UNCED contained Agenda 21, which was a global commitment by coastal nations to apply integrated management and sustainable development principles to coastal and marine environments under their jurisdiction. Agenda 21 declared that all coastal nations would have an integrated coastal management programme underway by the year 2000. Since UNCED, the Global Conference on Sustainable Development of Small Island Developing States (1994) and the International Coral Reef Initiative (1994) have also reinforced the need for coastal management (Cicin-Sain *et al.*, 1995).

Broadly stated, CZM is an attempt to integrate a number of often disparate ecological, political and socioeconomic methods to move away from a sectoral management model (e.g. managing fisheries and foreshore development separately). CZM reflects the complex and integrated ecological nature of coastal environments, while also reducing overlap in traditional management methods (Box 10.2). It creates a forum where coastal communities, and those that rely on the coastal environment, are provided with the best available information in order to make informed decisions on coastal use. Since the 1980s, CZM has alternatively been referred to integrated coastal zone management, coastal area management, integrated management, coastal resource management and other terms based on geographic location and the state(s) sponsoring the programmes. Fundamentally, however, all CZM-related programmes have similar intents and purposes.

CZM is more akin to a philosophy on how coastal environments can be managed than a specific set of directives that are applied in all instances. CZM has been termed 'governance beyond government', meaning that governments are just one of the four types of participant in CZM. The other participants are civil society, the private sector and the scientific community (Cicin-Sain and Knecht, 1998; Bremer and Glavovic, 2013). The application of CZM depends on the type of problems to be addressed (e.g. pollution, overharvesting), the characteristics of the environment under consideration (e.g. tropical, temperate), the social and political will and resources available to address these problems. There are, however, some common goals of all CZM initiatives, which are as follows:

- Achieve sustainable development of coastal and marine areas.
- Reduce vulnerability of coastal areas and their inhabitants to natural hazards.
- Maintain essential ecological processes, life support systems and biological diversity in coastal and marine areas.

Regardless of the definition of CZM, there are a number of widely accepted principles of

---

Box 10.2 Popular definitions of coastal zone management

...a dynamic process for the sustainable management and use of coastal zones, taking into account at the same time the fragility of coastal ecosystems and landscapes, the diversity of activities and uses, their interactions, the maritime orientation of certain activities and uses and their impact on both the marine and land parts.

...provide[s] for the best long-term and sustainable use of coastal natural resources and for perpetual maintenance of the most natural environment.

...to improve the quality of life of human communities who depend on coastal resources while maintaining the biological diversity and productivity of coastal ecosystems.

...an adaptive process of resource management for environmentally sustainable development in coastal areas. It is not a substitute for sectoral planning, but focuses on the linkages between sectoral activities to achieve more comprehensive goals.

Source: Clark (1992); FAO (1996); UNEP (1995, 2008)

CZM that must be adhered to in order for it to be effective. According to Cicin-Sain and Knecht, (1998) and Kay and Alder (2005) CZM must:

- be holistic, integrated and multi-sectoral in approach;
- be consistent with, and integrated into, development plans;
- be consistent with national environmental and fisheries policies;
- build on, and integrate into, existing institutionalized programmes;
- be participatory;
- be adaptive and rely on the 'learning by doing' model;
- build on local/community capacity for sustained implementation;
- build self-reliant financing mechanisms for sustained implementation;
- address quality of life issues of local communities as well as conservation issues;
- take a long-term perspective and consider the needs of present and future generations.

Success of CZM

Since the passage of the US Coastal Zone Management Act in 1972, more than 95 states have participated in over 150 CZM initiatives worldwide. In recent years, states have been cooperating with each other on CZM programmes. In addition, the UNEP Regional Seas Programme, with over 140 participating states, is an evolving venue for CZM discussions. Furthermore, the Mediterranean ICZM Protocol (adopted January 2008, entered into force March 2011) became the first international legal instrument mandating CZM. The Protocol is unique in that it creates binding legal obligations over disciplines normally under national jurisdiction including administrative law, economic policies and urban planning (Rochette and Billé, 2012).

Overall progress, however, on implementing CZM has been measured, if not sluggish. Only moderate headway has been achieved towards the UNCED goal of all coastal states having integrated coastal management programmes underway by the year 2000. A review of barriers to implementing CZM by Olsen (2003) concluded that the capacity of institutions (governance capacity) to implement CZM programmes is a larger barrier than limitations on funding or scientific knowledge.

A 2006 study commissioned by the EU determined that European CZM programmes, while not implemented by all member states, raised awareness and preparedness of issues facing coastal areas. Programmes could also improve the livelihood and employment in coastal areas, provide an opportunity to rethink approaches that attempt to reconcile socioeconomic and environmental interests, and provide a linkage between terrestrial and marine laws at the scale of 'regional seas'. Regional cooperation, stakeholder participation, science and information and monitoring CZM implementation were also singled out as areas that could be significantly bettered (Rupprecht Consult, 2006).

A 2007 study of the success of the US 1972 Coastal Zone Management Act concluded that the Act enabled voluntary partnerships between the federal and state government. Perhaps this is why 34 of 35 eligible states have participated in the programme. Under the programme, the federal government provides funding, oversight and a national vision and yet leaves states free to design tailored programmes that suit their needs (NOAA, 1972). The programme has generally been viewed as successful, with certain programmes (e.g. water quality) demonstrating improvements in coastal ecosystem health (NOAA, 2007).

Ecosystem approaches to management

The notion that the disparate parts of an ocean area can be managed concurrently is over a century old (Baird, 1873; Link, 2005). However, it was not until the concept of the 'ecosystem' was championed by Tansley (1935) that scientists and decision-makers began to debate the precise definition of an ecosystem, whether and how ecosystems should be 'managed', and what aspects of traditional management paradigms need to be tweaked to make them 'ecosystem-based' (Leopold, 1949; Larkin, 1996).

The central premise of what are generally termed ecosystem approaches to management is that a global and holistic approach to marine management is required that simultaneously considers multiple ecological and socioeconomic objectives in the management of either a geographic area (e.g. protected area) or ecosystem (however defined). These approaches strive to maintain the benefits humans derive from natural systems (ecological goods and services) while minimizing externalities associated with their use. There are over a dozen terms in use that fundamentally denote an approach to natural resource management that focus on sustaining ecosystems to meet both ecological and human needs for the future. Here, we will use the term 'ecosystem approaches to management' (EAM), since it is routinely applied in terrestrial and marine environments and is broader in scope than other terms in use (e.g. ecosystem-based management). See also Kidd et al. (2011) for further discussion of this issue.

The term EAM, however, lacks consensus on its definition. As such, confusion arises over what exactly EAM is, and what it entails for the marine environment. A myriad of definitions of the concept exist, including: ecosystem-based management; integrated management; integrated oceans management; coastal zone management; integrated coastal zone management; and sustainable development. Broadly, EAM incorporates socioeconomic, cultural and ecological inputs into management, conservation and decision-making processes. In this regard, COMPASS (2005), an American environmental non-government organization offers a pragmatic definition of EAM: 'An integrated approach to management that considers the entire ecosystem, including humans with the goal to maintain an ecosystem in a healthy, productive, and resilient condition so that it can provide the services we want and need'.

Many EAM principles have existed in international soft law for decades; however, it was not until the 1995 FAO Code of Conduct for Responsible Fisheries and the UN Fish Stocks Agreement that soft law or voluntary agreements began to outline the principles and procedures for what would become marine EAM. Specifically, the UN Fish Stocks Agreement states that there is a 'conscious need to avoid adverse impacts on the marine environment, preserve biodiversity, and maintain the integrity of marine ecosystems' (FAO, 2005).

Marine EAM concepts have a growing presence in the scientific, peer-reviewed literature and a significant presence in the 'grey' literature of the applied conservation and management disciplines. Marine EAM academic and application-related papers and reports are found in disciplines as broad as biology, economics, social science and indigenous study. Some papers are descriptive, some are model-based, but few follow the traditional empirical development and testing of hypotheses. Ecosystem approaches are also a global phenomenon, with over 100 states either participating in EAM efforts or hosting EAM programmes within their borders. Certain jurisdictions – most notably Australia, Canada and the US – have EAM principles enshrined in certain marine statutes. Nonetheless, the absence of a single authoritative guide to marine EAM continues to limit its application and use.

A more recent endeavour is to apply EAM as an alternative to 'hard' engineering solutions (e.g. shoreline armouring) to adapt to climate change. Termed 'ecosystem-based adaptation' (EbA), these approaches attempt to use natural coastal barriers, such as wetlands, mangroves and barrier beaches, to protect against aspects of climate change (e.g. sea level rise, increased storm frequency).

Marine protected areas

Humans have long desired to protect natural areas from their own activities. Indeed, evidence suggests that the first protected areas were established in India 2000 years ago in order to preserve natural resources (Holdgate, 1999). In more recent times, European nobility excluded their 'inferiors' from important hunting grounds. Based predominantly on resource value or ecological import, cultures worldwide have conferred sacred status upon certain terrestrial areas. In the marine environment, pre-Western, non-market economies established fishing closures in nearshore waters that displayed signs of overharvesting.

While a number of states introduced protected areas in the early twentieth century, the modern application of protected areas as a biodiversity management tool began in earnest with the inaugural World Conference on National Parks in 1962. This conference was probably the first time the need for protection of coastal and marine areas was internationally recognized. Prior to this, Beverton and Holt (1957) advocated for the use of closed areas in fisheries management and a number of states had established smaller protected areas for protection of sites of cultural and historic significance. However, the need for a systematic and representative approach to establishing protected areas in marine environments was first clearly articulated at the International Conference on Marine Parks and Protected Areas, convened by the IUCN in Tokyo in 1975 (Kenchington, 1996).

The concept of MPAs gained traction in the 1980s with the publication of the IUCN/WWF/UNEP *World Conservation Strategy* (1980). Following a number of workshops on the establishment of MPAs a 1984 IUCN report titled *Marine and Coastal Protected Areas: A Guide for Planners and Managers* was published (Salm and Clark, 1984). MPAs are now internationally recognized as an essential and fundamental component of marine conservation. They are referred to within many international agreements, including the Convention on Biological Diversity (Article 8) and the accompanying Jakarta Mandate, the Global Program of Action, MARPOL 73/78 and the more recent IMO Guidelines, and the World Heritage Convention. While the LOSC is silent on what conservation mechanisms (e.g. ecosystem approaches, MPAs) should be applied to meet the Convention's objectives to protect the marine environment, it is clear that states have specific duties. They are obliged to protect rare or fragile ecosystems and the habitats of depleted, threatened or endangered species and other forms of marine life (Articles 192 and 194).

Today, there are over 5880 MPAs worldwide in over 80 countries. Averaging a rate of expansion of 4.6 per cent per year (from 1984 to 2006) (see Figure 10.2), the current portfolio of MPAs covers 1.17 per cent (4.21 million square km) of the oceans's surface. Nevertheless, there continues to be some debate surrounding the definition of an MPA and, ergo, the purpose(s) of their existence (Toropova *et al.*, 2010; Jones, forthcoming). Two widely quoted definitions of an MPA are as follows: 'Any area of intertidal or subtidal terrain, together with its overlying water and associated flora, fauna, historical and cultural features which has been reserved by law or other effective means to protect part or all of the enclosed environment' (Kelleher and Kenchington, 1992); and:

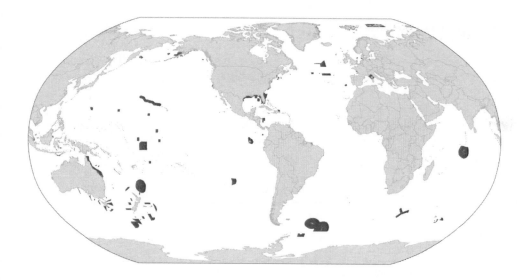

Figure 10.2 Global map of current marine protected areas
Source: protectedplanet.net

Any area of sea or ocean – where appropriate in combination with contiguous intertidal areas – together with associated natural and cultural features in the water column within, or on top of the seabed, for which measures have been taken for the purpose of protecting part or all of the enclosed environment.

(Nijkamp and Peet, 1994)

While minor disagreement continues to exist on the exact definition of an MPA, for the purposes of this book MPAs are defined as areas set aside under legislation or international agreement to protect marine values.

Whichever way they are defined, a plethora of regions throughout the world could be deemed MPAs. Ballantine (1991) cites some 40 names that are in common use around the globe for areas that are currently set aside for the protection of parts of the sea (see Table 10.1). Needless to say, this multitude of 'labels', definitions and terminologies has the potential to create misunderstanding and uncertainty. MPAs also range from small, highly protected 'no-take' reserves that sustain species and maintain natural resources, to very large multiple-use areas in which the use and removal of resources is permitted, yet controlled in order to ensure that conservation goals are achieved. There is a need, therefore, to ensure that the intention underlying the use of specific terminology is clarified.

There exists a wide variety of types of MPAs designed to accomplish various purposes (Box 10.3). MPAs may be created to satisfy a range of values and the goal or purpose of an MPA influences the 'type' of MPA. Consequently this affects its design and selection and, ultimately, its name. These different 'types' of MPAs are not necessarily autonomous: for example, within Australia's Great Barrier Reef Marine Park – which itself is regarded as a huge 'multiple-use' MPA – various other zones are defined to promote or restrict different human activities.

Another way to categorize MPAs is to consider them in terms of representative and distinctive areas (Box 10.4).

Marine protected areas in ABNJ

As discussed in previous chapters, the high seas and deep seabed (ABNJ) were, until the twentieth century, spared the impact of human activities since they were inaccessible and

Table 10.1 Various names in common use around the globe for areas that are currently set aside for the protection of parts of the sea and the 'values' they may protect

NAMES

Marine park	Marine protected area	Maritime park
Marine reserve	National seashore	Marine sanctuary
Marine nature reserve	Marine wildlife reserve	Marine life refuge
Marine habitat reserve	Marine wilderness area	Marine conservation area

VALUES

Conservation value	Recreational values
Commercial values	Scenery/aesthetics
Enhancement of special species	Unique features
Scientific importance	Cultural values
Historic features	Traditional uses

Source: 'Names' after Ballantine (1991)

Box 10.3 Some examples of types and purposes of MPAs

In general, an **MPA** is any marine area set aside under legislation to protect marine values.

A **representative MPA** is a special kind of MPA that has been designed to maximize representation based on an ecological or bio-geographical framework.

A **no-take MPA (or harvest refugia)** is a special type of MPA (or a zone within a multiple use MPA) where: any removal of marine species and modification or extraction of marine resources is prohibited (by such means as fishing, trawling, dredging, mining, drilling, etc.); and other human disturbance is restricted.

A **multiple-use MPA** is a particular type of MPA in which the use and removal of resources may be permitted, but where such use is controlled to ensure that long-term conservation goals are not compromised. Multiple-use MPAs generally have a spectrum of 'zones' within them, with some zones allowing greater use and removal of resources than other zones (e.g. no-take zones are commonly designated as one of the zones).

A **biosphere reserve** is a particular type of protected area, generally large, that has totally protected core area(s), surrounded by partially protected buffer areas, and an outer zone/transition area. Both the buffer and the outer zone may allow some ecologically sustainable resource use (e.g. low-key and traditional uses) and should also have facilities for research and monitoring; yet activities within these areas should not compromise the integrity of the core area. This concept was developed for terrestrial reserves but is considered by some to also have application for marine environments such as MPAs (Price and Humphrey, 1993).

Note: Various authors (Kenchington and Agardy, 1990; Brunckhorst *et al.*, 1997) suggest that, if the UNESCO Biosphere Reserve Programme were to be redesigned and properly planned and implemented, it would provide an efficacious tool for the wider aspects of marine conservation. See also Agardy (2010) on ocean zoning.

---

the continental shelves met human demands. However, with the mechanization of fisheries, interest in deep sea mining, ocean acidification and ocean warming have placed several high seas fish stocks (e.g. tunas, whales) and habitats (e.g. deep sea vents, feeding grounds) under threat (Drankier, 2012; Gjerde and Rulska-Domino, 2012).

Several international agreements provide a mandate for establishing MPAs in ABNJ. The LOSC envisions that a combination of regional seas organizations and sectoral international organizations should be responsible for the protection of high seas marine environments. Under this model, RFMOs (also enabled by the UN Fish Stocks Agreement – see Chapter 6) may enact fisheries closures (spatial and temporal) and the ISA has identified 'areas of environmental interest' for the Clarion-Clipperton Zone in the equatorial North Pacific Ocean.

Article 8 of the CBD (see Chapter 5) presented contracting parties with several commitments. They are required to: establish a system of protected areas to conserve biodiversity, develop guidelines for the selection, establishment and management of protected areas, and manage biological resources inside and outside of protected areas. The 1995 Jakarta Mandate on Marine and Coastal Biodiversity committed parties to institute a global network of MPAs by 2012. The 2008 CBD COP 9 (Decision IX/20) adopted seven

Box 10.4 Considering MPAs in terms of representative and distinctive habitats

Marine environments can be conceived as fundamentally comprising a set of habitat types each containing a particular type of biological community. Habitats and their communities can be viewed as either representative or distinctive where the term 'representative' means recurring or characteristic of its surroundings at a given scale and 'distinctive' means atypical of its surroundings at a given scale.

Management efforts have traditionally been directed to the rare, endangered, large animals and their associated habitats. This is especially true in the marine environment where 'focal species' (any species on which our attention is focused for some reason) such as 'charismatic megafauna' (whales, seals, migratory birds etc.) capture public attention and may be used as 'flagships' to enlist public support (Zacharias and Roff, 2001). Often these flagship species seasonally occupy distinctive habitats, whose communities are substantially different from those in surrounding representative habitats. However, a management strategy that addresses only individual species or only distinctive habitats leaves the majority of species and their habitats at risk. The representative or 'ordinary' parts of the biosphere remain unprotected.

Accordingly, emphasis in conservation is now directed both to 'species' and distinctive habitats, and to 'spaces' and representative habitats. In terrestrial environments this change in emphasis has emerged as the 'enduring features' approach, and in the marine environment as the 'geophysical' approach based on enduring features and recurrent processes (see Roff and Taylor, 2000, for review). This change in strategy has likely occurred for two main reasons: first, no species can survive without its habitat; and second, recognition that effort should be expended not only to protect the rare and endangered, but the ordinary, representative (and great majority of) species as well.

An analysis of representative marine habitats is vital for many reasons (see Roff *et al.*, 2003). If marine environments are to be systematically protected from the adverse effects of human activities, then identification of the types of marine habitats and the communities they contain, as well as delineation of their boundaries in a consistent classification, is required. Human impacts on defined communities can then be assessed, candidate MPAs can be designated and the 'health' of these communities can be monitored.

Within any region there will also be distinctive habitats containing special features that (by definition) are not revealed by any analysis of representative habitats. Distinctive habitats often appear to be distinguished because special oceanographic processes are occurring within them on a regional or local scale, whereas representative habitats are not notable in this way.

scientific criteria for the identification of EBSAs for open ocean and deep sea habitats. These include:

- Uniqueness of rarity.
- Special importance for life history of species.
- Importance for threatened, endangered or declining species and habitats.
- Vulnerability, fragility, sensitivity, slow recovery.

- Biological productivity.
- Biological diversity.
- Naturalness.

The IMO has a mandate to protect marine environments from shipping activities and may establish particularly sensitive sea areas (PSSAs) designated by the IMO Assembly. Special areas can also be introduced, which require an amendment to the MARPOL 73/87 Convention and respective Annex. While only two special areas and no PSSAs have been designated in ABNJ, the IMO is reviewing criteria for applying PSSAs and special areas in ABNJ. In matters related to fishing, the FAO is responsible for the FAO Code of Conduct for Responsible Fishing (see Chapter 6). This, in turn, led to the development of the FAO Technical Guidelines on MPAs and fishers (the MPA Guidelines) (FAO, 2006).

Article V of the ICRW (see Chapter 6) permits the designation of sanctuaries. In designating a sanctuary under the auspices of the Convention, the only regulatory measures that can be taken involve prohibiting the harvest of all whale species at any time from a specified geographic area, irrespective of their conservation status. A sanctuary in the South Pacific sector of the Southern Ocean was established in 1949 although six years later it was deregulated on the advice of the IWC Scientific Committee. Since then, two additional sanctuaries have been adopted: the Indian Ocean Sanctuary (IOS) in 1979 (renewed in 2002) and the Southern Ocean Sanctuary (SOS) in 1994. Additional sanctuary proposals in the South Atlantic and the South Pacific Oceans have been tabled at recent IWC annual meetings.

While a number of international agreements permit the establishment of spatially defined ocean areas for conservation purposes in specific sectors, there is a paucity of integrated MPAs that protect marine areas from multiple threats. To date, only the following four regional seas conventions (out of 13 or 18 depending on how 'regional seas' are defined) include provisions for the establishment of high seas MPAs:

- Convention for the protection of the Mediterranean Sea against Pollution (Barcelona Convention) and its Protocol on Biodiversity and Specially Protected Areas.
- Convention for the Protection of the Marine Environment of the North-East Atlantic of 1992 (OSPAR Convention) and its Annex V.
- Antarctic Treaty System Protocol on Environmental Protection.
- Convention for the Protection of the Marine Environment and Coastal Zones of the South-East Pacific (Lima Convention), Protocol for the Conservation and Management of Protected Marine and Coastal Areas of the South East Pacific.

The 2002 WSSD committed states to initiate MPAs in the high seas by 2012. This goal, unfortunately, has not been realized. Reasons for this failure are miscellaneous:

- The lack of a global process and institution for establishing MPAs.
- The current sector-by-sector approach of high seas management.
- The lack of environmental assessment procedures and understanding of cumulative effects in the high seas.
- The lack of regional governance bodies with a conservation mandate and the non-binding nature of international agreements and conventions.
- The lack of an international science and advisory body.

One success, however, has been in the OSPAR (North-East Atlantic) region where six MPAs were established in 2010, comprising 286,200 square km. The establishment was approved by 15 European states and all of the MPAs are located in ABNJ (Gjerde and Rulska-Domino, 2012).

In 2011, the UNGA agreed to develop a process to address the conservation and sustainable use of marine biodiversity in high seas areas.

Marine spatial planning

A recurrent theme throughout this book is the increasing conflict among the various communities and sectors (stakeholders) that depend on the oceans for their livelihood. As increased demands are placed on oceans to supply food, absorb wastes, provide energy and support transportation, the ongoing ability of certain ocean environments to provide ecosystem goods and services exceeds their capacity. Moreover, the concentration of multiple human activities, particularly in coastal environments, has the added risk of producing unexpected ecosystem responses (cumulative effects – see below).

The increasing multiple stresses on marine environments, combined with most of the world's oceans managed as a common property resource, is a twofold recipe for increased conflict between those that utilize the ocean and an overall decline in ocean health. While coastal management addresses balancing human activities with ocean health in nearshore areas, and while MPAs are a tool to protect certain geographic areas, marine management has, until recently, lacked a systematic approach to managing human activities to which everyone subscribes.

Marine spatial planning (MSP) has been adopted by a number of national and international organizations including the EU, UN and US. They envision optimizing the economic and social uses of oceans to enable them to continue to provide ecosystem goods and services (Young et al., 2007; Douvere, 2008). MSP builds on the principles of coastal management and MPA identification in order to establish a framework for informational gathering and decision-making that can be applied to all marine areas.

Inputs to MSP include the values and objectives of those that use a particular marine region (termed 'stakeholders') along with the cultural, political and socioeconomic context. Unique to MSP is that information inputs and decision outputs are managed spatially, meaning the MSP process occurs within a geographic (mapped/geospatial) framework.

MSP, based on the principles of ecosystem-based management (see above), is therefore founded on the premise that human activities and their associated stresses along with marine habitats can be spatially represented (mapped) in three dimensions. This permits the estimation of the location and severity of risks (both environmental and social). During the MSP process, different management alternatives (scenarios) are produced in order to assess their alignment with the overall vision for the marine region. MSP attempts to be inclusive of all marine stakeholders, including indigenous peoples, recreational users, environmental groups and traditional stakeholders – fishing and marine transportation interests being the foremost of these. Marine spatial plans are based on the best available science and information; they also exhort stakeholders to communicate and improve relationships with each other.

The tangible output from a MSP is an overall vision for a particular marine region, a set of objectives and related performance measures, and recommendations on how stakeholders must conduct their activities to achieve the overall vision. Less tangible benefits of MSP accompany this vision: they create forums for communication between stakeholder groups and often allow groups to understand each other's values and perspectives.

Key goals of MSP are as follows (adapted from the National Ocean Council, 2013):

- Support sustainable, safe, secure, efficient and productive uses of the ocean and ocean users.
- Protect, maintain and restore coastal and ocean resources and maintain resilient ecosystems and improve and sustain the provision of ecosystem services.
- Provide for and maintain public access to the ocean.
- Promote compatibility among users and reduce user conflicts and environmental impacts.
- Improve the consistency, transparency, efficiency and durability of decision-making and regulatory processes.
- Increase certainty and predictability in coastal and ocean planning.
- Enhance interagency, intergovernmental and international communication and collaboration.
- Decrease user conflict, improve planning and regulatory efficiencies, decrease associated costs and delays, engage affected stakeholders and preserve critical ecosystem functions.

MSP is not a replacement for traditional marine resource management, which includes sector plans for fishing, transportation, energy development, conservation and other uses. Marine spatial plans inform the development or amendment of sector plans and more broadly identify the opportunities for each sector within a marine region such that the overall health and productivity of the area can be maintained.

Despite its explicit advantages, a number of barriers to implementing MSP exist, including:

- Fear of changes to the status quo by user groups.
- A lack of objective, science-based evidence that MSP can successfully harness to reduce user conflicts and maintain improve ocean health.
- Quality and quantity of spatial data is often poor.
- Complex yet unknown interactions may occur even using the best available information.

Overall, however, MSP continues to be adopted by an increasing number of states. In 2008, the European Commission adopted the *Roadmap for Maritime Spatial Planning: Achieving Common Principles in the EU*, which proposed a set of key principles for MSP (European Commission, 2008). In July 2010 the White House announced the Ocean Stewardship Executive Order 13547 that commits the US to MSP (National Ocean Council, 2013). To date, MSP initiatives are operating in many nations and a selected list of MSP initiatives is provided in Table 10.2

Cumulative effects

The term 'cumulative effects' is widely used in natural resource management to connote situations where the seemingly rational regulation and management of individual sectors

Table 10.2 Selected list of MSP initiatives

Country	Agency	Project
Australia	Great Barrier Reef Marine Park Authority (GBRMPA)	GBRMPA zoning
	Department of the Environment, Water, Heritage and the Arts	South-east Regional Marine Plan
		National Marine Bioregionalisation of Australia
Canada	Fisheries and Oceans Canada	Eastern Scotian Shelf Integrated Management Plan
China	State Oceanic Administration	Territorial Sea zoning
Denmark, Germany and The Netherlands	Wadden Sea Secretariat	Trilateral Wadden Sea Cooperation Area
Germany	Federal Maritime and Hydrographic Agency	Spatial Plan for the North Sea Spatial Plan for the Baltic Sea
Norway	Ministry of the Environment	Integrated Management Plan of the Barents Sea
The Netherlands	Ministry of Transport, Public Works & Water Management – North Sea Directorate	Integrated Management Plan for the North Sea 2015/National Waterplan

Source: Modified from www.unesco-ioc-marinesp.be/msp_references?PHPSESSID=5jhoa8anh8b2o1jbocn2riaf14

(e.g. fishing, transportation) fails to consider the collective impacts of these activities on either the environment, public health or human use of the resource. This results in unintended consequences to ecological systems and human well-being. Cumulative effects then are changes to environmental, social or economic conditions (or values) caused by the combined effect of past, present and reasonably foreseeable actions or events. They may be localized (e.g. pollution discharge) or global (e.g. climate change) and may be acute (e.g. overharvesting a fishery) or enduring (e.g. disposed radionuclides).

The issue of cumulative effects in terrestrial natural resource management is well-known. It arises from the difficulty in demonstrating that small, seemingly benign incremental land-use changes may combine over time to produce unexpected deleterious results (Theobald *et al.*, 1997). Cumulative effects, furthermore, do not naturally emerge from sector- and project-based approaches to decision-making used in most jurisdictions. Thus, individual decisions, such as fishery allocations – despite being made with the best available information – may still have unintended consequences on other aspects of marine systems and vice versa.

Cumulative effects, in a mathematical sense, may be additive (CE = stress *a* + stress *b*) where two or more stresses (e.g. fishing, pollution) may affect a resource (e.g. the fishery) independent of one another. It can be antagonistic (cumulative effect < stress *a* + stress *b*) where two or more stresses interact to reduce the overall impacts on a particular value. Or it can be synergistic (cumulative effect > stress *a* + stress *b*) where two or more stresses interact to produce a negative response greater than the sum of their parts. A literature review by Crain et al. (2008) found that synergistic stresses were common in marine environments. Common examples of synergistic stresses included situations where the effects of toxins (pollution) on organisms were increased in the presence of changes in salinity, temperature or UV light exposure (as a result of climate change or ozone depletion).

Cumulative effects is a useful construct for considering how different activities interact to impact marine systems and the humans that depend on these systems, yet, in order for cumulative effects to inform management decisions, it must be nested within a broader consideration of scientific analysis, social choice and public policy development. Cumulative effects assessment or cumulative impacts analysis are approaches that seek to quantify the overall impacts to a particular value or set of values from different activities where the activities are typically managed independent of each other. Cumulative effects assessment therefore is part of the decision-making process and has been enshrined in domestic legislation in a number of jurisdictions (Canada, US), particularly as a component of environmental impact assessment of major projects (Parkins, 2011).

Depending on the jurisdiction and problem(s)/areas(s) under consideration, cumulative effects assessment uses different nomenclature and descriptions of approaches (Weber et al., 2012). Assessment programmes, however, have similar underlying methodologies that may include:

- *Identification of values*: Environmental and socioeconomic (including cultural) values are identified so as to focus the cumulative effects assessment on preventing changes to important aspects of ecological and human systems.
- *Identification of indicators*: Indicators measure the condition or 'health' of the values under consideration.
- *Monitoring of values*: Values are monitored spatially and temporally to support decision-making and to better understand ecological and human systems.
- *Identification of ecological and social thresholds*: Finding the 'tipping points' where small changes in resource use produce large non-linear responses to ecosystem condition or human well-being.
- *Understanding adaptive capacity and resilience*: Determining the capacity of a system to withstand and recover from disturbance and perturbations.
- *Understanding patterns of cause and effect*: Developing reliable generalizations about functional, measurable relationships between indicators and observed or forecast patterns of change.
- *Scenario analysis*: Used to evaluate options (often framed as trade-offs) and evaluate future outcomes.
- *Setting management direction*: Using the information gleaned in the above steps to inform management actions.

Historically, in marine environments the primary impacts of human activities included overexploitation of key species, pollution and nearshore habitat loss; however, invasive species and climate change are increasing becoming major concerns. At present, marine cumulative effects assessments are being piloted in areas such as the west coast of North America (Halpern et al., 2009; Ban et al., 2010).

Private ocean rights

A frequent theme throughout this text is the perils of open access (see Chapter 2) where individuals, corporations and states have few incentives to produce, regulate, manage or monitor the production of marine-derived goods and services efficiently in ocean areas that are shared with others. Open marine access leads to multifarious, undesirable effect: rapid depletion of resources due to 'gold rush' behaviour, low profits due to overcapitalization and excessive costs, dangerous working conditions due to low profit margins and competition, and poor product quality (Deacon, 2009). Most discussion related to the perils of open access centres on the rapid depletion of fisheries due to circumstances in which fishers have few economic incentives to consider the long-term health or productivity of fish stocks. Yet, other sectors – transportation, energy development, mining and others – often lack incentives to operate in a manner that minimizes environmental harm and maximizes long-term opportunities.

The proliferation of declarations of EEZs over the past several decades has resulted in 95 per cent of the world's fisheries and nearly all of the world's offshore hydrocarbon resources (found on continental shelves) vesting within state governments. The establishment of EEZs has been a significant step towards economic rationality and enclosing the commons. However, many of the most important global fisheries, including tuna, occur primarily outside of EEZs.

Introduction to marine property rights

Marine property is frequently conceived of as a tangible object (e.g. territory, fish, oil and gas reserve); but it can also be an intangible benefit or income stream (e.g. right to fish) recognized by some level of government (Edwards, 1994). Marine property rights span a continuum, ranging from weak usage rights – such as those granted to companies to explore for oil and gas deposits – to the full ownership of ocean resources found in some fisheries management systems (Table 10.3). Property rights have different meanings depending on whether the resource is found within a state's (or multiple state's) EEZs or ABNJ. With the exception of some privately held estuarine and intertidal areas in certain states, marine environments are held in trust by states on behalf of their citizens within territorial waters and EEZs. Outside of these areas, ABNJ are held in trust for the benefit of all states and managed cooperatively by the international organizations discussed in previous chapters.

Rights-based fishery management systems

As discussed in Chapters 5 and 6, throughout human history humans have managed to over-fish nearly any area or stock that was accessible with available technology. The fundamental reasons for the repeated and unchecked exploitation of fish stocks are open access (the rule of capture), weak governance (decision-making) and insufficient enforcement (Hannesson, 2011). These reasons are especially relevant to high seas fisheries where fish stocks are truly a global 'commons'. The problem is not limited to the high seas. This predicament also exists within many state's EEZs if the state cannot control or enforce access.

As early as 1954 it was understood that economic inefficiencies and overexploitation are inevitable consequences of fish as a 'commons' resource. Gordon (1954) defined the commons as '...free goods for the individual and scarce goods for society'. Under open

Table 10.3 Types of marine property rights

Types of rights	Characteristics	Examples
Proprietary (possessory)	Physical control of the property to use it (or not use it), to change its form and substance and to transfer all rights through the sale or some rights through the rental. Responsible for damages to other's interests and may claim damages if others harm or damage the property.	Collects rents, royalties from leases States have proprietary rights held in trust for their citizens
Use	Rights to use for specific purposes at specific times and places	Fishing rights, oil and gas exploration and development rights, offshore energy installations, aquaculture
Exclusionary	Right to exclude others from the property	Exclusive rights to explore and exploit (e.g. energy and minerals). May exclude others for safety concerns
Disposition	Right to transfer, sell, exchange	Individual transferrable fishing rights. Rights to resell use or exclusionary rights.

Source: Adapted from Young (2007)

access, fisheries would experience 'market failure', meaning fishing becomes unprofitable as fishing effort increases and fish stocks decrease (Gordon, 1954). Gordon was an early proponent of treating fish as a private resource; but it was not until 1979, when Iceland began limiting entry to its herring fishery using licence limitations, that modern rights-based fishing emerged (Chu, 2009).

Approaches to rights-based fishing can broadly be separated into approaches that:

* Limit entry to a fishery by determining who is allowed to fish and where they can fish (license limitation) (see Box 10.5)
* Limit who is allowed to utilize processing facilities (landing limitation).
* Limit who is allowed access to goods and services (e.g. fuel, supplies) that support fishing.
* Conferring ownership of a portion or all of a particular fish stock upon an individual, company or community (privatization) (see Box 10.5).

The most prominent and frequently used rights-based approaches are those that establish a TAC for a fish stock and then allocate a portion of the catch (known as 'catch shares') to individuals, vessels, corporations or communities. Types of catch share approaches include:

* Individual transferrable quotas (ITQs).
* Individual vessel quotas (IVQs).
* Community development quotas (CDQs).
* Enterprise allocations (allocated to fishing corporations).
* Territorial user rights in fisheries (TURFs – allocated to residents of a geographical area).

Box 10.5 Successful examples of catch share programmes

The Peruvian anchovy fishery

The Peruvian anchovy (*Engraulis ringens*) fishery comprises nearly 10 per cent of the global fisheries landings, supports approximately 1200 fishing vessels and 140 factories, employs over 18,000 people and represents a minimum of 7 per cent ($2 billion) of Peru's foreign exchange earnings. The fishery was also recently ranked as the world's most sustainable fishery in a study developed by the University of British Columbia Fisheries Centre.

The fishery operates under a TAC set annually by government with scientific advice from the Peruvian Marine Research Institute. Since 1992 the government has addressed overfishing by limiting the issuance of new licenses. In 2008 it adopted a system of IVQs to divide the TAC among vessels registered in the fishery. Each day during the fishing season the government publishes a list of vessels permitted to fish. In addition, fishing plants are prohibited from receiving fish from vessels not authorized to fish.

The Peruvian government has developed a legislated management scheme to control domestic and foreign access by designating geographical boundaries, limiting the number of boats and their fishing capacity, and maintaining biological productivity. These statutes are modified as necessary by ministerial resolutions relating to closed areas, periods or quotas.

The North-east Pacific halibut fishery

The Canadian and US populations of Pacific halibut (*Hippoglossus stenolepis*) have been managed by the IPHC since 1923. The Canadian fishery moved to an IVQ system in 1991 and the US to an IFQ (same as an ITQ) system in 1995.

The Canadian halibut fishing season was six days long in 1990. Today the season is ten months in duration, which has resulted in 95 per cent of the catch sold fresh (not frozen), yielding higher market prices. The longer season also produced stability for fish processors, greater crew safety (through shorter shifts), a smaller fleet fishing, more weeks per vessel and a product price increase of $3.30 per kg at the wholesale level (DFO, 2008).

The US halibut fishing season was reduced from 150 days in 1970 to 16 days in 1979. Overcapitalization in the fishery – characterized by an increase in the size and number of fishing vessels – occasioned this drastic measure. After the establishment of IFQs, the vessel fleet was reduced by 35 per cent through consolidation and the fishing season was increased to nine months. Additionally, the halibut price paid to fishers increased 10.5 per cent per year for the first eight years of the ITQ programme. The programme added further efficacious measures: it instituted vessel size quotas so that larger operators would be prohibited from dominating the fishery, it limited transferability to fishers with experience in commercial fishing and it introduced use caps on individual owners and vessels discourage consolidation.

Source: DFO (2008), Schreiber (2012), Carothers (2013)

ITQs are the most frequently used catch share approaches and the term 'ITQ' has become synonymous with all types of catch shares. Currently, ITQs have been introduced in over 121 fisheries in 18 states with some evidence that they can be successful. Hilborn *et al.* (2003) found that bycatch declined an average of 40 per cent in a study of ten fisheries in the US and Canada after they moved to management under catch shares. Chu (2009) found improvements in 12 of 20 stocks after ITQs were introduced. Despite these promising statistics, catch share programmes continue to be the exception in fisheries management. A paltry 1 per cent of the world's commercial fisheries and 15 per cent of the world's catch are managed under the scheme (Deacon, 2009).

It is important to note that catch share programmes do not bestow full property rights to fishers. Full property rights would require that the holder of a catch share have an exclusive, inalienable right in perpetuity that could be sold, leased or rented. While some catch share programmes, such as those practiced in some New Zealand fisheries, approach private rights, most programmes occupy a place between open access and full private rights. So long as the TAC is set correctly, catch shares can promote sustainability by providing fishers with an incentive to conserve the fish resource. Lest one forgets, the incentive to fish responsibly is based in part on the security, exclusivity and transferability of the fishing right.

There has been considerable debate as to whether catch shares are applicable to high seas fisheries such as tuna, swordfish and Patagonian toothfish. While certain ITQs and other catch share programmes have successfully established sustainable fisheries within EEZs or between EEZs or high seas areas (straddling stocks), there has been little progress on launching rights-based systems in high seas areas (Hannesson, 2011). As discussed previously, while nearly all the world's fisheries occur within EEZs, certain fisheries including cod (Grand Banks), tubot (Barents Sea) and Alaska pollock (North Pacific) occur outside of EEZs. RFMOs (Chapter 6) are responsible for managing high seas fisheries yet this remains problematic. The open access characteristic of the high seas makes developing and enforcing regulations difficult, especially when dealing with states that are not inclined to participate in RFMO processes. Two solutions are offered by Hannesson (2011) to establish private rights in high seas fisheries: first, states could extend their national EEZs to capture globally important fisheries that currently reside in the high seas and second, corporations could be established that would 'own' fish on the high seas where states are shareholders (Crothers and Nelson, 2006; Carothers, 2013). At this time, however, neither approach has much traction at the international level.

Chapter summary

- Integrated ocean management uses specific approaches and tools to assist with marine decision-making. These approaches may also be used to analyse existing policies and to formulate new policies. They are characteristically multi-sector, forward-looking, inclusive and attempt to inform durable decisions.
- Integrated approaches began with CZM in the 1970s followed by EAM, MPAs, MSP and cumulative effects.
- Rights-based approaches confer a private right to a public good. Marine property rights may be conceived of as tangible objects (e.g. territory, fish, oil and gas reserve) or intangible benefits or income streams (e.g. right to fish) recognized by some level of government. These approaches are used primarily in fisheries management and currently 15 per cent of the world's catch is managed under this scheme.

References

Agardy, T. (2010) *Ocean Zoning: Making Marine Management More Effective*, Earthscan, London.

Baird, S. F. (1873) 'Report on the condition of the sea fisheries of the south coast of New England in 1871 and 1872', *Report of the United States Fish Commission*, 1, GPO, Washington, DC.

Ballantine, W. J. (1991) 'Marine reserves for New Zealand', *Leigh Laboratory Bulletin*, 25, University of Auckland, New Zealand.

Ban, N. C., Alidina, H. M. and Ardron, J. A. (2010) 'Cumulative impact mapping: Advances, relevance and limitations to marine management and conservation, using Canada's Pacific waters as a case study', *Marine Policy*, 34, 876–886.

Beverton, R. J. H. and Holt, S. J. (1957) 'On the dynamics of exploited fish populations', *Fishery Investigations*, 19, Ministry of Agriculture, Fisheries and Food, London.

Bremer, S. and Glavovic, B. (2013) 'Mobilizing knowledge for coastal governance: Re-framing the science–policy interface for integrated coastal management', *Coastal Management*, 41 (1), 39–56.

Brunckhorst, D. J., Bridgewater, P. and Parker, P. (1997) 'The UNESCO Biosphere Reserve Program comes of age: Learning by doing, landscape models for a sustainable conservation and resource', in P. Hale and D. Lamb (eds) *Conservation Outside Reserves*, University of Queensland Press, Brisbane.

Carothers, C. (2013) 'A survey of US halibut IFQ holders: Market participation, attitudes, and impacts', *Marine Policy*, 38, 515–522.

Chu, C. (2009) 'Thirty years later: The global growth of ITQs and their influence on stock status in marine fisheries', *Fish and Fisheries*, 10, 217–230.

Cicin-Sain, B. and Knecht, R. W. (1998) *Integrated Coastal and Ocean Management: Concepts and Practices*, Island Press, Washington, DC.

Cicin-Sain, B., Knecht, R. W. and Fisk, G. W. (1995) 'Growth in capacity for integrated coastal management since UNCED: An international perspective', *Ocean and Coastal Management*, 29 (1–3), 93–123.

Clark, J. R. (1992) 'Integrated management of coastal zones', *FAO Fisheries Technical Paper No. 327*, FAO, Rome.

COMPASS (Communication Partnership for Science and the Sea) (2005) 'EBM consensus statement', http://www.compassonline.org/sites/all/files/document_files/EBM_Consensus_Statement_v12.pdf, accessed 23 December 2010.

Craig, R. K. (2012) *Comparative Ocean Governance: Place-Based Protections in an Era of Climate Change*, Edward Elgar, Cheltenham.

Crain, C. M., Kroeker, K. and Halpern, B. S. (2008) 'Interactive and cumulative effects of multiple human stressors in marine systems', *Ecology Letters*, 11, 1304–1313.

Crothers, G. T. and Nelson, L. (2006) 'High seas fisheries governance: A framework for the future?', *Marine Resource Economics*, 21, 341–353.

Deacon, R. T. (2009) *Creating Marine Assets: Property Rights in Ocean Fisheries*, PERC Policy Series, 43, Bozeman, Montana.

DFO (Department of Fisheries and Oceans) (2008) *Employment Impacts of ITQ Fisheries in Pacific Canada*, Fisheries and Oceans Canada, Ottawa, Canada.

Douvere, F. (2008) 'The importance of marine spatial planning in advancing ecosystem-based sea use management', *Marine Policy*, 32, 762–771.

Drankier, P. (2012) 'Marine protected areas beyond national jurisdiction', *The International Journal of Marine and Coastal Law*, 27, 291–350.

Edwards, S. F. (1994) 'Ownership of renewable ocean resources', *Marine Resource Economics*, 9, 253–273.

European Commission (2008) *Communication from the Commission: Roadmap for Maritime Spatial Planning: Achieving Common Principles in the EU*, COM(2008) 791 Final, http://eur-lex.europa.eu/LexUriServ/LexUriServ.do?uri=COM:2008:0791:FIN:EN:PDF, accessed 14 January 2013.

European Commission (2012) *Blue Growth: Opportunities for marine and maritime sustainable growth,*

COM(2012) 494 final, ec.europa.eu/maritimeaffairs/policy/blue_growth/documents/com_2012_494_en.pdf, accessed 12 March 2013.

FAO (Food and Agriculture Organization) (1996) 'The contributions of science to coastal zone management', *Reports and Studies No. 61*, GESAMP, Rome.

FAO (2005) *Progress in the Implementation of the Code of Conduct for Responsible Fisheries and Related Plans of Action*, FAO, Rome.

FAO (2006) 'Report and documentation of the expert workshop on marine protected areas and fisheries management: Review of issues and considerations', *FAO Fisheries Report No 825*, FAO, Rome.

Gjerde, K. M. and Rulska-Domino, A. (2012) 'Marine protected areas beyond national jurisdiction: Some practical perspectives and moving ahead', *The International Journal of Marine and Coastal Law*, 27, 351–373.

Gordon, H. S. (1954) 'The economic theory of a common-property resource: The fishery', *The Journal of Political Economy*, 62 (2), 124–142.

Halpern, B. S., Kappel, C. V., Sekloe, K. A., Micheli, F., Ebert, C. M., Kontgis, C., Crain, C., M., Martone, R. G., Shearer, R. and Teck, S. J. (2009) 'Mapping cumulative human impacts to California Current marine ecosystems', *Conservation Letters*, 2 (3), 138–150.

Hannesson, R (2011) 'Rights based fishing on the high seas: Is it possible?', *Marine Policy*, 35, 667–674.

Hilborn, R., Branch, T. A., Ernst, B., Magnusson A, Minte-Vera, C. V., Scheuerell, M. D. and Valero, J. L. (2003) 'State of the world's fisheries', *Annual Review of Environment and Resources*, 28, 359–399.

Holdgate, M. (1999) *The Green Web – A Union for World Conservation*, Earthscan, London.

IUCN/WWF/UNEP (International Union for the Conservation of Nature/World Wildlife Fund/United nations Environment Programme) (1980) *World Conservation Strategy: Living Resource Conservation for Sustainable Development*, IUCN, Nairobi, Kenya, data.iucn.org/dbtw-wpd/html/WCS-004/cover.html, accessed 25 October 2013.

Jones, P. (forthcoming) *Governing Marine Protected Areas*, Routledge, London.

Kay, R. and Alder, J. (2005) *Coastal Planning and Management*, Taylor & Francis, London.

Kelleher, G. and Kenchington, R. A. (1992) *Guidelines for Establishing Marine Protected Areas*, IUCN, Gland, Switzerland.

Kenchington, R. A. (1996) 'A global representative system of marine protected areas', in R. Thackway (ed.) *Developing Australia's Representative System of Marine Protected Areas*, Department of the Environment, Sport and Territories, Canberra, Australia.

Kenchington, R. A. and Agardy, M. T. (1990) 'Applying the biosphere reserve concept in marine conservation', *Environmental Conservation*, 17 (1), 39–44.

Kidd, S., Plater, A. and Frid, C. (eds) (2011) *The Ecosystem Approach to Marine Planning and Management*, Routledge, London.

Larkin, P. A. (1996) 'Concepts and issues in marine ecosystem management', *Reviews in Fish Biology and Fisheries*, 6, 139–164.

Leopold, A. (1949) *A Sand County Almanac and Sketches Here and There*, Oxford University Press, Oxford.

Link, J. S. (2005) 'Translating ecosystem indicators into decision criteria', *ICES Journal of Marine Science*, 62 (3), 569–576.

MA (Millennium Ecosystem Assessment) (2005) *Global Assessment Reports, 1: Current State and Trends*, Island Press, Washington, DC.

Martínez, M. L., Intralawan, A., Vázquez, G., Pérez-Maqueo, O., Sutton, P. and Landgrave, R. (2007) 'The coasts of our world: Ecological, economic and social importance', *Ecological Economics*, 63, 254–272.

NOAA (National Oceanic and Atmospheric Administration) (1972) *Coastal Zone Management Act of 1972*, coastalmanagement.noaa.gov/about/media/CZMA_10_11_06.pdf, accessed 1 April 2013.

NOAA (2007) *CZMA Section 312 Evaluation Summary Report – 2006*, http://coastalmanagement.noaa.gov/success/media/312summaryreport2006.pdf, accessed 4 April 2013.

National Ocean Council (2013) *CMSP National Goals and Principles*, www.whitehouse.gov/administration/eop/oceans/cmsp/goals-principles, accessed 13 January 2013.

Nijkamp, H. and Peet, G. (1994) *Marine Protected Areas in Europe*, Commission of European Communities, Amsterdam.

Olsen, S. B. (2003) 'Frameworks and indicators for assessing progress in integrated coastal management initiatives', *Ocean & Coastal Management*, 46, 347–361.

Parkins, J. R. (2011) 'Deliberative democracy, institution building, and the pragmatics of cumulative effects assessment', *Ecology and Society*, 16 (3).

Price, A. R. G. and Humphrey, S. L. (1993) *Application of the Biosphere Reserve Concept to Coastal Marine Areas*, IUCN, Gland, Switzerland.

Ray, G. C. and McCormick-Ray, J. (2004) *Coastal Marine Conservation: Science and Policy*, Blackwell, Malden, MA.

Rochette, J. and Billé, R. (2012) 'ICZM protocols to regional seas conventions: What? Why? How?', *Marine Policy*, 36, 977–984.

Roff, J. C. and Taylor, M. E. (2000) 'National frameworks for marine conservation: A hierarchical geophysical approach', *Aquatic Conservation: Marine and Freshwater Ecosystems*, 10, 209–223.

Roff, J. C., Taylor, M. E. and Laughren, J. (2003) 'Geophysical approaches to the classification, delineation and monitoring of marine habitats and their communities', *Aquatic Conservation, Marine and Freshwater Ecosystems*, 13, 77–90.

Rupprecht Consult (2006) *Evaluation of Integrated Coastal Zone (ICZM) Management in Europe: Final Report*, Rupprecht Consult, Cologne, Germany, ec.europa.eu/environment/iczm/pdf/evaluation_iczm_report.pdf, accessed 10 April 2013.

Salm, R. V. and Clark, J. R. (1984) *Marine and Coastal Protected Areas: A Guide for Planners and Managers*, IUCN, Gland, Switzerland.

Schreiber, M. A. (2012) 'The evolution of legal instruments and the sustainability of the Peruvian anchovy fishery', *Marine Policy*, 36, 667–674.

Sklandany, M., Clausen, R. and Belton, B. (2007) 'Offshore aquaculture: The frontier of redefining oceanic property', *Society & Natural Resources: An International Journal*, 20 (2), 169–176.

Sobel, J. (1996) 'Marine reserves: Necessary tools for biodiversity conservation?', *Global Biodiversity*, 6 (1), Canadian Museum of Nature, Ottawa, Canada.

Tansley, A. G. (1935) 'The use and abuse of vegetational terms and concepts', *Ecology*, 16, 284–307.

Theobald, D. M., Miller, J. R. and Hobbs, N. T. (1997) 'Estimating the cumulative effects of development on wildlife habitats', *Landscape and Urban Planning*, 39, 25–36.

Toropova, C., Meliane, I., Laffoley, D., Matthews, E. and Spalding, M. (2010) *Global Ocean Protection: Present Status and Future Possibilities*, IUCN, Gland, Switzerland, data.iucn.org/dbtw-wpd/edocs/2010-053.pdf, accessed 12 April 2013.

UNEP (United Nations Environment Programme) (1995) *Guidelines for Integrated Management of Coastal and Marine Areas – With Special Reference to the Mediterranean Basin*, UNEP Regional Seas Reports and Studies No. 161, Nairobi, Kenya.

UNEP (2008) *Protocol on Integrated Coastal Zone Management in the Mediterranean*, www.pap-thecoast-centre.org/razno/PROTOCOL%20ENG%20IN%20FINAL%20FORMAT.pdf, accessed 11 April 2013.

Weber, M., Krogman, N. and Antoniuk, T. (2012) 'Cumulative effects assessment: Linking social, ecological and governance dimensions', *Ecology and Society*, 17 (2), article 22.

Young, O. (2007) 'Rights, rules, and common pools: Solving problems arising in human/environment relations', *Natural Resources Journal*, 47 (1), 1–16.

Young, O., Oshrenko, G., Ekstron, J. Crowder, L., Ogden, J., Wilson, J., Day, J., Douvere, F., Ehler, C., McLeod, K., Halpern, B. and Peach, R. (2007) 'Solving the crisis in ocean governance: Place-based management of marine ecosystems', *Environment*, 49, 21–30.

Zacharias, M. A. and Roff, J. C. (2001) 'Use of focal species in marine conservation and management: A review and critique', *Aquatic Conservation: Marine and Freshwater Ecosystems*, 11, 59–76.

Conclusion

A report card on ocean management

Albert Einstein, with a conjecture analogous to Bertrand Russell's contention that 'wisdom' is a prerequisite to peacefully harness human innovation, famously surmised that 'The significant problems we have cannot be solved at the same level of thinking with which we created them'. In terms of this book, the first 'significant problem' addressed by states was the overharvesting of fish and marine mammals in the early twentieth century. Although there was substantial denial and foot-dragging concerning the problem, the agreements that were eventually negotiated prevented the near certain extirpation of many important fish and marine mammal populations.

In 1951, Rachel Carson brought attention to the vulnerability of the world's oceans to pursuits extending beyond fishing (Carson, 1951). The recognition that the management of coastal oceans should be more coordinated spawned coastal zone management as a discipline, which has since been adopted world-wide. In addition, it this recognition that has been responsible for the foundation of other integrated management tools such as marine protected areas and marine spatial planning. The mid-twentieth century also gave rise to the Antarctic Treaty and the need to zero in on the Antarctic as a place of scientific research and international concord. It has been over 20 years since the publication of seminal 'calls to action' (e.g. Thorne-Miller and Catena, 1991; Norse, 1993) that alerted us to the unceasing degradation of the ocean environment and the increasing complexity of prickly issues (e.g. climate change) that demanded our immediate attention.

Extensive progress has since been made towards rectifying many of these thorny problems – establishing marine protected areas, reducing greenhouse gases from shipping and steadfastly regulating pollution are merely three examples of this headway in managing the oceans benevolently and efficaciously. The Law of the Sea Convention, Antarctic Treaty and major pollution conventions (LDC, MARPOL) have improved both the health of the oceans and the plight of people that depend on them. Marine transportation is governed by a widely accepted set of rules that have fostered growth in international trade. Nearly all international conventions recognize the needs of less developed, geographically disadvantaged and landlocked states. Finally, most states have fully embraced the need to cooperate on a regional basis to address shared ocean issues.

Significant challenges persist. IUU fishing, piracy, poverty, worker safety, flags of convenience, invasive alien species and an incomplete comprehension of the biological and physical functions of the ocean are but some. The effects of climate change – ocean acidification, warming and sea level rise – on the oceans are poorly understood, but likely to be severe. Likewise are the environmental and social consequences of rapid human population growth along the world's coastlines. Certain regions, most notably the South China Sea, continue to be at the nexus of jurisdictional disagreements, leading to political and economic friction. Yet pessimism need not prevail; states have demonstrated that cooperation with each other is not only possible, but can lead to improved economic, environmental and political outcomes.

References

Carson, R. (1951) *The Sea Around Us*, Oxford University Press, Oxford.

Norse, E. A. (ed.) (1993) *Global Marine Biodiversity: A Strategy for Building Conservation into Decision Making*, Island Press, Washington, DC.

Thorne-Miller, B. and Catena, J. (1991) *The Living Ocean: Understanding and Protecting Marine Biodiversity*, Island Press, Washington, DC.

Index